T0305960

CIRCULAR STORAGE TANKS AND SILOS

THIRD EDITION

CIRCULAR STORAGE TANKS AND SILOS

THIRD EDITION

A M I N G H A L I

CRC Press is an imprint of the
Taylor & Francis Group, an **informa** business

CRC Press
Taylor & Francis Group
6000 Broken Sound Parkway NW, Suite 300
Boca Raton, FL 33487-2742

First issued in paperback 2017

Version Date: 20140127

ISBN 13: 978-1-4665-7104-4 (hbk)
ISBN 13: 978-1-138-07336-4 (pbk)

Library of Congress Cataloging-in-Publication Data

Ghali, A. (Amin)
Circular storage tanks and silos / Amin Ghali. -- Third edition.
pages cm
Summary: "This book presents the most relevant practical methods for the analysis and design of circular concrete tanks. The methods can also be used for silos, pipes, or any circular shells subjected to arbitrary axially symmetrical loading, and also deal with the more general problem of beams on elastic foundations. A new chapter is presented with guidance construction design of circular tanks. Examples of satisfactory designs are presented; including posttensioned concrete walls, footings, floors and roofs, and liquid-tight connections between these components"-- Provided by publisher.
Includes bibliographical references and index.
ISBN 978-1-4665-7104-4 (hardback)

1. Storage tanks--Design and construction. 2. Silos--Design and construction. 3. Cylinders. I. Title.
TA660.T34G47 2014 690'.53--dc23

Visit the Taylor & Francis Web site at
http://www.taylorandfrancis.com

and the CRC Press Web site at
http://www.crcpress.com

Contents

Preface to third edition

This book, now in its third edition, is a guide for analysis and design of circular storage tanks and silos. Recommended practice for design and construction of concrete water tanks and liquefied natural gas tanks is presented in a new chapter. A Web site companion of the third edition provides the computer program CTW (Cylindrical Tank Walls) to perform the analysis for the load combinations anticipated in the design of cylindrical prestressed walls. The analysis in the design of tanks or silos having the general shape of an axisymmetrical shell can be performed by the finite-element computer program SOR (Shells of Revolution), which is available from the same Web site (http://www.crcpress.com/product/isbn/9781466571044).

Like the earlier editions, the new edition is suitable for use by practicing engineers, students, and researchers in any country. No specific system of units is used in most solved examples. However, it is advantageous to use actual dimensions and forces on the structure in a small number of examples. These problems are set in SI units and Imperial units (still used in the United States); the answers and the graphs related to these examples are given in the two systems. The presented methods of analysis and design procedures are independent of codes. However, occasional reference is made to North American and European codes. Periodic revisions of the codes should not lessen the relevance of the presented material.

For the analysis of a symmetrically loaded cylindrical wall, it is sufficient to consider an elemental strip parallel to the cylinder axis. The radial displacement of the strip is accompanied by hoop forces, which make the strip behave as a beam on an elastic foundation, receiving at every point transverse reaction forces proportional to the magnitude of the displacement. Solution of a differential equation relating the deflection to the applied load provides the analysis of the internal forces for the beam or the cylinder.

Cylindrical walls are often monolithic with other members in the form of annular plates or other shapes of shells of revolution. When subjected to axisymmetrical loading, these structures can be analyzed by the general force or displacement methods in the same way as for plane frames. For this purpose, the flexibility or the stiffness parameters are derived for circular cylindrical walls and annular plates. The finite element method presented in Chapter 5 and the computer program SOR are suitable for domes, cones, and other forms of shells of revolution.

Tanks have to sustain high-quality serviceability over a long lifespan; for this purpose, control of cracking is of paramount importance. Computing the stresses in service is treated in several chapters. Thermal stresses and the time-dependent stresses produced by creep and shrinkage of concrete and relaxation of prestressed steel are considered. The effects of cracking and means for its control are treated in Chapters 9 and 10.

To avoid large bending moments in the walls of circular-cylindrical prestressed concrete tanks, some designs allow sliding of the wall at their base during prestressing and filling of the tank. This has to be accompanied by the means, discussed in Chapter 11, to prevent leakage at the sliding joint. The distribution of prestressing that minimizes the bending moment in a wall having no sliding joint is explained in Chapter 8 with design examples.

As a design aid, a set of tables intended mainly for analysis of circular-cylindrical walls of linearly variable thickness are presented in Chapter 12; examples for their use are given in Chapters 4, 6, and 7. The tables can also be employed in the analyses of silos, pipes, or other cylinders subjected to arbitrary axisymmetric loading and support conditions. All tabulated parameters are dimensionless, thus usable with any system of units.

In Chapter 11, the sections relevant to water tanks were written in collaboration with Robert Bates and the sections on liquefied natural gas tanks were written in collaboration with Dr. Josef Roetzer. Dr. Ramez Gayed reviewed the new material of the third edition. The author is grateful to these individuals for their valued help.

Author

Amin Ghali, Ph.D., is a professor emeritus, civil engineering, University of Calgary, Canada. His books include Ghali, Neville, and Brown, *Structural Analysis,* 6th ed. (CRC Press/Spon Press, 2009), which has been translated into six languages, is of considerable international renown, and is widely adopted in engineering education and practice; Ghali, Favre, and Elbadry, *Concrete Structures: Stresses and Deformations: Analysis and Design for Serviceability*, 4th ed. (CRC Press/Spon Press, 2011), translated into Japanese, the book provides an emphasis on designing structures for high-quality service for a longer lifespan. Dr. Ghali has published more than 250 papers on the analysis and design of concrete structures including the strength of flat slabs, deflections, time-dependent deformations of reinforced and prestressed structures, partially prestressed structures, composite members, bridges, storage tanks, thermal effects, seismic effects, and serviceability. Dr. Ghali's joint invention "Headed Stud Shear Reinforcement" is recognized worldwide as the most effective practical type for punching resistance of concrete slabs, footings, rafts, and walls.

Dr. Ghali's structural consulting on major projects has included three offshore platforms in the North Sea; major bridges in Canada, France, and Bahrain; and the Northumberland Strait Bridge connecting Prince Edward Island and New Brunswick.

List of Examples

SI system of units and British equivalents

Length	
meter (m)	1 m = 39.37 in
	1 m = 3.281 ft
Area	
square meter (m^2)	1 m^2 = 1550 in^2
	1 m^2 = 10.76 ft^2
Volume	
cubic meter (m^3)	1 m^3 = 35.32 ft^3
Moment of inertia	
meter to the power four (m^4)	1 m^4 = 2403 × 10^3 in^4
Force	
newton (N)	1 N = 0.2248 lb
Load intensity	
newton per meter (N/m)	1 N/m = 0.068 52 lb/ft
newton per square meter (N/m^2)	1 N/m^2 = 20.88 × 10^{-3} lb/ft^2
Moment	
newton meter (N m)	1 N m = 8.851 lb in
	1 N m = 0.7376 × 10^{-3} kip ft
	1 kN m = 8.851 kip in
Stress	
newton per square meter (pascal)	1 Pa = 145.0 × 10^{-6} lb/in^2
	1 MPa = 0.1450 ksi
Curvature	
$(meter)^{-1}$	1 m^{-1} = 0.0254 in^{-1}
Temperature change	
Celsius degree (°C)	1°C = (5/9)°F (Fahrenheit degree)
Energy and power	
joule (J) = 1 N m	1 J = 0.7376 lb ft
watt (W) = 1 J/s	1 W = 0.7376 lb ft/s
	1 W = 3.416 Btu/h
Nomenclature for decimal multiples in the SI system	
	10^9 giga (G)
	10^6 mega (M)
	10^3 kilo (k)
	10^{-3} milli (m)

Notation

The following symbols are common in various chapters of the book. All the symbols are defined where they first appear in each chapter.

A Any action, which may be a reaction or a stress resultant. A stress resultant is an internal force bending moment, shearing, or axial force.

a Cross-sectional area.

D_i Displacement (rotational or translational) at coordinate i. When a second subscript j is used, it indicates the coordinate at which the force causing the displacement acts.

d Diameter of the middle surface of a circular wall.

E Modulus of elasticity.

EI Flexural rigidity. For a beam, its "units" are force × length^2; for a shell or plate, it indicates flexural rigidity per unit width with "units" force × length (see Equation 2.5).

F A generalized force: a couple or a concentrated load. For an axisymmetric shell or plate, the symbol indicates the intensity of force or moment uniformly distributed over the periphery.

f_{ij} Element of flexibility matrix; that is, displacement at coordinate i caused by a unit force at coordinate j. For an axisymmetric shell or circular plate, the symbol represents displacement at coordinate i due to a distributed force of unit intensity uniformly distributed over the periphery at coordinate j.

h Thickness: when subscripts b and t are used, they refer to the bottom and top edges of a wall, respectively.

I Moment of inertia.

i, j, m, n Integers.

J_i Ratio of the moment of inertia at a section i to a reference value (usually moment of inertia at one edge).

k Modulus of elastic foundation ("units" are force/length^2 for a beam and force/length^3 for an axisymmetrical cylindrical shell; see Equation 2.3).

l Length of beam on elastic foundation or height of a circular-cylindrical wall.

M Bending moment. For a beam, its units are force × length; for a shell or plate the symbol indicates the bending moment per unit width and has the "units" force × length/length = force. For a circular-cylindrical wall the symbol is used without a subscript to represent the bending moment in the direction of a generatrix.

Mr, M_t	Bending moment in radial and tangential direction in a circular plate.
M_{AB}	Moment on end A of element AB.
N	Hoop force. For a circular-cylindrical wall, the symbol indicates the ring force per unit height of the cylinder (force/length).
P, Q	Concentrated load (force). For an axisymmetrical shell, the symbol represents the intensity of a line load on the periphery (force/length).
q	Load intensity (force/length for a beam and force/length^2 for a shell or a plate).
R	Reaction (force for a beam and force/length for an axisymmetrical shell).
r	Radius.
S_{ij}	Element of stiffness matrix; that is, force at coordinate i caused by a unit displacement at coordinate j. For an axisymmetric shell or circular plate, the symbol represents the force intensity of a uniformly distributed load over the periphery at coordinate i when a unit displacement is introduced at j.
S_{AB}	End-rotational stiffness of member AB.
t	Carry-over moment.
V	Shearing force. For a beam its "unit" is force; for a shell or plate, the symbol indicates the shearing force per unit width and has "units" force/length.
w	Deflection.
Y	Function of the dimensionless quantity βx (see Equation 2.14).
y	Value of the function Y when $x = l$.
Z	Function of the dimensionless quantity βx (see Equation 2.57).
α	Characteristic dimensionless parameter for a beam on elastic foundation (Equation 2.18a,b).
β	$= \alpha/l$.
γ	Specific weight.
η	Characteristic dimensionless parameter for circular-cylindrical walls; $= l^2/dh$ when thickness is constant, and $= l^2/(dh_b)$, where h_b is the thickness at the bottom edge, when h varies.
λ	Interval between nodes (in the finite-difference method).
μ	Coefficient of thermal expansion.
ν	Poisson's ratio.
{ }	Curly brackets indicate a vector; that is, a matrix of one column. To save space, the elements of a vector are sometimes listed in a row between two curly brackets.
$[\ ]_{m \times n}$	Square brackets indicate a rectangular (or square) matrix of m rows and n columns.

Part I

Analysis and recommended practice

Chapter 1

Introduction to the analysis of circular tanks

1.1 SCOPE

Cylindrical walls of circular tanks and other containers are usually subjected to radial pressure from the contained material or from externally retained earth. This pressure is assumed here to have an intensity that is constant at any one level but varies in the vertical direction. Other sources of such axisymmetrical loading on walls are circumferential prestressing, weight of overhanging circular platforms, or peripheral channels.

This type of loading produces axisymmetrical radial displacement. The wall edges at the top or bottom may be free to rotate or translate, and may be restrained by the base or the cover. Thus, the edges may receive axisymmetrical radial shear or bending moment. Such end forces will also develop at a restrained edge due to the effects of axisymmetrical temperature variation, shrinkage, or creep of concrete.

For the analysis of a wall of this type it is sufficient to consider the forces and the deformations of a typical elemental strip parallel to the cylinder axis. The radial displacement of the strip must be accompanied by hoop forces. As will be discussed later, the elemental strip behaves as a beam on elastic foundation, which receives transverse reaction forces proportional at every point to the deflection of the beam. The analysis constitutes a solution of one governing differential equation relating the deflection to the applied load.

The objective of this book is to provide a solution of the aforementioned differential equation to obtain the reactions on the edges and the internal forces in circular-cylindrical walls. For the sake of simplicity in practical application, design tables are provided and their use illustrated by examples. Although the tables are mainly intended for use in the design of concrete tanks, they can also be utilized in the analysis of silos, pipes, or any circular-cylindrical shell when subjected to axisymmetrical loading and support conditions. The tables are also applicable for the more general problem often met in practice of a beam on elastic foundation.

1.2 BASIC ASSUMPTIONS

The methods of structural analysis presented in this book are based on the assumption that the material is linearly elastic. Besides, circular walls are considered as cylindrical shells for which the thickness is small compared to the radius. While bending, the normal to the middle surface of a wall is assumed to remain straight and normal to the deformed middle surface. The governing differential equation based on the aforementioned assumptions is solved in a closed form for walls of constant thickness. When the thickness varies in an arbitrary fashion, closed solutions become difficult, so that the analysis is best achieved by numerical procedures.

The time-dependent effect of creep of concrete or of the bearings supporting the wall and relaxation of steel in prestressed concrete tanks, and the nonlinear behavior due to cracking of reinforced concrete tanks are important problems that are treated in this book. For the analysis of these problems the linear elastic analysis with the assumptions adopted from earlier constitutes a basic solution from which (by iterative or step-by-step procedures or other approaches) the time-dependent or the nonlinear behavior may be derived (see Chapters 6 to 10).

1.3 GENERAL METHODS OF STRUCTURAL ANALYSIS

Figure 1.1a,b represents two cross-sections of tanks formed by the intersection of shells of revolution. When subjected to axisymmetrical loading, these structures can be analyzed by the general displacement or force methods of analysis[1] in the same way as for plane frames. For this, consider a strip obtained by cutting the shell by two radial vertical planes with an arbitrary small angle between them. This strip is then treated as an assemblage of elements for which the stiffness coefficients in the displacement method or the flexibility coefficients in the force method need to be calculated. Derivation of these coefficients for cylindrical walls as well as for circular and annular

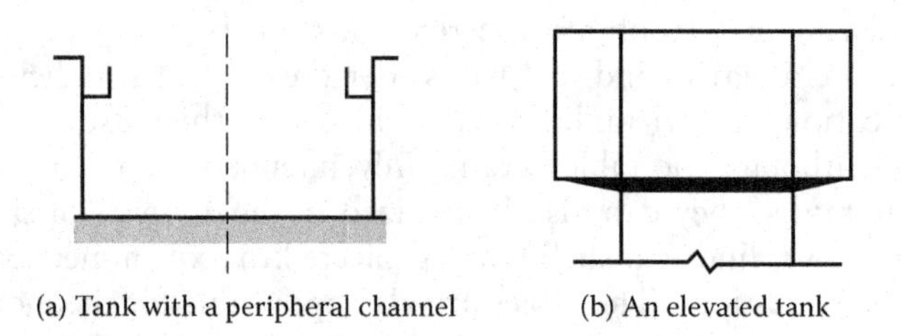

(a) Tank with a peripheral channel (b) An elevated tank

Figure 1.1 Cross-sections of tanks made up of shells of revolution.

plates will be discussed in other chapters. In this chapter, it is assumed that these coefficients are known quantities.

Domes and cones or other forms of shells of revolutions that are used in tank construction can be analyzed by the same general methods, but the stiffness or the flexibility coefficients needed for the analysis are beyond the scope of this book. However, the finite element method presented in Chapter 5 can be used for the analysis of this type of structure.

1.4 THE DISPLACEMENT METHOD

To explain the displacement method, consider, for example, the wall of a tank (shown in Figure 1.2a) subjected to axisymmetrical loading (not shown in the figure). The wall is assumed to be free at the top edge A but continuous with an annular plate BC at the base. The structure is supported by a roller support at B and is totally fixed at C. The analysis by the displacement method involves the following five steps.

1. A coordinate system is established to identify the location and the positive directions of the joint displacements (Figure 1.2b). The number of coordinates n is equal to the number of possible independent joint displacements (degrees of freedom). There are generally two translations and a rotation at a free (unsupported) joint. The number of unknown displacements may be reduced by ignoring the extension of a straight generatrix, which is equivalent to ignoring the axial deformation of a plane frame. For example, by considering that lengths AB and BC remain unchanged, the degrees of freedom are reduced to the three shown in Figure 1.2c.
2. Restraining forces $\{F\}_{n\times 1}$ are introduced at the n coordinates to prevent the joint displacements. The forces $\{F\}$ are calculated by summing the fixed-end forces for the elements meeting at the joints. The internal forces at various positions within the elements are also determined with the joints in the restrained position.
3. The structure is now assumed to be deformed by the displacement at coordinate j, $D_j = 1$, with the displacements prevented at all the other coordinates. The forces S_{1j}, S_{2j}, ..., S_{nj} required to hold the shell in this configuration are determined at the n coordinates. This process is repeated for unit values of displacement at each of the coordinates, respectively. Thus, a set of $n \times n$ stiffness coefficients is calculated, which forms the stiffness matrix $[S]_{n\times n}$ of the structure; a general element S_{ij} is the force required at coordinate i due to a unit displacement at coordinate j.
4. The displacements $\{D\}_{n\times 1}$ in the actual (unrestrained) structure must be of such a magnitude that the restraining forces vanish, which is expressed by the superposition equations:

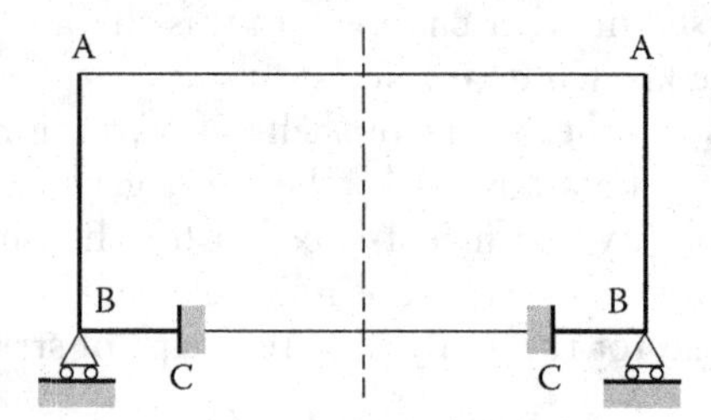

(a) A circular-cylindrical wall continuous with an annular slab subjected to axisymmetrical loading (not shown)

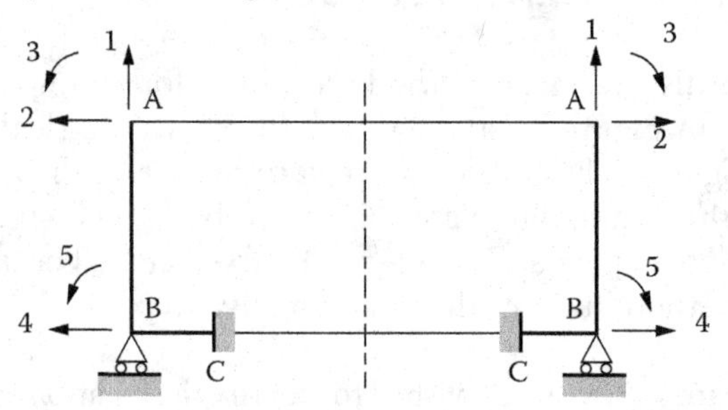

(b) Coordinate system representing forces $\{F\}$ or displacements $\{D\}$

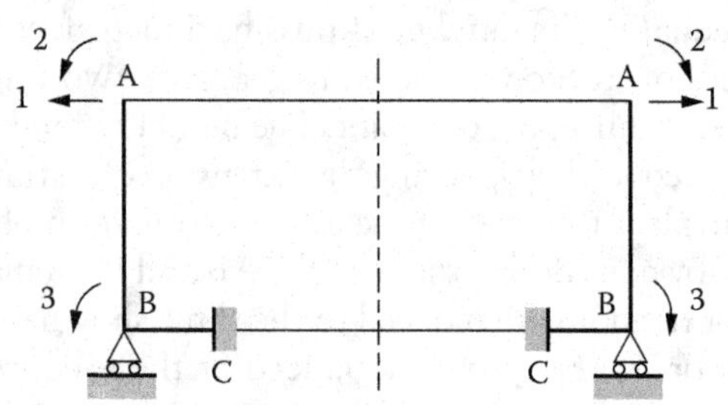

(c) Reduced degrees of freedom (compared to part [A] of Figure) by assuming that the lengths AB and BC do not change

Figure 1.2 Analysis of an axisymmetrical shell by the displacement method.

$$F_1 + S_{11}D_1 + S_{12}D_2 + \cdots + S_{1n}D_n = 0,$$

$$F_1 + S_{21}D_1 + S_{22}D_2 + \cdots + S_{2n}D_n = 0,$$

$$\vdots$$

$$F_n + S_{n1}D_1 + S_{n2}D_2 + \cdots + S_{nn}D_n = 0,$$

or, in matrix form,

$$[S]\{D\} = \{F\}. \tag{1.1}$$

The solution of this group of simultaneous equations gives the n unknown displacements $\{D\}$.

5. Finally, the reactions and the internal forces at any position of the actual shell structure are obtained by adding the values in the restrained structure (calculated in step 2) to the values caused by the joint displacements. This is expressed by the superposition equation:

$$A_i = A_{ri} + [A_{ui1} \quad A_{uin} \quad \cdots \quad A_{uin}]\{D\}, \tag{1.2}$$

where A_i is the value of any action, an internal force or reaction in the actual structure; A_{ri} is the value of the same action in the restrained condition; and A_{uij} is the value of the same action corresponding to a displacement $D_j = 1$. When there are m values to be determined, Equation (1.2) can be used m times as follows:

$$\{A\}_{m\times1} = \{A\}_{m\times1} + [A_u]_{m\times n}\{D\}_{n\times1}. \tag{1.3}$$

1.5 THE FORCE METHOD

The analysis by the force method also involves five steps:

1. Select a number of releases n by the removal of external or internal forces ("redundants"). These redundant forces $\{F\}$ should be chosen in such a way that the remaining released structure is stable and easy to analyze. As an example, releases for the structure in Figure 1.2a are shown in Figure 1.3, where a system of n coordinates indicates arbitrarily chosen positive directions for the released forces and the corresponding displacements.
2. With the given external loads applied on the released structure, calculate the displacements $\{D\}$ at the n coordinates. These represent inconsistencies to be eliminated by the redundant forces. The reactions and internal forces are also determined at various positions of the released structure.

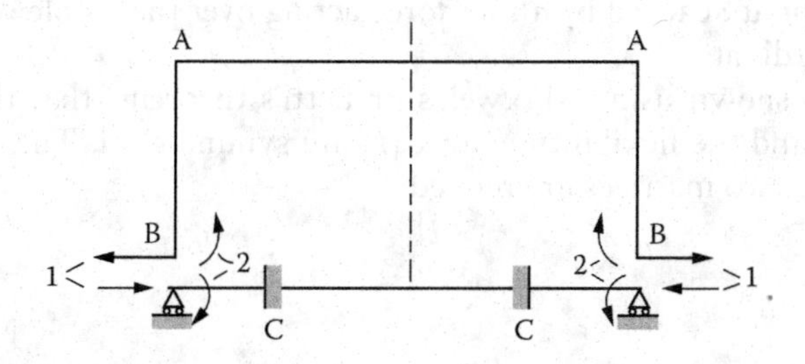

Figure 1.3 Released structure and coordinate system for the analysis of the axisymmetrical shell in Figure 1.2a by the force method.

3. The released structure is subjected to a force $F_1 = 1$ and the displacement f_{11}, f_{21}, f_{n1} at each of the n coordinates is determined. The process is repeated for unit values of the forces at each of the n coordinates, respectively. Thus, a set of $n \times n$ flexibility coefficients is generated, which forms the flexibility matrix $[f]_{n \times n}$; a general element f_{ij} is the displacement of the released structure at coordinate i when subjected to a force $F_j = 1$.
4. The values of the redundant forces $\{F\}$ necessary to eliminate the inconsistencies in the displacements are determined by solving the superposition equation

$$[f]\{F\} = -\{D\}. \tag{1.4}$$

5. The reactions and the internal forces in the actual shell structure are obtained by adding the values in the released structure calculated in step 2 to the values caused by the redundants. This is expressed by the superposition equation

$$\{A\}_{m\times1} = \{A_s\}_{m\times1} + [A_u]_{m\times1}\{F\}_{n\times1}, \tag{1.5}$$

where $\{A\}$ is a vector of m actions to be determined; $\{A_s\}$ are their values in the released structure; and any column j of the matrix $[A_u]$ has m values of the actions due to $F_j = 1$ acting separately on the released structure.

1.6 FLEXIBILITY AND STIFFNESS MATRICES

As mentioned earlier (Section 1.3), the analysis by the force or displacement method may be applied to a strip obtained by cutting the shell of revolution by two radial vertical planes with an arbitrary small angle between them. The width at any section of such a strip is directly proportional to the radius of revolution of the strip at this section. A force F_i or a stiffness coefficient S_{ij} represents the value of the force acting over the whole width of the strip at coordinate i. Similarly, a flexibility coefficient f_i represents a displacement at i caused by a unit force acting over the whole width of the strip at coordinate j.

It can be shown using Maxwell's or Betti's theorem[2] that the stiffness matrix $[S]$ and the flexibility matrix $[f]$ are symmetrical. The elements of either of the two matrices are related:

$$f_{ij} = f_{ji} \tag{1.6}$$

and

$$S_{ij} = S_{ji}. \tag{1.7}$$

An alternative approach that is adopted in this book is to consider for an axisymmetrical shell of revolution that the value F_i or S_{ij} represents the intensity of a uniformly distributed force or couple applied on the periphery of the shell at coordinate i. Similarly, a flexibility coefficient f represents the displacement at coordinate i when a uniformly distributed load of unit intensity is applied on the periphery of the shell at coordinate j. The stiffness and flexibility matrices in this case will be symmetrical only when all the coordinates are at one radius as in the case of a circular-cylindrical wall.

In the general case, Maxwell's or Betti's theorem leads to

$$r_i = S_{ij} = r_j S_{ji}, \tag{1.8}$$

$$r_i f_{ij} = r_j f_{ji}, \tag{1.9}$$

where r_i and r_j are the radii of the shell of revolution at coordinates i and j, respectively. (This is discussed further in Appendix A, Section A.3.)

With any of the two aforementioned approaches, it can be shown that for the same system of coordinates the stiffness and the flexibility matrices are related:

$$[S] = [f]^{-1} \tag{1.10}$$

or

$$[f] = [S]^{-1}. \tag{1.11}$$

1.7 MOMENT DISTRIBUTION

In some cases, it may be advantageous to use the moment distribution in order to avoid or reduce the number of equations to be solved (step 4 of the displacement method). This is particularly the case when the joints between the elements rotate without translation. The axisymmetrical circular structure in Figure 1.4 is an example; by ignoring the changes of length of EA, AB, AD, and BC, the degrees of freedom will be the rotations at A and B.

With moment distribution, the joints are first assumed to be fixed. The effect of joint displacements is then introduced by successive corrections. Thus, moment distribution is a displacement method, but different from that described earlier, since no equations are usually required to find the joint displacements. Instead, the displacements are allowed to take place in steps and their effects are seen as a series of successive converging corrections to the moments at the ends of the various members.

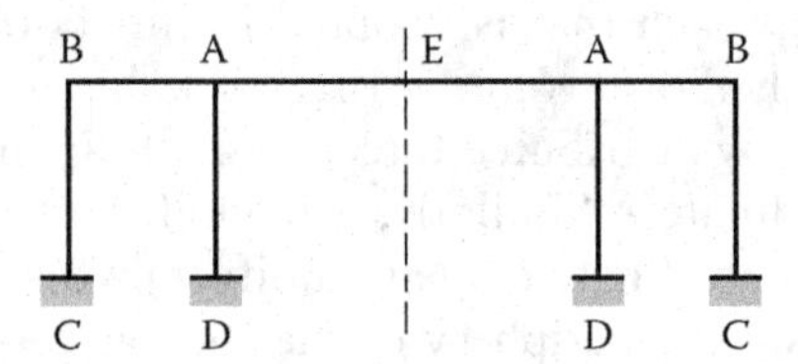

Figure 1.4 An axisymmetrical circular structure to be analyzed by the moment distribution method.

1.7.1 Notation and sign convention

Before expounding further details about analysis by moment distribution, it is important to define the notation and sign convention adopted. Consider the right-hand half of the circular-cylindrical wall shown in Figure 1.5a. A clockwise moment or rotation of either end is considered positive. The following notation is also adopted.

S_{AB} = rotational stiffness at edge A of element AB; that is, the axisymmetric moment per unit length corresponding to a unit rotation at A while the displacements of B are restrained (Figure 1.5a).

t_{AB} = moment carried over from A to B; that is, the moment per unit length at the fixed edge B caused by a unit rotation of edge A (Figure 1.5a).

Similarly, S_{BA} and t_{BA} are represented in Figure 1.5b.

In general, for a cylinder with variable thickness, S_{AB} is not equal to S_{BA} but

$$t_{AB} = t_{BA} = t. \tag{1.12}$$

When the radii at the two ends r_A and r_B are different (such as for the annular plate AB in Figure A.2; see Appendix A), then

$$r_A t_{BA} = r_B t_{AB}. \tag{1.13}$$

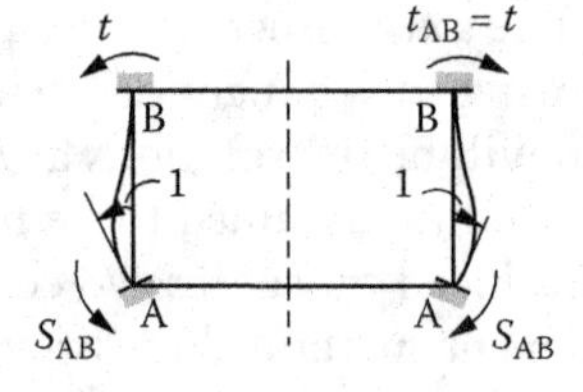

(a) End moments caused by a unit rotation at edge A

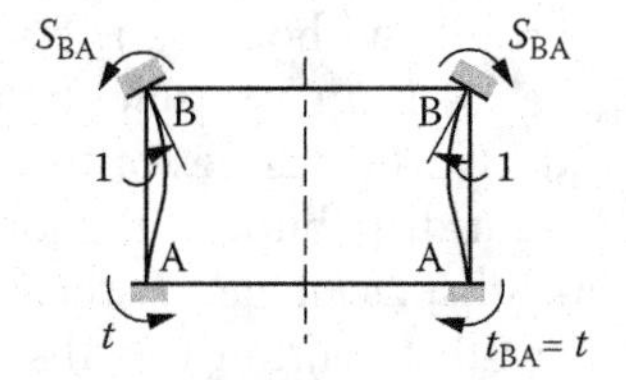

(b) End moments caused by a unit rotation at edge B

Figure 1.5 Sign convention and terminology used in moment distribution.

Equation (1.12) and Equation (1.13) follow from application of Betti's theorem.[3] The carry-over factor from A to B is defined as

$$C_{AB} = \frac{t_{AB}}{S_{AB}}. \tag{1.14}$$

Similarly,

$$C_{BA} = \frac{t_{BA}}{S_{BA}}. \tag{1.15}$$

1.7.2 Steps of analysis

The analysis by moment distribution is done in five steps, which are discussed next for the structure in Figure 1.4.

1. With the rotations restrained at all joints, determine the fixed-end moments (in the radial plane) due to the applied loads.
2. At the joints where rotations occur (joints A and B), calculate the relative end-rotational stiffnesses as well as the carry-over factors. Determine the distribution factors as usually done for plane frames.
3. Perform cycles of moment distribution, carrying over in the usual way, then sum the member end moments, which are the final radial moments at the edges.
4. Calculate the changes in edge moments caused by the rotations of the joints; these are equal to the differences of the final moments and the fixed-end moments.
5. The final end reactions or the internal forces at any point in the shell are then obtained by adding the values for the restrained structure to the values due to the radial moment changes calculated in step 4.

The procedure is similar to what engineers usually apply for the analysis of plane frames. However, the values of the fixed-end moments, the end-rotational stiffnesses, and the carry-over factors are of course different, as will be discussed in later chapters.

1.8 ADJUSTED STIFFNESS AND FLEXIBILITY COEFFICIENTS

In later chapters, tables or equations are presented to give elements of the stiffness matrix [S] or its inverse, the flexibility matrix [f], for cylindrical

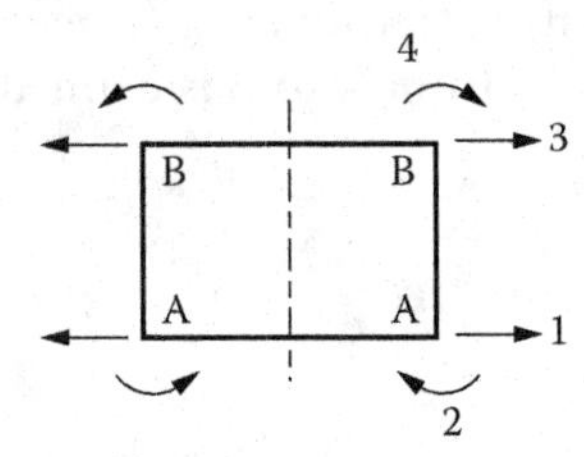

(a) Coordinate system

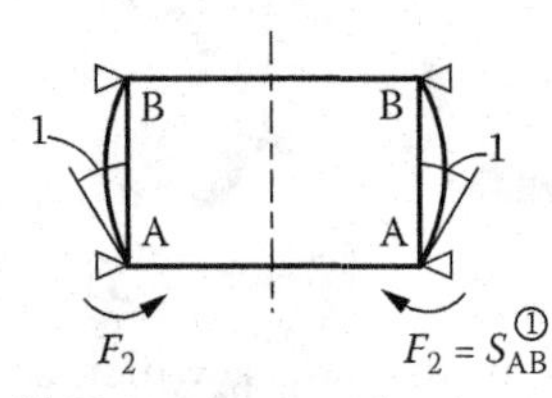

(b) Unit rotation at edge A, with edge B hinged

Given $D_2 = 1; D_1 = 0; D_3 = 0;$ $F_4 = 0$, relations derived from Equation 1.16:

$$D_4 = -\frac{S_{24}}{S_{44}} \tag{1.18}$$

$$F_1 = S_{12} - \frac{S_{14}S_{24}}{S_{44}} \tag{1.19}$$

$$F_2 = S_{22} - \frac{S_{24}^2}{S_{44}} \tag{1.20}$$

$$F_3 = S_{23} - \frac{S_{34}S_{24}}{S_{44}} \tag{1.21}$$

Rotational stiffness of edge A, with edge B hinged:

$$S_{AB}^{①} = S_{22} - \frac{S_{24}^2}{S_{44}} \tag{1.22}$$

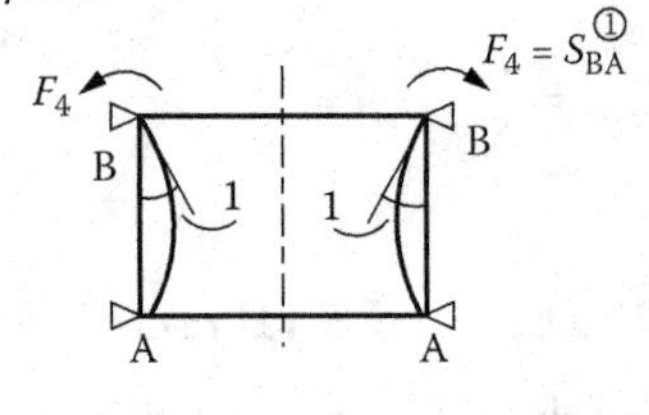

(c) Unit rotation at edge B, with edge A hinged

Given $D_4 = 1; D_1 = 0; D_3 = 0;$ $F_2 = 0$, relations derived from Equation 1.16:

$$D_2 = -\frac{S_{24}}{S_{22}} \tag{1.23}$$

$$F_1 = S_{14} - \frac{S_{12}S_{24}}{S_{22}} \tag{1.24}$$

$$F_3 = S_{34} - \frac{S_{23}S_{24}}{S_{22}} \tag{1.25}$$

$$F_4 = S_{44} - \frac{S_{24}^2}{S_{22}} \tag{1.26}$$

Rotational stiffness of edge B, with edge A hinged:

$$S_{BA}^{①} = S_{44} - \frac{S_{24}^2}{S_{22}} \tag{1.27}$$

Figure 1.6 Effect of rotation (or distributed moment) applied at one edge of an axisymmetrical shell with hinged edges. The coefficients S_{ij} are elements of the stiffness matrix corresponding to the coordinates in (a).

walls and for annular slabs. For any of these axisymmetrical shells, the matrices correspond to four coordinates as shown in Figure 1.6a. The elements of $[S]$ and $[f]$, the stiffness and flexibility coefficients, are given in terms of the geometrical and material properties of the shell (see Table A.1 in Appendix A and Tables 12.16 and 12.17 in Chapter 12).

The matrix $[S]$ or $[f]$ relates forces to displacements at the four coordinates by the equation

$$[S]\{D\} = \{F\} \qquad (1.16)$$

or its inverse

$$[f]\{F\} = \{D\}. \qquad (1.17)$$

Equation (1.16) must not be confused with Equation (1.1), $[S]\{D\} = -\{F\}$, used in the displacement method of analysis where unknown displacements $\{D\}$ are introduced of such a magnitude as will produce forces $-\{F\}$ to eliminate the artificial restraining forces $\{F\}$. Similarly, Equation (1.17) must not be confused with Equation (1.4) of the force method.

It is often expedient to reduce the number of simultaneous equations by using adjusted stiffness or flexibility coefficients, for example, to allow directly for a hinged or a free end. Likewise, in moment distribution, it may be expedient to use adjusted or modified rotational stiffness of one end with the other end hinged or free as shown in Figure 1.6b,c and Figure 1.7b,c. In general, when four of the eight end forces and displacements are known to be either zero or unity, the other four can be derived by solving Equation (1.16) or (1.17), with preference given to the equation requiring an easier solution.

For example, to find the adjusted end-rotational stiffness $S_{AB}^{①}$ of the cylindrical wall in Figure 1.6b (with end B hinged instead of fixed), set $D_2 = 1$, $D_1 = D_3 = 0$, and $F_4 = 0$ and solve Equation (1.16) for the remaining forces or displacements to obtain the relations listed in Figure 1.6b. The relations given in Figure 1.6c and $S_{AB}^{①}$ are derived in a similar way. In Figure 1.7, two other adjusted end-rotational stiffnesses are derived, $S_{BA}^{②}$ and $S_{BA}^{②}$, which are the intensities of the moments required to introduce unit rotation at ends A and B, respectively, of a cylinder with the other end free.

1.9 GENERAL

The decision to use the displacement or the force method for the analysis depends in general on the amount of computation required to generate the matrices $[S]$ and $\{F\}$ in the displacement method and $[f]$ and $\{D\}$ in the force method and in solving Equations (1.1) or (1.4).

The design tables presented in Part II and the appendices of this book are prepared mainly to help with analysis by the force method, or by moment distribution. However, some of the tables are also helpful in analysis by the displacement method in its general form.

Numerical examples involving the use of the force and displacement methods are presented in Chapters 2 to 4. Moment distribution is used in Chapter 4, Example 4.3.

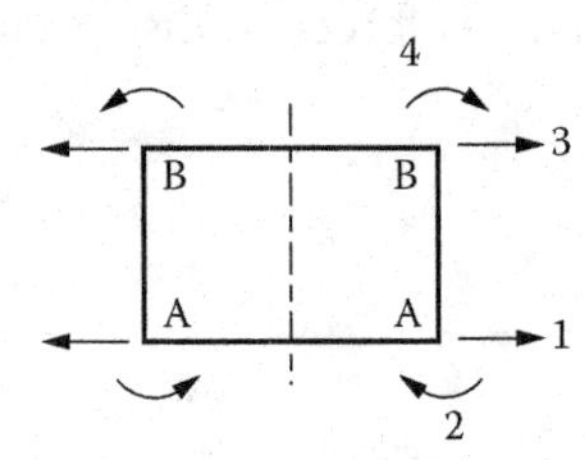

(a) Coordinate system

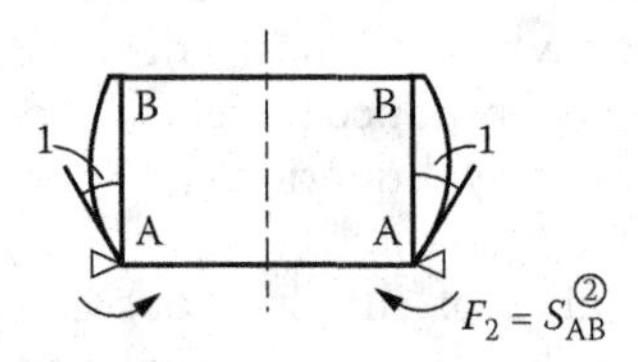

(b) Unit rotation at edge A, with edge B free

Given $D_2 = 1; D_1 = 0; F_3 = 0;$ $F_4 = 0$, relations derived from Equation 1.17:

$$D_3 = \frac{f_{11}f_{23} - f_{12}f_{13}}{f_{11}f_{22} - f_{12}^2} \tag{1.28}$$

$$D_4 = \frac{f_{11}f_{24} - f_{12}f_{14}}{f_{11}f_{22} - f_{12}^2} \tag{1.29}$$

$$F_1 = \frac{-f_{12}}{f_{11}f_{22} - f_{12}^2} \tag{1.30}$$

$$F_2 = \frac{f_{11}}{f_{11}f_{22} - f_{12}^2} \tag{1.31}$$

Rotational stiffness of edge A, with edge B free

$$S_{AB}^{②} = \frac{f_{11}}{f_{11}f_{22} - f_{12}^2} \tag{1.32}$$

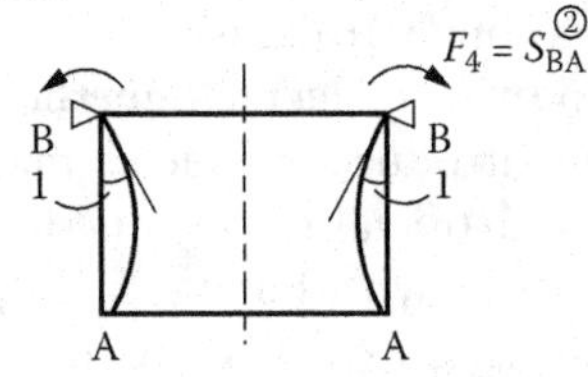

(c) Unit rotation at edge B, with edge A free

Given $D_4 = 1; D_3 = 0; F_1 = 0;$ $F_2 = 0$, relations derived from Equation 1.17:

$$D_1 = \frac{f_{33}f_{14} - f_{13}f_{34}}{f_{33}f_{44} - f_{34}^2} \tag{1.33}$$

$$D_2 = \frac{f_{33}f_{24} - f_{23}f_{34}}{f_{33}f_{44} - f_{34}^2} \tag{1.34}$$

$$F_3 = \frac{-f_{34}}{f_{33}f_{44} - f_{34}^2} \tag{1.35}$$

$$F_4 = \frac{f_{33}}{f_{33}f_{44} - f_{34}^2} \tag{1.36}$$

Rotational stiffness of edge B, with edge A free

$$S_{BA}^{②} = \frac{f_{33}}{f_{33}f_{44} - f_{34}^2} \tag{1.37}$$

Figure 1.7 Effect of rotation (or distributed moment) applied at one edge of an axisymmetrical shell with one edge hinged and the other edge free. The coefficients f are elements of the flexibility matrix corresponding to the coordinates in (a).

NOTES

1. Ghali, A., Neville, A. M., and Brown, T. G. (2009). *Structural Analysis: A Unified Classical and Matrix Approach*, 6th ed., Spon Press, London.
2. Ibid.
3. Ibid.

Chapter 2

Circular walls of constant thickness

2.1 INTRODUCTION

As mentioned earlier, to analyze a circular-cylindrical shell subjected to axisymmetrical loading, it is sufficient to consider a typical elemental strip parallel to the axis of the cylinder. In this chapter, it will be shown that such an element deflects as a beam on elastic foundation, and the differential equation relating the load to the deflection will be derived. The solution of the equation in a closed form for a constant-thickness wall will also be discussed here. When the thickness of the wall varies in an arbitrary fashion, the solution is obtained by numerical procedures discussed in Chapter 3.

2.2 BEAM-ON-ELASTIC-FOUNDATION ANALOGY

An axisymmetrical outward radial loading of intensity q per unit area on a thin cylindrical shell (Figure 2.1a) causes an outward radial deflection, w, which represents an increase in the radius of the cylinder, r. A circumferential strain of magnitude w/r is developed accompanied by hoop force per unit length

$$N = \frac{Eh}{r} w, \tag{2.1}$$

where r is the radius of the middle surface of the cylinder, h its thickness, and E the modulus of elasticity. A tensile hoop force is considered positive; q and w are positive when outward.

Consider a strip of unit width along a generatrix of the cylinder. The hoop forces on the two sides of the strip have a radial resultant of magnitude (Figure 2.1b)

$$\bar{q} = -\frac{N}{r} = -\frac{Eh}{r^2} w. \tag{2.2}$$

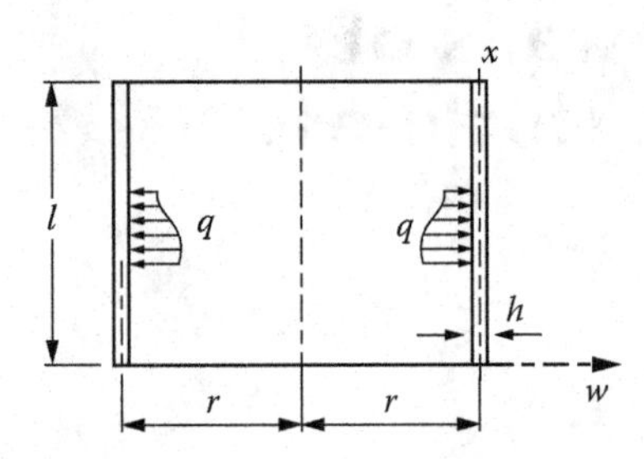

(a) Section through axis of cylinder showing positive directions of load and deflection

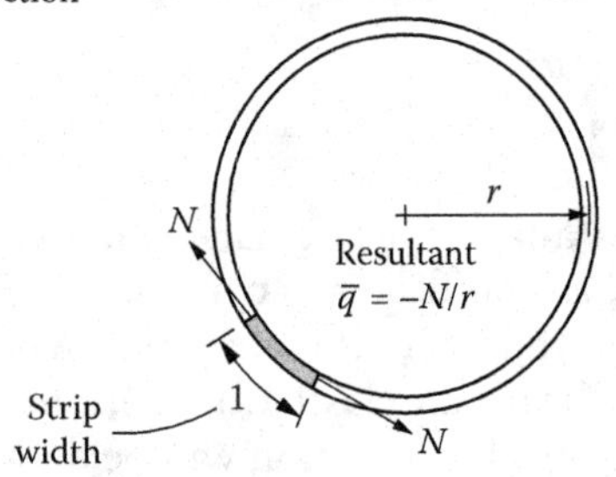

(b) Section normal to the cylinder axis

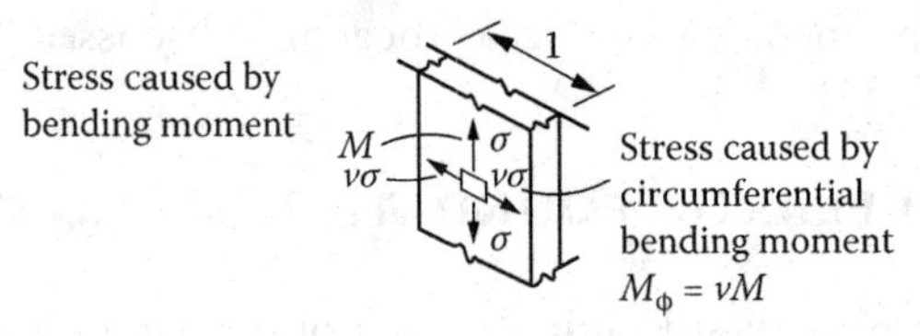

(c) Stresses at any lamina on a vertical strip

Figure 2.1 Circular-cylindrical wall subjected to axisymmetrical loading.

The minus sign indicates that the resultant acts in the inward direction. The strip thus behaves as a beam on elastic foundation subjected to external applied load of intensity q and receives in the opposite direction a reaction of intensity $\bar{q}$ proportional at any point to the deflection, such that $\bar{q} = -kw$. The value k is the modulus of the foundation (force/length^3)

$$k = \frac{Eh}{r^2}. \tag{2.3}$$

When the strip is subjected to positive bending moment M, tensile and compressive stresses are produced parallel to a generatrix on the outer and inner faces, respectively. Owing to Poisson's effect, the two edges of the strip tend to rotate out from their original radial plane. Because of symmetry, this rotation cannot occur as the sides of any strip must remain in radial planes; thus as the strip bends the lateral extension is prevented. The restraining influence is produced by a bending moment M_ϕ in the circumferential direction

$$M_\phi = \upsilon M, \tag{2.4}$$

where υ is Poisson's ratio. The stresses at any point on the outer surface of the shell, or at any lamina, will be as shown in Figure 2.1c, and the strain parallel to a generatrix will be $(\sigma/E) - \upsilon(\upsilon\sigma/E) = \sigma(1 - \upsilon^2)/E$. Thus, the effect of M_ϕ is equivalent to an increase in the modulus of elasticity by the ratio $1/(1 - \upsilon^2)$. It follows that a strip of unit width along the generatrix of a circular cylinder subjected to axisymmetrical load has the same deflection as that of a beam on elastic foundation for which the foundation modulus is given by Equation (2.3) and the flexural rigidity of the beam is

$$EI = \frac{Eh^3}{12(1-\upsilon^2)}. \tag{2.5}$$

With the usual assumption in the bending of beams, that is, that plane cross-sections remain plane, the moment M and the deflection w at any point x (Figure 2.1a) are related:

$$M = -EI\frac{d^2w}{dx^2}. \tag{2.6}$$

The intensity q^* of the resultant transverse load at any position is equal to the algebraic sum of the external applied load and the elastic foundation reaction:

$$q^* = q + \bar{q} = q - kw. \tag{2.7}$$

The shearing force V, the bending moment, and the load intensity q^* are related:

$$V = \frac{dM}{dx} = -\frac{d}{dx}\left(EI\frac{d^2w}{dx^2}\right), \tag{2.8}$$

$$q^* = -\frac{dV}{dx} = -\frac{d^2M}{dx^2}. \tag{2.9}$$

Substituting Equations (2.6) and (2.7) into Equation (2.9) gives the differential equation relating the deflection of a beam on elastic foundation to the intensity of the external applied load:

$$\frac{d^2}{dx^2}\left(EI\frac{d^2w}{dx^2}\right)+kw=q. \tag{2.10}$$

When the beam has a constant flexural rigidity, the differential equation of the elastic line is

$$EI\frac{d^4w}{dx^4}+kw=q. \tag{2.11}$$

2.3 GENERAL SOLUTION OF THE DIFFERENTIAL EQUATION OF THE DEFLECTION OF A WALL OF CONSTANT THICKNESS

Consider a beam on elastic foundation, with EI and k constants, subjected to a concentrated load P, a couple C, and a distributed load of intensity q (Figure 2.2).

It can be shown[1] that the deflection at any point of the portion AG (between the left end and the first external applied load) is

$$w=w_0Y_1+\frac{\theta_0}{\beta}Y_2-\frac{M_0}{\beta^2EI}Y_3-\frac{V_0}{\beta^3EI}Y_4. \tag{2.12}$$

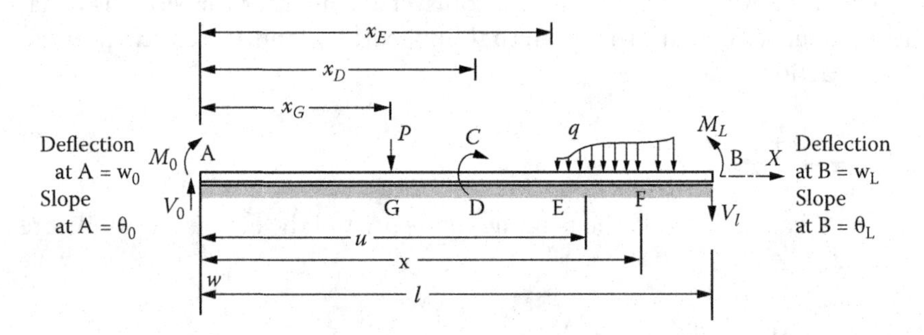

Figure 2.2 Derivation of Equation (2.16) for the deflection at any point in terms of the loads and the conditions at the left end of a beam on elastic foundation. M_0, V_0, M_l, and V_l are the bending moments and shearing forces at the two ends, indicated in their positive directions.

This equation, which satisfies the differential Equation (2.11), expresses w in terms of w_0, θ_0, M_0, and V_0, which are respectively the deflection, the rotation (dw/dx), the bending moment, and the shearing force at the left end of the beam. The Y's are functions of the dimensionless variable (βx), where β is called the characteristic factor of the beam:

$$\beta = \sqrt[4]{\left(\frac{k}{4EI}\right)}. \tag{2.13}$$

Since β has the dimension length^{-1}, the term ($1/\beta$) is called the characteristic length. The Y-functions are used by several authors and their values given in tables for different values of the variable.[2] The functions are defined next:

$$\begin{aligned}
Y_1 &= Y_1(\beta x) = \cosh\beta x \cos\beta x, \\
Y_2 &= Y_2(\beta x) = \tfrac{1}{2}(\cosh\beta x \sin\beta x + \sinh\beta x \cos\beta x), \\
Y_3 &= Y_3(\beta x) = \tfrac{1}{2}\sinh\beta x \sin\beta x, \\
Y_4 &= Y_4(\beta x) = \tfrac{1}{4}(\cosh\beta x \sin\beta x - \sinh\beta x \cos\beta x).
\end{aligned} \tag{2.14}$$

The Y-functions and their first derivatives are related:

$$Y_1' = -4\beta Y_4, \qquad Y_2' = \beta Y_1, \qquad Y_3' = \beta Y_2, \qquad Y_4' = \beta Y_3. \tag{2.15}$$

Successive differentiation results in similar sets of equations for higher derivatives of the function Y.

The third and fourth terms in Equation (2.12) represent, respectively, the effect of a couple M_0 and an upward force V_0 on the deflection at any point to the right of A. The couple C and the load P produce similar effects at sections to the right of their respective points of application. For example, noting that the factor of M_0 in Equation (2.12) is $(-1/\beta^2 EI)Y_3(\beta x)$, the couple C applied at D should have a modifying effect of $(-1/\beta^2 EI)Y_3[\beta(x - x_D)]$ on the deflection at points where $x \geq x_D$. The distributed load q can be treated as if consisting of infinitesimal concentrated forces. The effect of the edge forces and displacements at edge A together with the applied forces P, C, and q can be obtained by superposition.

For instance, at any point F, where $x_E \leq x \leq l$ (l = the total length of beam), the deflection is

$$w = w_0 Y_1(\beta x) + \frac{\theta_0}{\beta} Y_2(\beta x) - \frac{M_0}{\beta^2 EI} Y_3(\beta x) - \frac{V_0}{\beta^3 EI} Y_4(\beta x)$$

$$- \frac{C}{\beta^2 EI} Y_3[\beta(x - x_D)] + \frac{P}{\beta^3 EI} Y_4[\beta(x - x_G)] \tag{2.16}$$

$$+ \frac{1}{\beta^3 EI} \int_{x_E}^{x} q Y_4[\beta(x - u)]\, du,$$

where u is the distance from the left end to a variable point between E and F (Figure 2.2). By differentiation of Equation (2.16), the slope, the bending moment, and the shear at any section can be obtained (see Equations 2.6 and 2.8). For the use of Equation (2.16), the values w_0, θ_0, M_0, and V_0 at the left end A must be known. In practical application, two of these values are known as well as two of the four quantities—w_l, θ_l, M_l, and V_l—at the right end B. Substituting for $x = l$ in Equation (2.16) and using the two known quantities at B will give two equations from which the two unknown values at A can be determined.

2.4 CHARACTERISTIC PARAMETERS

When the equations in Section 2.3 are used for a circular-cylindrical wall of constant thickness, the characteristic factor β is obtained by substituting Equations (2.3) and (2.5) into Equation (2.13):

$$\beta = \frac{\sqrt[4]{[3(1-v^2)]}}{\sqrt{(rh)}}. \tag{2.17}$$

As discussed earlier, w, θ, M, and V will be the deflection, the slope, the bending moment, and the shear of a strip of unit width along a generatrix of the cylinder. The hoop force N can be calculated from w using Equation (2.1).

For any end conditions and loading, the variation of w, θ, M, and V along a generatrix of the wall is characterized by the dimensionless parameter

$$\alpha = \beta l \tag{2.18a}$$

or

$$\alpha = l \frac{\sqrt[4]{[3(1-v^2)]}}{\sqrt{(rh)}}, \tag{2.18b}$$

where l is the length of the cylinder. For concrete structures, Poisson's ratio is between 1/6 and 1/5; using the first of these two values,

$$\alpha = 1.3068 \frac{l}{\sqrt{(rh)}} \tag{2.18c}$$

or

$$\alpha = 1.848\sqrt{\eta}, \tag{2.19}$$

where

$$\eta = l^2/(dh) \tag{2.20}$$

and $d = 2r$ is the diameter of the middle surface of the wall.

All cylinders or beams on elastic foundations with the same value of α and loading have a similar variation of w, θ, M, and V, such that each of these quantities can be expressed as the product of a coefficient depending on α and an appropriate multiplier. This fact enabled the preparation of the design tables in Part II of this book in which the parameter η is used instead of α. The equivalent values of the two parameters can be calculated by Equation (2.19), which makes the tables usable for cylinders and more generally for beams on elastic foundation. This will be discussed further in Chapter 4, Section 4.8. The tables are prepared using a value of Poisson's ratio $v = 1/6$; varying this value to 1/5 has negligible effect. For a larger variation in the value of v, the tables are usable but with simple corrections discussed in Chapter 4, Section 4.7.

2.5 STIFFNESS AND FLEXIBILITY MATRICES

The solution discussed in the previous section is used to derive the stiffness and the flexibility matrices for a cylindrical wall or a beam on elastic foundation corresponding to the four coordinates in Figure 2.3a or Figure 2.3b. The displacement at the four coordinates can be expressed in terms of w; using Equation (2.12) and noting that $Y_1(0) = 1$ and $Y_2(0) = Y_3(0) = Y_4(0) = 0$,

$$\begin{Bmatrix} D_1 \\ D_2 \\ D_3 \\ D_4 \end{Bmatrix} = \begin{Bmatrix} (w)_{x=0} \\ \left(\dfrac{dw}{dx}\right)_{x=0} \\ (w)_{x=l} \\ \left(\dfrac{dw}{dx}\right)_{x=l} \end{Bmatrix} = \begin{bmatrix} 1 & 0 & 0 & 0 \\ 0 & \alpha/l & 0 & 0 \\ y_1 & y_2 & y_3 & y_4 \\ -4\alpha y_4/l & \alpha y_1/l & \alpha y_2/l & \alpha y_3/l \end{bmatrix} \begin{Bmatrix} w_0 \\ -\dfrac{\theta_0 l}{\alpha} \\ -\dfrac{M_0 l^2}{\alpha^2 EI} \\ -\dfrac{V_0 l^3}{\alpha^3 EI} \end{Bmatrix}, \tag{2.21}$$

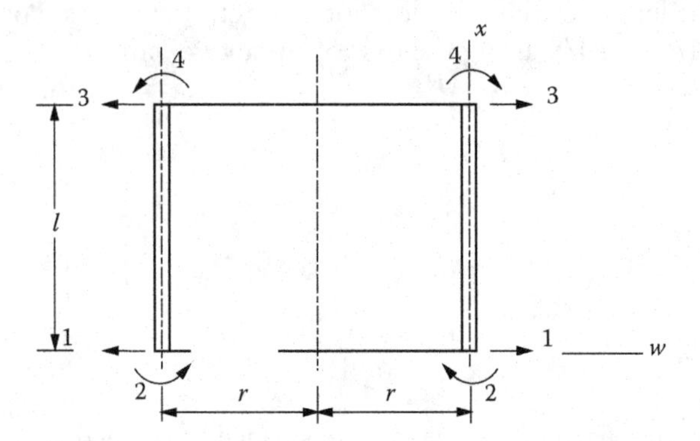

(a) Circular wall: The four arrows represent the positive directions of uniformly distributed forces {F} or displacements {D} at the edges

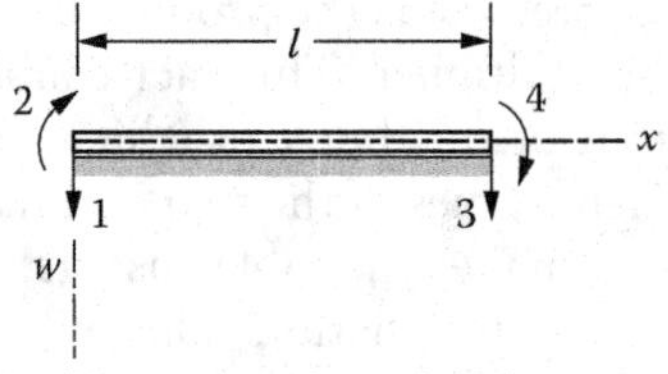

(b) Beam on elastic foundation

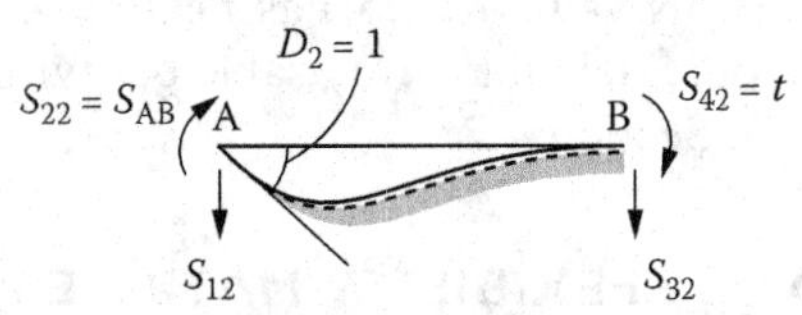

(c) End forces corresponding to the deflected configuration with $D_2 = 1$ and $D_1 = D_3 = D_4 = 0$

Figure 2.3 Coordinate system corresponding to the stiffness and flexibility matrices in Equations (2.27) and (2.28), and Tables 12.16 and 12.17.

where $y_i = Y_i(\alpha)$ are the values of the Y-functions at $x = l$ (see Equation 2.14). Equation (2.21) may be written in short form:

$$\{D\} = [C]\{B\}, \tag{2.22}$$

where [C] and {B} are the 4 × 4 and 4 × 1 matrices on the right-hand side of Equation (2.21).

Similarly, the forces $\{F\}$ at the four coordinates are related to the shearing force and the bending moment at the two ends; thus using Equations (2.12), (2.8), and (2.6),

$$\begin{Bmatrix} F_1 \\ F_2 \\ F_3 \\ F_4 \end{Bmatrix} = \begin{bmatrix} -(V)_{x=0} \\ (M)_{x=0} \\ (V)_{x=l} \\ -(M)_{x=l} \end{bmatrix} = EI \begin{bmatrix} 0 & 0 & 0 & \alpha^3/l^3 \\ 0 & 0 & -\alpha^2/l^2 & 0 \\ 4\alpha^3 y_2/l^3 & 4\alpha^3 y_3/l^3 & 4\alpha^3 y_4/l^3 & -\alpha^3 y_1/l^3 \\ 4\alpha^2 y_3/l^2 & -4\alpha^3 y_4/l^2 & \alpha^3 y_1/l^2 & \alpha^2 y_2/l^2 \end{bmatrix} \{B\} \quad (2.23)$$

or

$$\{F\} = [G]\{B\}, \quad (2.23a)$$

where $[G]$ is the product of EI and the square matrix in Equation (2.23).

Solving for $\{B\}$ from Equation (2.22) and substituting into Equation (2.23a),

$$\{F\} = [G][C]^{-1}\{D\}. \quad (2.24)$$

Putting

$$[S] = [G][C]^{-1}, \quad (2.25)$$

Equation (2.24) takes the form $[S]\{D\} = \{F\}$, where $[S]$ is the required stiffness matrix. Likewise, solving for $\{B\}$ from Equation (2.23a) and substituting into Equation (2.22), the latter equation takes the form $[f]\{F\} = \{D\}$, where

$$[f] = [C][G]^{-1} \quad (2.26)$$

is the flexibility matrix. Performing the matrix operations in Equations (2.25) and (2.26) gives the stiffness matrix of a circular wall or a beam on elastic foundation corresponding to the coordinate system in Figure 2.3a or Figure 2.3b:

$$[S] = EI \times$$

$$\begin{bmatrix} \left[\dfrac{\alpha^3(y_1y_2+4y_3y_4)}{12(y_3^2-y_2y_4)}\right]\dfrac{12}{l^3} & & & \text{symmetrical} \\ \left[\dfrac{\alpha^2(y_2^2-y_1y_3)}{6(y_3^2-y_2y_4)}\right]\dfrac{6}{l^2} & \left[\dfrac{\alpha(y_2y_3-y_1y_4)}{4(y_3^2-y_2y_4)}\right]\dfrac{4}{l} & & \\ -\left[\dfrac{\alpha^3y_2}{12(y_3^2-y_2y_4)}\right]\dfrac{12}{l^3} & \left[\dfrac{\alpha^2y_3}{6(y_3^2-y_2y_4)}\right]\dfrac{6}{l^2} & \left[\dfrac{\alpha^3(y_1y_2+4y_3y_4)}{12(y_3^2-y_2y_4)}\right]\dfrac{12}{l^3} & \\ \left[\dfrac{\alpha^2y_3}{6(y_3^2-y_2y_4)}\right]\dfrac{6}{l^2} & \left[\dfrac{\alpha y_4}{2(y_3^2-y_2y_4)}\right]\dfrac{2}{l} & \left[\dfrac{\alpha^2(y_2^2-y_1y_3)}{6(y_3^2-y_2y_4)}\right]\dfrac{6}{l^2} & \left[\dfrac{\alpha(y_2y_3-y_1y_4)}{4(y_3^2-y_2y_4)}\right]\dfrac{4}{l} \end{bmatrix} \tag{2.27}$$

and the flexibility matrix

$$[f] = \frac{l^3}{4\alpha^3EI(y_3^2-y_2y_4)} \times$$

$$\begin{bmatrix} y_2y_3-y_1y & & & \text{symmetrical} \\ -\alpha(y_2^2-y_1y_3)/l & \alpha^2(y_1y_2+4y_3y_4)/l^2 & & \\ -y_4 & \alpha y_3/l & y_2y_3-y_1y_4 & \\ -\alpha y_3/l & \alpha^2 y_2/l^2 & \alpha(y_2^2-y_1y_3)/l & \alpha^2(y_1y_2+4y_3y_4) \end{bmatrix}. \tag{2.28}$$

Each element of the stiffness matrix in Equation (2.27) includes a dimensionless term between square brackets; its value tends to unity when the foundation modulus k tends to zero (or each of β and $\alpha \to 0$). The stiffness matrix will then be reduced to that of an unsupported prismatic beam; hence $[S]$ becomes singular and cannot be inverted to obtain $[f]$.

The equation $[S]\{D\} = \{F\}$, or its inverse, $[f]\{F\} = \{D\}$, relates the values of four end forces to the corresponding four displacements. Usually, four of these eight values are known and one of these two equations can be solved to find the other four (e.g., see Figure 1.6 and Figure 1.7). When D_1, D_2, F, and F_2 are determined, the variation of w, θ, M, and V can be found by using Equation (2.12) and its derivatives, noting that $D_1 = w_0$, $D_2 = \theta_0$, $F_1 = -V_0$, and $F_2 = M_0$:

$$\begin{Bmatrix} w \\ \theta \\ M \\ V \end{Bmatrix} = \begin{bmatrix} Y_1 & (Y_2/\alpha)l & (Y_4/\alpha^3)(l^3/EI) & -(Y_3/\alpha^2)(l^2/EI) \\ -4\alpha Y_4/l & Y_1 & (Y_3/\alpha^2)(l^2/EI) & -(Y_2/\alpha)(l/EI) \\ 4\alpha^2 Y_3 EI/l^2 & 4\alpha Y_4 EI/l & -(Y_2/\alpha)l & Y_1 \\ 4\alpha^3 Y_2 EI/l^3 & 4\alpha^2 Y_3 EI/l^2 & -Y_1 & -4\alpha Y_4/l \end{bmatrix} \begin{Bmatrix} D_1 \\ D_2 \\ F_1 \\ F_2 \end{Bmatrix}. \tag{2.29}$$

2.6 END-ROTATIONAL STIFFNESS AND CARRY-OVER FACTOR

The end-rotational stiffness and the carry-over moments used in the method of moment distribution were defined in Chapter 1, Section 1.7.1. Figure 2.3c shows a beam on elastic foundation for which a unit angular displacement is introduced at end A while end B is fixed. The magnitudes of the end forces to hold the beam in this deflected shape are equal to the elements of the second column of the stiffness matrix corresponding to the coordinates in Figure 2.3b. The two end moments at A and B are the end-rotational stiffness and the carry-over moment; thus

$$S_{AB} = S_{22}, \tag{2.30}$$

$$t = S_{42}, \tag{2.31}$$

where S_{22} and S_{42} are elements of the stiffness matrix (Equation 2.27). The carry-over factor

$$C_{AB} = S_{42}/S_{22}. \tag{2.32}$$

The values of C_{AB} are given in Table 2.1 for various values of the characteristic parameter $\eta = l^2/(dh)$. It can be seen that C_{AB} becomes very small when $\eta \geq 3.0$ (or $\beta l \geq 3.2$).

Table 2.1 Carry-over factor for circular-cylindrical wall of constant thickness

η	0.0	0.4	0.8	1.2	1.6	2.0	3.0	4.0
$C_{AB} = C_{BA}$	0.5000	0.4785	0.4214	0.3459	0.2681	0.1983	0.0763	0.0160

2.7 FIXED-END FORCES

The solution discussed in Section 2.3 is used next to derive equations for the fixed-end forces, the variation of the deflection w [or the hoop force $N = (Eh/r)w$], and the bending moment M for a circular-cylindrical wall or a beam on elastic foundation subjected to cases of loading frequently used in practice.

2.7.1 Uniform or linearly varying load

The four fixed-end forces (Figure 2.4a or Figure 2.4b) are given by

$$\{F\} = [H]\begin{Bmatrix} q_1 \\ q_0 - q_1 \end{Bmatrix}, \tag{2.33}$$

where q_0 and q_1 are the load intensities at $x = 0$ and $x = l$, respectively (force per unit area for the cylinder or force per unit length for the beam); and $[H]$ is a 4 × 2 matrix for which the elements are

$$H_{11} = H_{31} = \left[\frac{y_1y_2 - y_2 + 4y_3y_4}{2\alpha(y_3^2 - y_2y_4)}\right]\left(-\frac{l}{2}\right) \tag{2.34}$$

$$H_{21} = -H_{41} = \left[\frac{3(y_1y_3 - y_3 + 4y_4^2)}{\alpha^2(y_3^2 - y_2y_4)}\right]\left(-\frac{l^2}{2}\right) \tag{2.35}$$

$$H_{12} = \left[\frac{\alpha^{-1}(y_1y_3 - y_2^2 - y_3) + y_1y_2 + 4y_3y_4}{1.4\alpha(y_3^2 - y_2y_4)}\right]\left(-\frac{7l}{20}\right) \tag{2.36}$$

$$H_{22} = \left[\frac{\alpha^{-1}(y_1y_4 - y_2y_3 - y_4) + y_1y_3 + 4y_4^2}{0.2\alpha^2(y_3^2 - y_2y_4)}\right]\left(-\frac{l^2}{20}\right) \tag{2.37}$$

$$H_{32} = H_{11} - H_{12} \tag{2.38}$$

$$H_{42} = H_{22} - H_{21}. \tag{2.39}$$

Each element of $[H]$ includes a dimensionless term written between square brackets in Equations (2.34) to (2.37); each of these terms tends to unity when the foundation modulus k tends to zero (or when β or α tends to zero). The elements in the first and second columns of $[H]$ will then

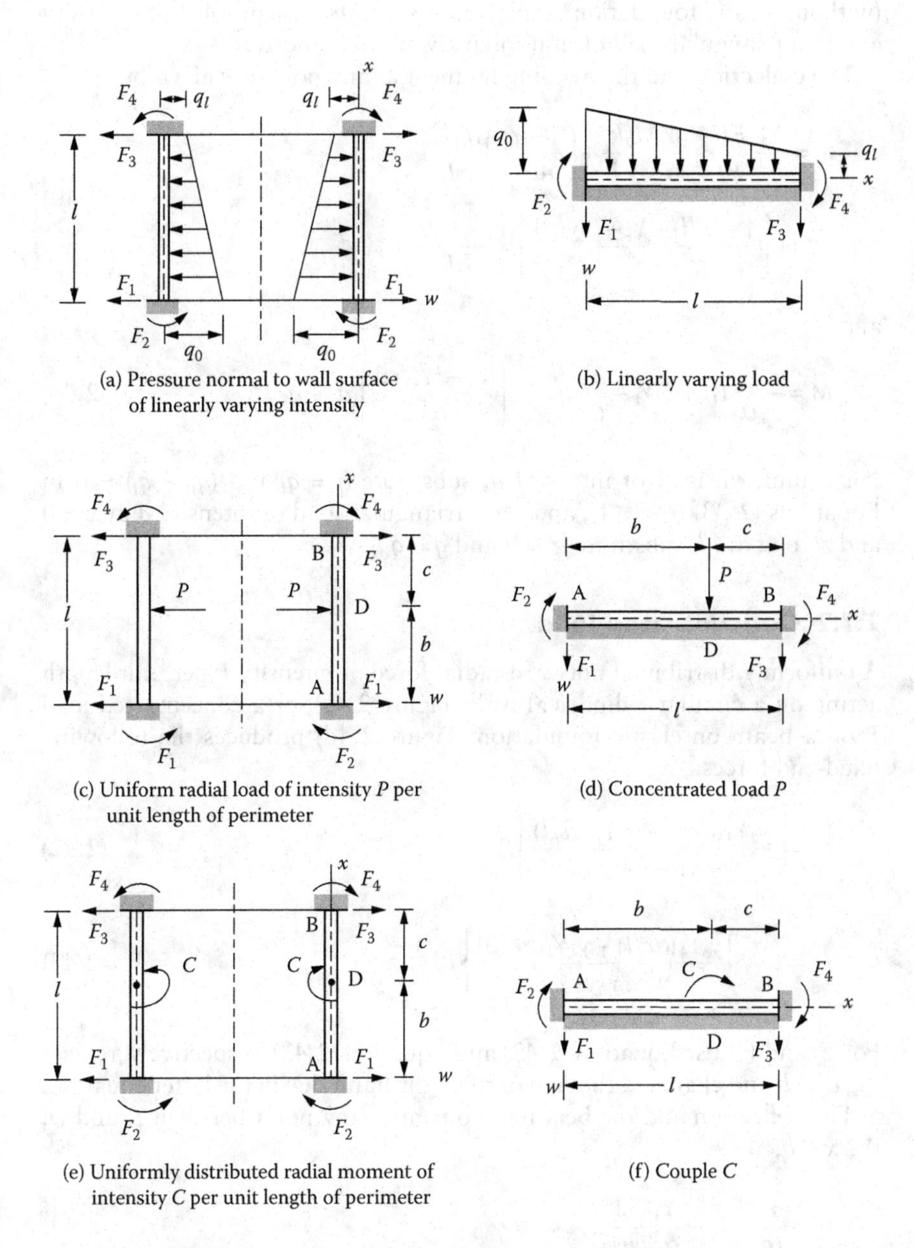

Figure 2.4 Fixed-end forces for a circular-cylindrical wall and for a beam on elastic foundation.

respectively be equal to the fixed-end forces for an ordinary prismatic beam (without elastic foundation) subjected to a uniform load of unit intensity and to a triangular load of unit intensity at $x = 0$ and zero at $x = l$.

The deflection and the bending moment at any point are given by

$$w = \frac{Y_4}{\alpha^3}\frac{F_1 l^3}{EI} - \frac{Y_3}{\alpha^2}\frac{F_2 l^2}{EI} + \left(\frac{1 - Y_1}{4\alpha^4}\right)\frac{q_l l^4}{EI} + \left[\frac{1 - (x/l) - Y_1 + (Y_2/\alpha)}{4\alpha^4}\right]\frac{(q_0 - q_l)l^4}{EI} \tag{2.40}$$

and

$$M = -\frac{Y_2}{\alpha}F_1 l + F_2 Y_1 - \frac{Y_3}{\alpha^2}q_l l^2 - \left[\frac{Y_3 - (Y_4/\alpha)}{\alpha^2}\right](q_0 - q_1)l^2. \tag{2.41}$$

For a uniform load of intensity q, substitute $q = q_l$ and $(q_0 - q_l) = 0$ in Equations (2.33) to (2.41), and for a triangular load of intensity q at $x = 0$ and zero at $x = l$, substitute $q_1 = 0$ and $q = q_0 - q_1$.

2.7.2 Concentrated load

A uniformly distributed outward radial force of intensity P per unit length acting on a circular-cylindrical wall (Figure 2.4c), or a concentrated load P on a beam on elastic foundation (Figure 2.4d) produces the following fixed-end forces:

$$F_1 = \left[\frac{y_2 Y_4(\alpha c/l) - y_3 Y_3(\alpha c/l)}{y_3^2 - y_2 y_4}\right]P, \tag{2.42}$$

$$F_2 = \left\{\frac{\alpha^{-1}[y_3 Y_4(\alpha c/l) - y_4 Y_3(\alpha c/l)]}{y_3^2 - y_2 y_4}\right\}Pl. \tag{2.43}$$

For F_3 and F_4, use Equation (2.42) and Equation (2.43), respectively, replacing c by b and changing the sign of the right-hand side in the latter equation.

The deflection and the bending moment at any point between A and D, $0 \le x \le b$, are

$$w = \frac{Y_4}{\alpha^3}\frac{F_1 l^3}{EI} - \frac{Y_3}{\alpha^2}\frac{F_2 l^2}{EI}, \tag{2.44}$$

$$M = -\frac{Y_2}{\alpha}F_1 l + F_2 Y_1, \tag{2.45}$$

and between D and B, $b \le x \le l$, are

$$w = \frac{Y_4}{\alpha^3}\frac{F_1 l^3}{EI} - \frac{Y_3}{\alpha^2}\frac{F_2 l^2}{EI} + \frac{Y_4[\alpha(x-b)/l]}{\alpha^3}\frac{Pl^3}{EI}, \tag{2.46}$$

$$M = -\frac{Y_2}{\alpha}F_1 l + F_2 Y_1 - \frac{Y_2[\alpha(x-b)/l]}{\alpha}Pl. \tag{2.47}$$

2.7.3 Couple

An external applied moment of intensity C per unit length applied in radial planes on a circular-cylindrical wall (Figure 2.4e), or a couple C on a beam on elastic foundation (Figure 2.4f) produces the following fixed-end forces:

$$F_1 = \frac{\alpha[y_3 Y_2(\alpha c/l) - y_2 Y_3(\alpha c/l)]}{y_3^2 - y_2 y_4}\frac{C}{l}, \tag{2.48}$$

$$F_2 = \frac{y_4 Y_2(\alpha c/l) - y_3 Y_3(\alpha c/l)]}{y_3^2 - y_2 y_4}C, \tag{2.49}$$

$$F_3 = \frac{\alpha[y_2 Y_3(\alpha b/l) - y_3 Y_2(\alpha b/l)]}{y_3^2 - y_2 y_4}\frac{C}{l}. \tag{2.50}$$

For F_4, use Equation (2.49), replacing b by c. The deflection and the bending moment at any point between A and D, $0 \le x \le b$, are given by Equations (2.44) and (2.45) with F_1 and F_2 as in Equations (2.48) and (2.49). For part DB (Figure 2.4f or Figure 2.4e),

$$w = \frac{Y_4}{\alpha^3}\frac{F_1 l^3}{EI} - \frac{Y_3}{\alpha^2}\frac{F_2 l^2}{EI} - \frac{Y_3[\alpha(x-b)/l]}{\alpha^2}\frac{Cl^2}{EI}, \tag{2.51}$$

$$M = -\frac{Y_2}{\alpha}F_1 l + F_2 Y_1 + \{Y_1[\alpha(x-b)/l]\}C. \tag{2.52}$$

2.8 SEMI-INFINITE BEAM ON ELASTIC FOUNDATION AND SIMPLIFIED EQUATIONS FOR LONG CYLINDERS

The analysis of the effect of the edge forces or displacements of a circular-cylindrical wall or of a beam on elastic foundation becomes considerably

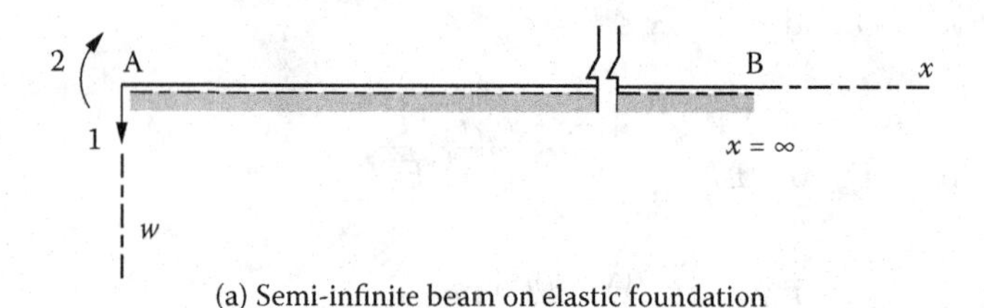

(a) Semi-infinite beam on elastic foundation

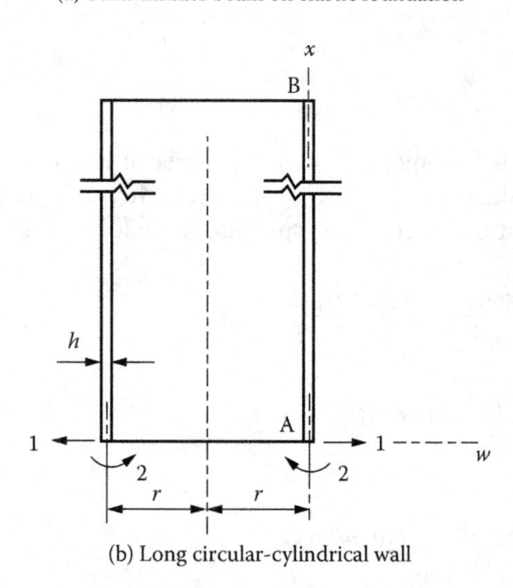

(b) Long circular-cylindrical wall

Figure 2.5 Coordinate system used for the derivation of Equations (2.53) to (2.58).

simpler when the value of the characteristic parameter $\alpha = \beta l$ becomes relatively high. For derivation of simplified equations, consider the semi-infinite beam or cylinder in Figure 2.5a or Figure 2.5b. It is reasonable to assume that the effect of forces F_1 and F_2 at edge A dies out gradually as x increases; in other words as $x \to \infty$, $w \to 0$ and $(dw/dx) \to 0$. Thus, it is possible to relate the forces and the displacements at the two coordinates (Figure 2.5) by the equation $[f]\{F\} = \{D\}$ or its inverse $[S]\{D\} = \{F\}$, where

$$[S] = \beta EI \begin{bmatrix} 4\beta^2 & 2\beta \\ 2\beta & 2 \end{bmatrix} \tag{2.53}$$

and

$$[f] = \frac{1}{4\beta^3 EI} \begin{bmatrix} 2 & -2\beta \\ -2\beta & 4\beta^2 \end{bmatrix}. \tag{2.54}$$

The last two equations can be derived by partitioning each of the 4 × 4 matrices in Equations (2.27) and (2.28) into four 2 × 2 submatrices. As α tends to ∞ the off-diagonal submatrices become null, indicating uncoupling of the forces (or of the displacements at the two ends), while the diagonal submatrices tend to the values in Equations (2.53) and (2.54). (For this derivation, it should be noted that for large values of α, sinh $\alpha \simeq$ cosh $\alpha \simeq e^{\alpha}/2$, $y_1 = (e^{\alpha} \cos^{\alpha})/2$ and so on.)

Substitution for F_1 and F_2 in terms of the displacements D_1 and D_2 (or vice versa) into Equation (2.29) results in the following two equations, which give the variation of w, θ, M, and V caused by displacements or forces introduced at the edge of a semi-infinite beam on elastic foundation or cylinder (Figure 2.5):

$$\begin{Bmatrix} w \\ \theta \\ M \\ V \end{Bmatrix} = \begin{bmatrix} Z_1 & Z_2/\beta \\ -2\beta Z_2 & Z_3 \\ 2\beta^2 EIZ_3 & 2\beta EIZ_4 \\ -4\beta^3 EIZ_4 & -2\beta^2 EIZ_1 \end{bmatrix} \begin{Bmatrix} D_1 \\ D_2 \end{Bmatrix} \tag{2.55}$$

or

$$\begin{Bmatrix} w \\ \theta \\ M \\ V \end{Bmatrix} = \begin{bmatrix} Z_4/(2\beta^3 EI) & -Z_3/(2\beta^3 EI) \\ -Z_1/(2\beta^2 EI) & Z_4/(\beta EI) \\ -Z_2/\beta & Z_1 \\ -Z_3 & -2\beta Z_2 \end{bmatrix} \begin{Bmatrix} F_1 \\ F_2 \end{Bmatrix}, \tag{2.56}$$

where the Z's are functions[3] of the variable (βx), defined as follows:

$$\begin{aligned} Z_1 &= Z_1(\beta x) = e^{-\beta x}(\cos\beta x + \sin\beta x), \\ Z_2 &= Z_2(\beta x) = e^{-\beta x}\sin\beta x, \\ Z_3 &= Z_3(\beta x) = e^{-\beta x}(\cos\beta x - \sin\beta x), \\ Z_4 &= Z_4(\beta x) = e^{-\beta x}\cos\beta x. \end{aligned} \tag{2.57}$$

The four functions are shown graphically in Figure 2.6, which also represent the shape in which w, θ, M, and V vary with x when a unit force

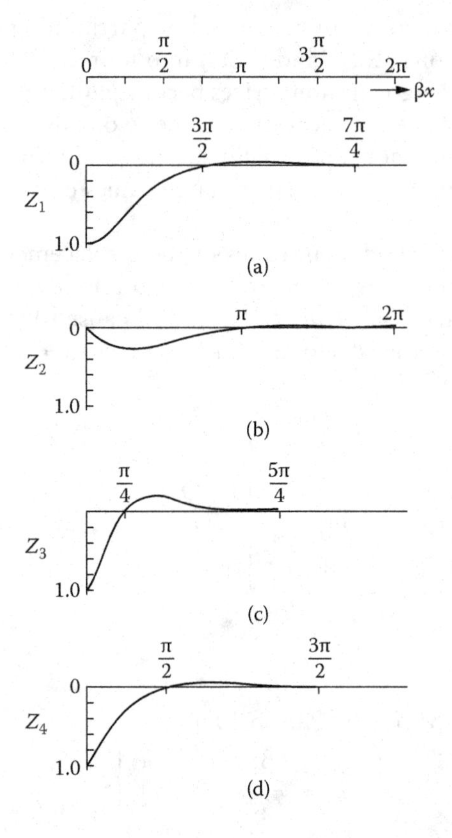

Figure 2.6 The Z-functions (defined by Equation 2.57) representing the shape of variation of *w*, θ, *M*, or *V* of a semi-infinite beam or cylinder due to forces or imposed displacements at the left-hand edge.

or a unit displacement is introduced at coordinates 1 or 2 (Figure 2.5). For example, Figure 2.6a,c have the same shapes as the deflection and the bending moment, respectively, of a semi-infinite beam with the imposed displacement $D_1 = 1$ while $D_2 = 0$.

The end-rotational stiffness S_{AB} of a semi-infinite beam is

$$S_{AB} = 2\beta EI. \tag{2.58}$$

This is the value of F_2 when $D_2 = 1$ and $D_1 = 0$, which is elements S_{22} of the stiffness matrix, Equation (2.53). The carry-over moment t and the carry-over factor C_{AB} are, of course, zero because the effect of $D_2 = 1$ dies out before the far end B is reached at infinity.

2.9 CLASSIFICATION OF BEAMS ON ELASTIC FOUNDATION AND CIRCULAR WALLS AS LONG OR SHORT

Any of the functions plotted in Figure 2.6 has a value at $\beta x = \pi$ much smaller than the maximum value of the function. It can, therefore, be suggested that when a beam on elastic foundation or a cylinder has a length $\geq \pi/\beta$, forces or displacements at one edge have a negligible effect on the other end. Such a beam or cylinder is considered here to be "long," and for its analysis the equations derived for a semi-infinite beam in Section 2.8 can be applied with negligible error. For "short" beams or cylinders having the characteristic parameter $\alpha = \beta l \leq \pi$, the deflection and the internal forces due to forces or displacements introduced at one end are affected by the conditions at the other end, so that the equations in Section 2.8 should not be used.

Hetényi[4] considers beams having the characteristic parameter $\alpha > \pi$ as "long" and suggests that, if greater accuracy is required, the limit may be changed to 5. The corresponding values of $\eta = l^2/(dh)$ for cylinders are 2.9 and 7.3, respectively (see Equation 2.19).

2.10 EXAMPLES

Examples of the use of the force and displacement methods for the analysis of circular walls are presented here. The equations in this chapter are used to calculate stiffness and flexibility coefficients and fixed-end forces, although some of these values are available from the design tables in Part II of this book.

EXAMPLE 2.1
Circular Wall, Hinged at the Bottom

A circular wall of constant thickness hinged at the bottom edge and free at the top is subjected to hydrostatic pressure of intensity γl at the bottom and zero at the top (Figure 2.7a); wall height = l, radius = $3l$, and thickness = $l/15$. Find the reaction at the bottom edge and derive an equation for the hoop force and the bending moment at any point. Poisson's ratio = 1/6.

This example is conveniently solved by the force method as follows:

1. The released structure is shown in Figure 2.7b, with one coordinate representing one unknown redundant force, the radial reaction at the base.
2. With both edges free, the wall deflects in a straight line,

$$w_s = \gamma l r^2 (l - x)/(lEh), \tag{2.59}$$

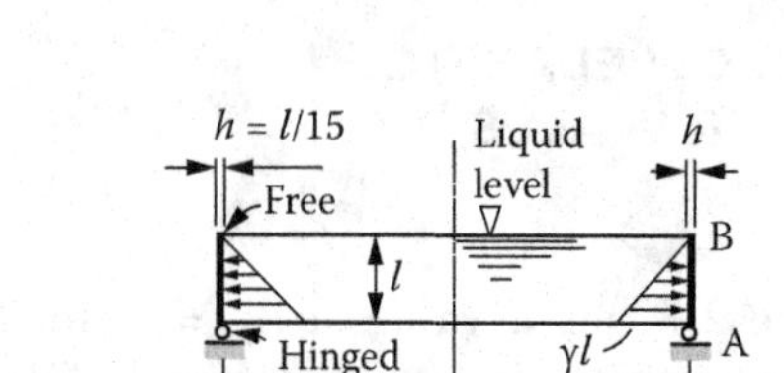

(a) Cross-section of the circular wall to be analyzed

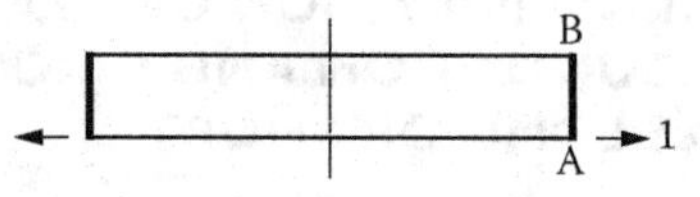

(b) Released structure and coordinate system (Example 2.1)

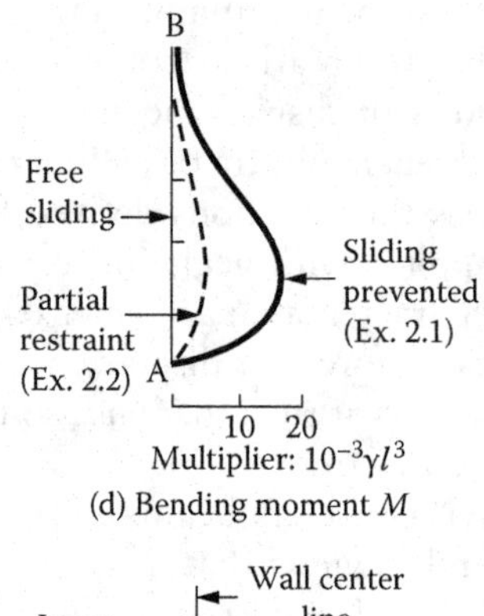

(d) Bending moment M

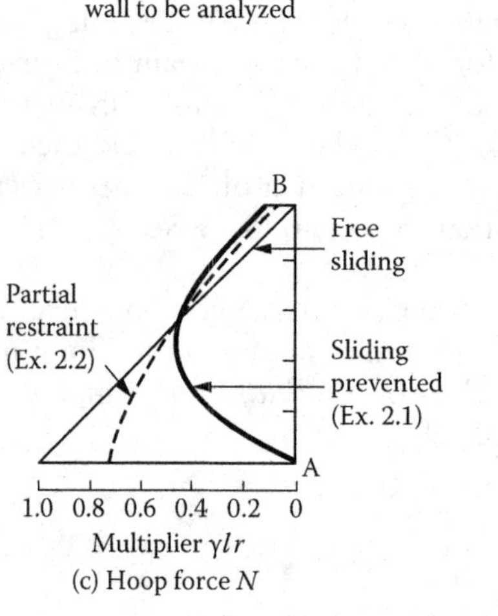

(c) Hoop force N

Wall center line
Inner face
A
Bearing pad

(e) Detail at A for wall of Example 2.2

Figure 2.7 Circular wall analyzed in Examples 2.1 and 2.2.

where x is the vertical distance from edge A to any point on the wall. The values of the required actions in the released structure are the hoop force and the bending moment. Using Equation (2.1),

$$\{A_s\} = \begin{Bmatrix} N_s \\ M_s \end{Bmatrix} = \begin{Bmatrix} \gamma lr(l-x)/l \\ 0 \end{Bmatrix}.$$

The displacement at coordinate 1 is

$$D_1 = \gamma l \frac{r^2}{Eh} = \gamma l \frac{(3l)^2}{E(l/15)} = 135 \frac{\gamma l^2}{E}.$$

3. The characteristic parameter (Equation 2.18c) is $\alpha = 2.9221$ and the flexural rigidity (Equation 2.5) is $EI = 25.40 \times 10^{-6} El^3$. One flexibility coefficient is needed; its value is the same as element f_{11} of the matrix in Equation (2.28):

$$f_{11} = 796.03/E$$

The deflection and moment at any point caused by $F_1 = 1$ are given by Equation (2.29), substituting for $D_1 = f_{11} = 796.03/E$, $D_2 = -2308.24/El$ (= element f_{21} of matrix in Equation 2.28), $F_1 = 1$, and $F_2 = 0$. Thus, using Equation (2.29) in this manner and noting that $N = (Eh/r)w$, the following equation is obtained:

$$\{A_u\} = \begin{Bmatrix} N_u \\ M_u \end{Bmatrix} = \begin{Bmatrix} \frac{1}{45}(796.03Y_1 - 789.92Y_2 + 1578.10Y_4) \\ l(0.6905Y_3 - 0.6852Y_4 - 0.3422Y_2) \end{Bmatrix}.$$

4. The value of the redundant force (Equation 1.4) is

$$F_1 f_{11}^{-1}(-D_1) = -0.169\gamma l^2 = -0.0565\gamma lr.$$

5. By superposition, Equation (1.5) gives

$$\{A\} = \begin{Bmatrix} N \\ M \end{Bmatrix} = \begin{Bmatrix} \gamma lr\left[\left(\frac{l-x}{l}\right) - Y_1 + 0.9924Y_2 - 1.9826Y_4\right] \\ \gamma l^3[-0.1171Y_3 + 0.1162Y_4 + 0.0580Y_2] \end{Bmatrix}.$$

The graphs in Figure 2.7c,d show the variation of N and M.

EXAMPLE 2.2
Circular Wall, on Bearing Pads

Solve Example 2.1 assuming that the wall is supported on bearing pads that behave as continuous elastic support providing uniform radial horizontal reaction in the direction opposite to the deflection. The stiffness of the elastic support is $0.5 \times 10^{-3}E$ (force per unit length of the perimeter per unit deflection = force/length^2), where E is the modulus of elasticity of the wall material.

The solution is based on the same steps as in Example 2.1 modified as follows:

The redundant force is represented by a pair of opposite arrows (Figure 2.7e), which also indicate the relative movement of the wall and the top of the pads in the released structure. The flexibility coefficient in this case is equal to the sum of the deflection of the wall and that of the pads due to two equal and opposite radial forces of intensity unity applied at coordinates 1:

$$f_{11} = \frac{796.03}{E} + \frac{1}{0.5\times 10^{-3}E} = \frac{2796.03}{E}.$$

The value of the redundant force (Equation 1.4) is

$$F_1 = f_{11}^{-1}(-D_1) = \frac{F}{2796.03}\left(-\frac{135\gamma l^2}{E}\right)$$

$$= 0.0483\gamma l^2 = -0.0161\gamma lr.$$

Using superposition as in step 5 of Example 2.1 gives the values of N and M, which are plotted in Figure 2.7c,d. The figures indicate the variation of N and M when the base of the wall is free to slide, or when sliding is prevented (Example 2.1) or partially restrained by the pads.

EXAMPLE 2.3
Circular Wall, with Thickness Change

Figure 2.8a is a cross-section of a circular wall with sudden change in thickness at midheight. The top edge is free while the bottom edge is totally fixed. Find the values of the bending moment at A and B and the hoop force at B and C due to axisymmetrical triangular loading as

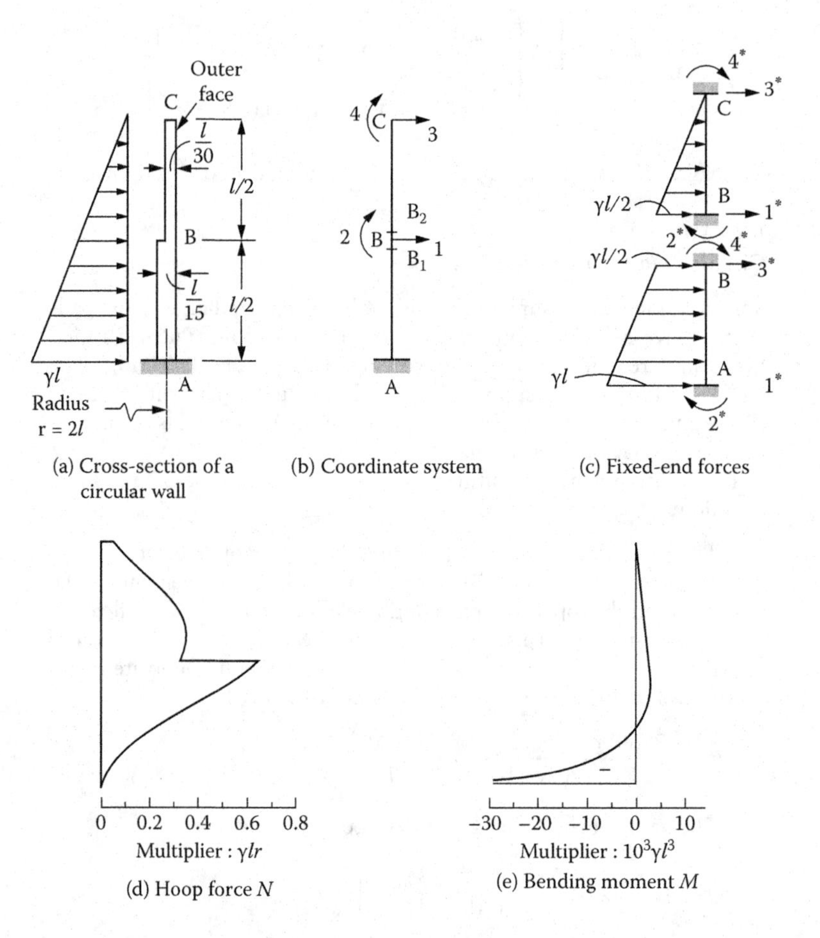

Figure 2.8 Circular wall with sudden change in thickness analyzed in Example 2.3.

shown. (Wall height = l; radius = $2l$; wall thicknesses = $l/15$ and $l/30$; Poisson's ratio = 1/6.)

The solution is obtained by using the five steps of the displacement method as follows:

1. The degrees of freedom are represented by the four coordinates in Figure 2.8b. Five actions are to be calculated:

$$\{A\} = \{M_A, M_{B1}, N_{B1}, N_{B2}, N_C\}.$$

The subscripts B1 and B2 refer to two points just below and above B; the bending moments at these two points are equal but the hoop forces are of different values because of the change in thickness (see Equation 2.1).

2. The characteristic values for the two cylinders AB and BC are (using Equation 2.18c): α_{AB} = 1.7894 and α_{BC} = 2.5306. The fixed-end forces for the two parts are (see Equation 2.33 and Figure 2.8c)

$$\{F_r^*\}_{AB} = \frac{\gamma}{1000}\begin{Bmatrix} -202.49l^2 \\ -15.60l^3 \\ -152.76l^2 \\ +13.53l^3 \end{Bmatrix} \qquad \{F_r^*\}_{BC} = \frac{\gamma}{1000}\begin{Bmatrix} -76.23l^2 \\ -5.06l^3 \\ -27.28l^2 \\ +3.05l^3 \end{Bmatrix}.$$

The subscript r is used to indicate restrained elements. These end forces are assembled by inspection to give the four forces necessary to prevent the displacement at the four coordinates in Figure 2.8b:

$$\{F\} = 10^{-3}\gamma\left\{-228.99l^2, 8.47l^3, -27.28l^2, 3.05l^3\right\}.$$

With the displacements artificially restrained, the values of the five actions to be calculated are

$$\{A_r\} = 10^{-3}\gamma\left\{-15.60l^3, -13.53l^3, 0,\ 0,\ 0\right\}.$$

As usual, the bending moment is considered positive when it produces tension at the outer face of the wall.

3. Using Equation (2.5),

$$(EI)_{AB} = 25.40\times10^{-6}El^3 \quad \text{and} \quad (EI)_{BC} = 3.175\times10^{-6}El^3,$$

and the stiffness matrices for the individual cylinders (given by Equation 2.27) are

$$[S^*]_{AB} = 10^{-6}E\begin{bmatrix} 5417.3 & & \text{symmetrical} & \\ 815.6l & 221.7l^2 & & \\ -1470.5 & -491.9l & 5417.3 & \\ 491.9l & 87.9l^2 & -815.6l & 221.7l^2 \end{bmatrix}$$

$$[S^*]_{BC} = 10^{-6}E\begin{bmatrix} 1661.5 & & \text{symmetrical} & \\ 165.4l & 33.2l^2 & & \\ 63.3 & -30.2l & 1661.5 & \\ 30.2l & 7.3l^2 & -165.4l & 33.2l^2 \end{bmatrix}.$$

By inspection, the last two matrices are assembled[5] to give the stiffness matrix of the wall corresponding to the coordinates in Figure 2.8b:

$$[S] = 10^{-6}E\begin{bmatrix} 7078.8 & & & \\ -650.2l & 254.9l^2 & & \\ 63.3 & -30.2l & 1661.5 & \\ 30.2 & 7.3l & -165.4l & 33.2l^2 \end{bmatrix}.$$

The values of the actions due to unit displacements are listed in the matrix:

$$[A_u] = 10^{-6}E\begin{bmatrix} -491.9l & 87.9l^2 & 0 & 0 \\ 815.6l & -221.7l^2 & 0 & 0 \\ 33333.3 & 0 & 0 & 0 \\ 16666.7 & 0 & 0 & 0 \\ 0 & 0 & 16666.7 & 0 \end{bmatrix}.$$

The first two rows representing M_A and M_B are elements of the matrix $[S^*]_{AB}$ with the sign adjusted to follow the convention adopted for bending moment. The last three rows of $[A_u]$ representing the hoop forces are calculated from the deflection w, using Equation (2.1).

4. The solution of Equation (1.1) gives

$$\{D\} = (\gamma/E)\left\{39.38l^2,\ 71.18l,\ 3.89l^2,\ -123.78l\right\}.$$

5. The superposition Equation (1.3) gives

$$\{A\} = 10^{-3}\gamma\left\{-28.71l^3,\ 2.81l^3,\ 656.33lr,\ 328.17lr,\ 32.42lr\right\}.$$

The variations of N and M are shown in Figure 2.8d,e. These are obtained by adding the values for each of parts AB and BC subjected to the external load with its ends free and the effect of edge forces and displacements of known magnitudes using Equation (2.29).

2.11 GENERAL

The analytical solutions presented involve the use of the Y-functions and the Z-functions for short and long beams (or cylinders), respectively (Equations 2.14 and 2.57). With the use of programmable calculators or computers, the use of the functions can be considered for practical computations with or without employment of tabulated values of the functions.

In the following two chapters numerical solution using finite differences is used to prepare tables for the analysis of cylinders of varying thickness (see Part II of this book), with the special case of constant thickness treated in the same way and included in the tables. Some engineers may prefer to use the tables even for constant-thickness walls rather than use the analytical solutions in this chapter. Although the values given in the tables are accurate enough for all practical applications, they involve the approximation inevitable with numerical solutions. The closed-form mathematical solutions in this chapter may be used by the reader who wishes to verify or assess the error involved in the tables. They are also useful to the engineer who wants to incorporate the equations in his or her own computer programs, and, of course, to derive values or solutions for cases not included in the tables.

NOTES

1. Hetényi, M. (1946). *Beams on Elastic Foundation*, University of Michigan Press, Ann Arbor, p. 6.
2. See, for example, Vlasov, V. Z., 1964, *General Theory of Shells and its Applications in Engineering*, translation from Russian, NASA, Washington, DC, pp. 860–77.
3. Tables for values of the Z-functions are given by Hetényi, pp. 217–39.
4. See Hetényi, pp. 46–7.
5. See Ghali, A., Neville, A. M., and Brown, T. G., 2009, *Structural Analysis: A Unified Classical and Matrix Approach*, 6th ed., Spon Press, London.

Chapter 3

Circular walls of varying thickness

3.1 INTRODUCTION

The differential equation relating the deflection of a circular-cylindrical wall of variable thickness to the intensity of a radial axisymmetrical loading is derived in Chapter 2, Section 2.2 (Equation 2.10). A closed-form solution to the equation is available[1] for the case of a wall with linearly varying thickness subjected to an axisymmetrical distributed load of linearly varying intensity. The solution is usually given as the sum of a particular integral and complementary solution of the homogeneous equation obtained by setting the right-hand side of Equation (2.10) equal to zero. It is assumed that only the second part of the solution implies bending deformation, while the contribution of the particular solution to the bending moment is ignored. The error induced by this approximation may not be negligible in very short cylinders.

For arbitrary variation in thickness or loading, numerical methods of analysis are more suitable. The finite-element method, discussed in Chapter 5, can be adopted for the analysis of shells of revolution of any shape. The finite-difference method is less general but adequate for the solution of Equation (2.10). Both methods involve the solution of a set of linear simultaneous equations, but for the same degree of accuracy the number of equations is smaller and their preparation simpler when the finite-difference method is used. For these reasons, the method of finite differences is selected for the preparation of a set of tables (presented in Part II of this book) to be used in the design of circular tank walls of variable thickness and also for the more general problem of a beam on elastic foundation.

In this chapter, the method of finite differences is discussed in sufficient detail for the sake of practical use in cases that are not covered by the tables.

It is often important in the design of tanks to consider the internal forces caused by temperature variation. Shrinkage of concrete walls has a similar effect to drop of temperature and can be analyzed in a similar way. In Section 3.8 the analysis for thermal effects is discussed for circular-cylindrical walls of varying thickness.

3.2 FINITE-DIFFERENCE EQUATIONS

The procedure for the analysis of circular walls of variable thickness or beams on elastic foundations of variable flexural rigidity and foundation modulus by finite differences is presented here.[2] The governing differential Equation (2.10) (see Chapter 2) is put in a finite-difference form, which, when applied at a node, relates the deflections w at this node and at equally spaced nodes in its vicinity to the external applied load. Application of the finite-difference equation at each node where the deflection is unknown provides a sufficient number of simultaneous linear equations whose solution gives the deflection at the nodes. At a general node i away from the edges, the finite-difference form of Equation (2.10) is (Figure 3.1a or Figure 3.1b):

$$\frac{EI_b}{\lambda^3}\left[J_{i-1} \quad -2(J_{i-1}+J_i) \quad (J_{i-1}+4J_i+J_{i+1}+C_i) \quad -2(J_i+J_{i+1}) \quad J_{i+1}\right]$$

$$\begin{Bmatrix} w_{i-2} \\ w_{i-1} \\ w_i \\ w_{i+1} \\ w_{i+2} \end{Bmatrix} \cong Q_i \tag{3.1}$$

where

$$C_i = k_i\lambda^4/(EI_b). \tag{3.2}$$

Here, λ is the spacing between the nodes; k_i is the value of the modulus of the elastic foundation at node i; for a circular wall (see Equation 2.3)

$$k_i = Eh_i/r^2 \tag{3.3}$$

where h_i is the thickness at node i and r is the radius of the cylinder. The J_s in the one-row matrix in Equation (3.1) are the ratios of the flexural rigidity at individual nodes to an arbitrarily chosen flexural rigidity EI_b; thus for node i:

$$J_i = EI_i/EI_b. \tag{3.4}$$

For a circular wall, EI_b may be conveniently chosen at the bottom edge; using Equation (2.5),

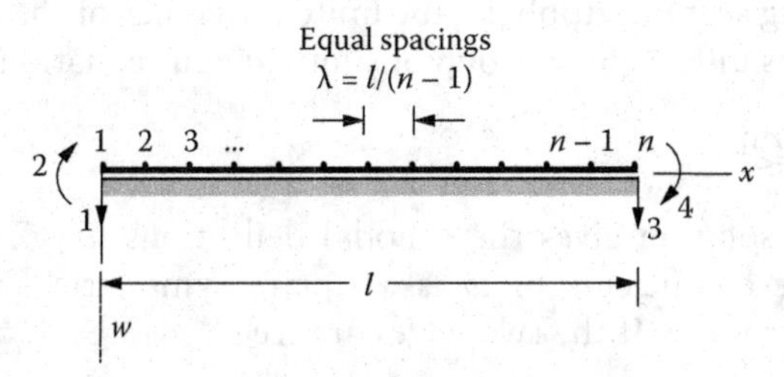

(a) Beam on an elastic foundation of variable flexural rigidity and foundati modulus

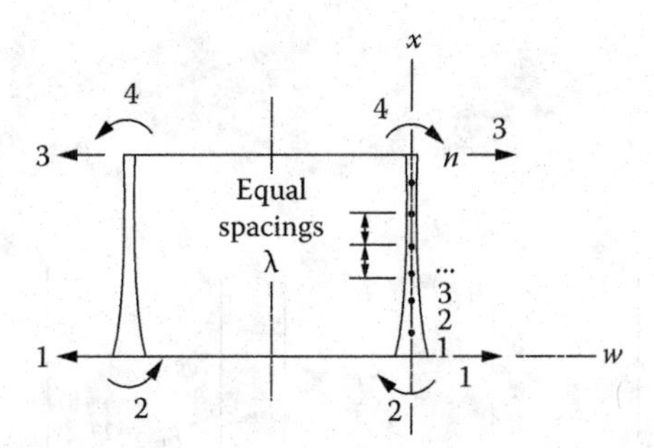

(b) Circular wall of variable thickness

Figure 3.1 Equally spaced nodes for the analysis of circular walls or beams on elastic foundation using the finite-difference method. Arrows at the edges represent a coordinate system corresponding to the flexibility and stiffness matrices derived in Section 3.7.

$$EI_b = Eh_b^3 / \left[12(1-\nu^2)\right] \tag{3.5}$$

and

$$J_i = h_i^3 / h_b^3 \tag{3.6}$$

where h_b is the thickness at the bottom edge.

Q_i is an equivalent concentrated load at node i, which may be simply taken equal to $q_i\lambda$, where q_i is the external load intensity at node i, or more accurately Q_i equals the sum of the reactions[3] at node i of two simple beams of span λ loaded by the external loading q.

Equation (3.1) relates the equivalent external applied load Q_i to the deflections at node i and at two adjacent nodes on either side of i. When i is on or adjacent to the edge, the coefficients of w in the finite-difference equation are adjusted according to the boundary conditions, as will be discussed in the following section. Applying the finite difference of the n nodes where the deflection is unknown, a group of simultaneous equations is generated:

$$[K]\{w\} \simeq \{Q\} \tag{3.7}$$

for which the solution gives the n nodal deflections $\{w\}$. The matrix $[K]$ formed by the coefficients of w is a square symmetrical pentadiagonal matrix of order $n \times n$. If the two edges are free,

$$[K] = \frac{EI_b}{\lambda^3} \times$$

$$\begin{bmatrix}
J_2 + \frac{C_1}{2} & & & & & & & \text{symmetrical} \\
-2J_2 & 4J_2+J_3+C_2 & & & & & & \\
\cdots & \cdots & \cdots & & & & & \\
 & & J_{i-1} & -2J_{i-1}-2J_i & J_{i-1}+4J_i+J_{i+1}+C_i & & & \\
 & & & \cdots & \cdots & \cdots & & \\
 & & & & & J_{n-2} & -2J_{n-2}-2J_{n-1} & J_{n-2}+4J_{n-1}+C_{n-1} \\
\text{elements not shown are a zero} & & & & & & J_{n-1} & -2J_{n-1} & J_{n-1}+\frac{C_n}{2}
\end{bmatrix} \tag{3.8}$$

The elements of a typical row of $[K]$ in Equation (3.8) are the coefficients of w in the finite-difference Equation (3.1) applied at node i away from the edges. The elements in the first and last two rows are coefficients of w in the adjusted forms of this equation to account for the boundary conditions, as will be discussed in the following section.

When the values of w are determined, the hoop force at any node i is calculated by Equation (2.1):

$$N_i = (Eh_i/r)w_i \tag{3.9}$$

and the bending moment is calculated by using the finite-difference form of Equation (2.6):

$$M_i \cong (EI_b/\lambda)J_i(-w_{i-1} + 2w_i - w_{i+1}). \tag{3.10}$$

The shear is more conveniently calculated midway between the nodes. The finite-difference form of Equation (2.8) applied at $i+\frac{1}{2}$ midway between nodes i and $i + 1$ is

$$V_{i+1/2} \cong (EI_b/\lambda^3)[J_i w_{i-1} - (2J_i + J_{i+1})w_i + (J_i + 2J_{i+1})w_{i+1} - J_{i+1}w_{i+2}]. \tag{3.11}$$

The reactions at a supported node can also be calculated from the deflections, using equations from the following section.

3.3 BOUNDARY CONDITIONS AND REACTIONS

As mentioned earlier, application of Equation (3.1) at a node i includes the deflection at two adjacent nodes on either side of i (see Table 3.1a). When i is at or near an edge, fictitious nodes outside the wall (or the beam) are included. The boundary conditions are used to eliminate the fictitious deflections by expressing their values in terms of the deflections at other nodes. Three end conditions need to be considered.

At a simply supported end, the deflection and the bending moment are zero. These two conditions are satisfied if the beam is considered to be continuous with another similar beam subjected to similar loading acting in the opposite direction. Thus, with reference to Table 3.1b, the deflection at a fictitious node $i - 2$ (not shown in the table) is $w_{i-2} = -w_i$. Thus, the finite-difference Equation (3.1) when applied at node i takes the form of Equation (3.12) written in the table.

The conditions for a fixed end, that the deflection and the slope are zero, are satisfied if the beam is considered to be continuous with a similar beam having the same loading as the actual beam. Thus, $w_{i-2} = w_i$ and the finite-difference Equation (3.1) when applied at node i adjacent to a fixed end takes the form of Equation (3.13) in the table.

The bending moment at node $i + 1$ near a free edge (Table 3.1d) is $M_{i+1} = -Q_i\lambda + k_i w_i(\lambda^2/2)$. Expressing M_{i+1} in terms of the deflection using Equation (3.10) gives the finite-difference Equation (3.14) applied at node i, the free end. In Table 3.1e, the bending moment at the free end is zero; $M_{i-1} = 0$ which, when put in finite-difference form, gives the deflection at a fictitious point $w_{i-2} = 2w_{i-1} - w_i$. This relation is then used to eliminate w_{i-2} from Equation (3.1), thus giving the finite-difference Equation (3.15) applied at node i adjacent to a free end (Table 3.1e).

If there is a spring support at node i of stiffness K_i (force/length^2 for circular wall or force/length for a beam on elastic foundation), add K_i to the coefficient of w in any of the equations in Table 3.1. An example of this type of support that occurs in practice is in circular prestressed concrete tanks where the wall is supported by elastomeric bearing pad providing a radial reaction proportional to the deflection w.

The equations derived above and listed in Table 3.1 express the equivalent concentrated load Q_i at node i (force/length for circular walls or force for a beam) to the deflections at adjacent nodes. In a similar way, the reaction R_i at a supported node can be expressed in terms of the nodal deflections. Three equations[4] are presented in Table 3.2 for the reaction at an intermediate, a hinged, and a totally fixed support. When the deflection

Table 3.1 Finite-difference equations relating deflection of a circular wall or beam on elastic foundation to the applied force

Position of node i	*Coefficients of the deflection in terms of* EI_b/λ^3					*Right-hand side*	*Equation number*
	w_{i-2}	w_{i-1}	w_i[a]	w_{i+1}	w_{i+2}		
(a)	J_{i-1}	$-2(J_{i-1} + J_i)$	$J_{i-1} + 4J_i + J_{i+1} + C_i$	$-2(J_i + J_{i+1})$	J_{i+1}	$\cong Q_i$	(3.1)
(b) Hinged support	—	—	$4J_i + J_{i+1} + C_i$	$-2(J_i + J_{i+1})$	J_{i+1}	$\cong Q_i$	(3.12)
(c) Fixed support	—	—	$2J_{i-1} + 4J_i + J_{i+1} + C_i$	$-2(J_i + J_{i+1})$	J_{i+1}	$\cong Q_i$	(3.13)
(d) Free end	—	—	$J_{i+1} + (C_i/2)$	$-2J_{i+1}$	J_{i+1}	$\cong Q_i$	(3.14)
(e) Free end	—		$-2J_i\ 4J_i + J_{i+1} + C_i$	$-2(J_i + J_{i+1})$	J_{i+1}	$\cong Q_i$	(3.15)

[a] If there is an elastic spring at i of stiffness K_i (force/length^2 for circular wall or force/length for a beam), add K_i to the coefficient of W_i in all equations.

Table 3.2 Finite-difference equations relating deflection of a circular wall or beam on elastic foundation to the reaction

Type of support	Coefficients of the deflection in terms of EI_b/λ^3					Right-hand side	Equation number
	w_{i-2}	w_{i-1}	w_i	w_{i+1}	w_{i+2}		
(a) Intermediate support	$-J_{i-1}$	$2(J_{i-1}+J_i)$	$-(J_{i-1}+4J_i+J_{i+1}+C_i)$	$2(J_i+J_{i+1})$	$-J_{i+1}$	$\cong (R_i - Q_i)$	(3.16)
(b) Hinged end	—	—	$-\left(J_{i+1}+\frac{C_i}{2}\right)$	$2J_{i+1}$	$-J_{i+1}$	$\cong (R_i - Q_i)$	(3.17)
(c) Fixed-end moment (given by Equation 3.19) Rotation at *i* is prevented	—	—	$-\left(2J_i+J_{i+1}+\frac{C_i}{2}\right)$	$2(J_i+J_{i+1})$	$-J_{i+1}$	$\cong (R_i - Q_i)$	(3.18)

at the support is zero, set $w_i = 0$ in each of the three equations. The fixed-end moment at end i in Table 3.2(c) can be calculated by Equation (3.10) substituting $w_{i-1} = w_{i+1}$,

$$M_i \cong \frac{EI_b}{\lambda^2} J_i(2w_i - 2w_{i+1}). \quad (3.19)$$

3.4 GENERATION OF SIMULTANEOUS EQUATIONS

Table 3.1 is intended to simplify the generation of the matrix $[K]$ in Equation (3.7) for which the solution provides the nodal deflections. As an example, when the two edges are free (see Figure 3.1 and Equation 3.8), the coefficients in the first and last two rows of $[K]$ are the coefficients of w in Equations (3.14) and (3.15). If the edge at node 1 (or n) is simply supported, the first (or last) row and column of the matrix in Equation (3.8) are to be deleted (resulting in a smaller matrix). Similarly, if the deflection is prevented by the introduction of a support at any intermediate node j, the order of $[K]$ is reduced by the deletion of the jth row and column. Examples 3.1 and 3.2 are chosen because all the conditions discussed here and in the previous section are applied.

As mentioned before, the elements of the vector $\{Q\}$ in Equation (3.3) are equal to the sum of the reactions at the nodes of simple beams of span λ loaded by external loading q (force/length^2 for a circular wall and force/length for a beam). For any continuous variation of the load intensity, the following equation may be used:

$$\{Q\} = \lambda \left\{ \frac{7q_1 + 6q_2 - q_3}{24}, \ldots, \frac{q_{i-1} + 10q_i + q_{i+1}}{12}, \ldots \frac{-q_{n-2} + 6q_{n-1} + 7q_n}{24} \right\} \quad (3.20)$$

which gives Q_1 and Q_n at the edge nodes and Q_i at a typical node i. This involves the assumption of parabolic variation of q over any three consecutive nodes.

3.5 SUDDEN CHANGE IN THICKNESS

When a circular wall or a beam on elastic foundation has a sudden change in flexural rigidity at a node i, an effective value EI_i is to be used in the finite-difference equations in Tables 3.1 and 3.2. The effective flexural rigidity is given by the equation[5]

$$EI_i = \frac{2EI_{iL}E_{iR}}{EI_{iL} + EI_{iR}} \quad (3.21)$$

in which the subscripts L and R refer to the two sections just to the left and right of node i. A sudden change in thickness in a circular wall corresponds also to a sudden change in k value; at node i where the change occurs, k is to be taken as the average of the values just before and after i.

3.6 EXAMPLES

EXAMPLE 3.1
Circular Wall, on Bearing Pads, with Linearly Varying Load

A circular concrete wall (Figure 3.2a) is subjected to circumferential prestressing providing an inward load normal to the surface of linearly varying intensity covering half the wall height. The top edge of the wall is free and the bottom edge is supported on bearing pads providing an elastic support of stiffness K = $0.6 \times 10^{-3}E$, where E is the modulus of elasticity of the wall material. Find the bending moment and the hoop force variation over the wall height as well as the value of the shearing force at the bottom edge. The wall dimensions are shown in the figure; Poisson's ratio, $\nu = 0.2$.

The loading covering half the wall only represents the condition during the application of the prestress. In some cases, loading with such partial prestress may produce a larger moment than when the prestress is completed covering the whole height.

Nine equally spaced nodes are shown in Figure 3.2b, with $\lambda = l/8$. At each node the thickness h_i and the values EI_i, J_i, and C_i are calculated, using Equations (3.2) to (3.6). The equivalent concentrated loads Q_i are listed in the last column of the table in Figure 3.2b (see Equation 3.20).

Applying the appropriate finite-difference equation from Table 3.1 at each of the nodes 1, 2, ..., 9 gives a group of simultaneous equations $[K]\{w\} = \{Q\}$ in which

$$[K] = \frac{EI_b}{\lambda^3} \times$$

$$\begin{bmatrix}
2.0546 & & & & & & & & \\
-1.7602 & 5.8432 & & & & & & & \\
0.8801 & -3.3008 & 6.1162 & & & \text{symmetrical} & & & \\
 & 0.7703 & -2.8804 & 5.4461 & & & & & \\
 & & 0.6699 & -2.4972 & 4.8309 & & & & \\
 & & & 0.5787 & -2.1498 & 4.2679 & & & \\
 & & & & 0.4962 & -1.8362 & 3.7542 & & \\
\text{elements not shown are zero} & & & & & 0.4219 & -1.5546 & 2.9910 & \\
 & & & & & & 0.3554 & -0.7108 & 0.8954
\end{bmatrix}$$

It can be seen that this matrix is calculated by substitution in Equation (3.8), which is applicable when the two edges of the wall

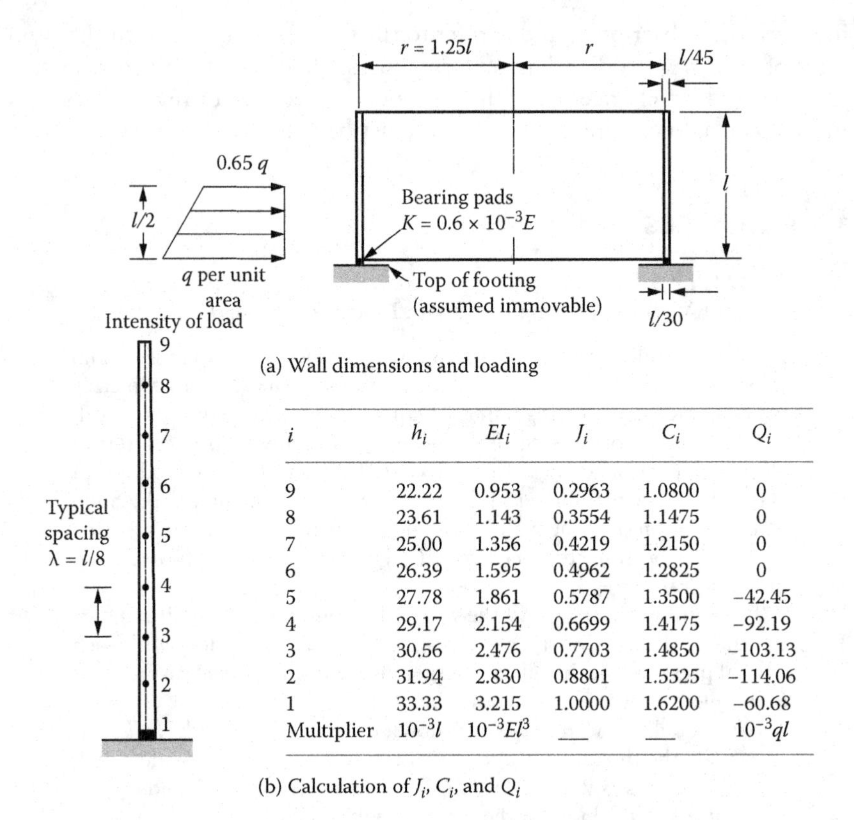

i	h_i	EI_i	J_i	C_i	Q_i
9	22.22	0.953	0.2963	1.0800	0
8	23.61	1.143	0.3554	1.1475	0
7	25.00	1.356	0.4219	1.2150	0
6	26.39	1.595	0.4962	1.2825	0
5	27.78	1.861	0.5787	1.3500	−42.45
4	29.17	2.154	0.6699	1.4175	−92.19
3	30.56	2.476	0.7703	1.4850	−103.13
2	31.94	2.830	0.8801	1.5525	−114.06
1	33.33	3.215	1.0000	1.6200	−60.68
Multiplier	$10^{-3}l$	$10^{-3}El^3$	—	—	$10^{-3}ql$

(b) Calculation of J_i, C_i, and Q_i

Figure 3.2 **Prestressed concrete circular wall with loading representing condition before completion of the application of the prestress (Example 3.1).**

are free. To account for the effect of the bearing pad at node 1, the value $K = 0.6 \times 10^{-3}E$ is added to element (1,1) of the matrix. (Note that $K = 0.3645EI_1/\lambda^3$.)

Solution of the simultaneous Equation (3.7) gives the deflections at the nodes:

$$\{w\} = (-10^{-3}q/\lambda^3/EI_1)\{58.6872, 68.6176, 69.1763, 58.0965, 33.1659, 8.6600, -0.8389, -1.9454, -1.2114\}.$$

Substitution in Equations (3.9) and (3.10) gives the hoop force and the bending moment at each of the nine nodes:

$$\{N\} = qr\{-0.761, -0.852, -0.822, -0.659, -0.358, -0.089, 0.008, 0.018, 0.010\}$$

$$\{M\} = (ql^2/1000)\{0, -1.03, -1.12, -1.16, 0.03, 0.93, 0.44, 0.08, 0\}.$$

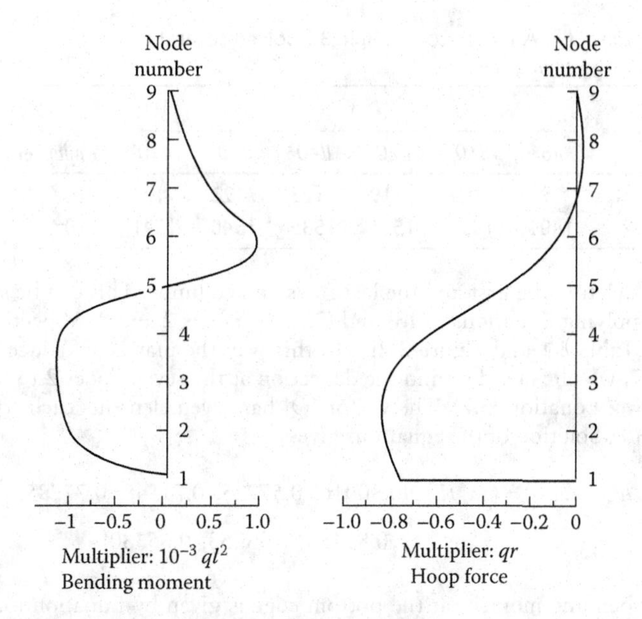

Figure 3.3 **Bending moment and hoop force in a prestressed concrete circular wall in a construction stage when prestressing is applied only on the lower half of the height (see Figure 3.2a).**

Applying Equation (3.17) (in Table 3.2) at node i = 1 and using the values of Q, J, and C listed in Figure 3.2b gives the reaction at the bearing pad, which is equal to the required shearing force at the bottom edge

$$R_1 = Q_i + \frac{EI_1}{\lambda^3}\left[-\left(J_2 + \frac{C_1}{2}\right)w_1 + 2J_2 w_2 - J_2 w_2\right] = -21.4 \times 10^{-3} ql.$$

This value should, of course, be equal to the product k_{w1}. The variation of the hoop force and the bending moment over the wall height are depicted in Figure 3.3.

EXAMPLE 3.2
Circular Wall, with Bottom Edge Encastré and Top Edge Hinged

Consider a circular wall with the same dimensions as in Figure 3.2a but which has the bottom edge encastré and the top edge hinged. Find the bending moment and the shear at the bottom edge due to a uniform outward load of intensity q. Use the same spacing λ as in the previous example. Assume Poisson's ratio v = 0.2.

The matrix $[K]$ is obtained simply by adding $2J_1 = 2 \times 1.0 = 2.0$ to element (2,2) of the corresponding matrix in the preceding example

Table 3.3 Answers to Example 3.2 solved several times with varying λ

	λ						
	l/8	*l/10*	*l/20*	*l/40*	*l/60*	*l/70*	*Multiplier*
M_1	94	104	119	123	123	124	$10^{-4}ql^2$
V_1	1401	1436	1511	1535	1540	1541	$10^{-4}ql$

and deleting the first and the last rows and columns. This can be seen by applying Equations (3.13) and (3.12) at nodes 2 and 8, respectively (see Table 3.1 and Figure 3.2b). In this way the matrix is reduced to 7 × 7, which is used to find the deflection at the seven nodes 2 to 8 by solving Equation (3.7). The vector $\{Q\}$ has seven elements each equal to $ql/8$. Solution of the equations gives

$$\{w_2, w_3, \cdots, w_8\} = (q\lambda^4/EI_1)\{0.30016,\ 0.57725,\ 0.71548,\ 0.77785,\ 0.82486,\ 0.83753,\ 0.65330\}$$

The bending moment at the bottom edge is given by Equation (3.19) (with $i = 1$ and $w_1 = 0$):

$$M_1 \frac{EI}{\lambda^2} J_1(-2w_2) = -0.60032q\lambda^2 = -0.00938ql^2.$$

For the shear at the bottom edge, use Equation (3.18) in Table 3.2c:

$$(R_1 - Q_1) = \frac{EI_1}{\lambda^3}\left[2(J_1 + J_2)w_2 - J_2w_3\right] = 0.62062q\lambda = 0.07758ql$$

with

$$Q_1 = ql/16, \quad R_1 = V_1 = 0.14088ql.$$

The accuracy of the finite-difference solution is improved by reducing λ and increasing the number of nodes. To study the effect of varying λ on the results, the above example is solved with $\lambda = l/10$, $l/20$, $l/40$, $l/60$, and $l/70$, giving the results in Table 3.3.

Table 3.3 indicates that in this particular example little gain in accuracy is achieved by choosing λ smaller than $l/40$.

3.7 FLEXIBILITY AND STIFFNESS OF CIRCULAR WALLS OF VARIABLE THICKNESS

In Section 2.4 (Chapter 2), the flexibility and stiffness matrices are derived in a closed form for a circular wall or beam on elastic foundation with *EI* and k constant. In this section, finite differences are used for the same purpose when EI and k are variable.

The deflections $\{w\}$ at the n nodes in Figure 3.1a or Figure 3.1b are related to the nodal forces $\{Q\}$ by Equation (3.7), which when inverted gives

$$\{w\} \cong [K]^{-1}\{Q\}. \tag{3.22}$$

$[K]^{-1}$ is a flexibility matrix corresponding to coordinates $\{w\}$. The elements of $[K]^{-1}$ are influence coefficients of the deflections. The deflection $\{w\}$ and the displacements at the four coordinates indicated at the edges in Figure 3.1a,b are related:

$$\{D\} \cong [C]\{w\}, \tag{3.23}$$

where

$$[C]_{4\times n} = \begin{bmatrix} 1 & 0 & & 0 & 0 \\ -1/\lambda & 1/\lambda & \text{elements not shown are zero} & 0 & 0 \\ 0 & 0 & & 0 & 1 \\ 0 & 0 & & -1/\lambda & 1/\lambda \end{bmatrix}. \tag{3.24}$$

Equation (3.23) approximates the rotations D_2 and D_4 by a finite-difference equation of the first derivative of the deflection at the middle of the two edge intervals.

It can be proved[6] that the flexibility matrix in the $\{D\}$ coordinates is

$$[f] \cong [C][K]^{-1}[C]^{T}. \tag{3.25}$$

The preceding approximation of the end rotation is acceptable when the interval is small or when the bending moment (and hence the rate of change of slope) is zero at the end. The use of $[C]$ as in Equation (3.24) to calculate $[f]$ results in values of the elements f_{22} and f_{44} that are less accurate than the other elements of the matrix. A more accurate value of f_{22} is given by

$$f_{22} \cong \frac{1}{6\lambda}[-11 \quad 18 \quad -9 \quad 2 \quad 0 \quad \cdots \quad 0][K]^{-1}\begin{Bmatrix} -1/\lambda \\ 1/\lambda \\ 0 \\ \vdots \\ 0 \end{Bmatrix} \tag{3.26}$$

in which the slope D_2 is approximated by the finite-difference relation[7]

$$D_2 = \left(\frac{dw}{dx}\right)_1 \cong \frac{1}{6\lambda}(-11w_1 + 18w_2 - 9w_3 + 2w_4), \tag{3.27}$$

while a unit couple at coordinate 2 is replaced by a pair of equal and opposite forces normal to the beam axis, and of magnitude $1/\lambda$ at nodes 1 and 2. An equation similar to Equation (3.26) is to be used for f_{44}.

When the flexibility matrix is generated, its inverse gives the stiffness matrix corresponding to the four edge coordinates in Figure 3.1a or Figure 3.1b.

3.8 EFFECT OF TEMPERATURE

Two cases of axisymmetrical temperature-induced loads are considered here for circular walls with linearly varying thickness as shown in Figure 3.4a:

Case (a)—The wall undergoes a uniform temperature rise T.
Case (b)—Temperature gradient is assumed to occur in the radial direction, as, for example, due to heating of the stored material. The temperature rises of the inner and outer surfaces are assumed T_i and T_o, respectively, with linear variation through the thickness.

The expansion of the walls of storage tanks and silos is seldom restrained in the vertical direction, such that no axial forces develop in this direction for the two cases—(a) and (b).

If the edges in Case (a) are free to expand in the radial directions, no forces develop and the edge displacements at the four coordinates in Figure 3.4b will be

$$\{D\} = \mu r T \begin{Bmatrix} 1 \\ 0 \\ 1 \\ 0 \end{Bmatrix}, \tag{3.28}$$

where μ is the coefficient of thermal expansion.

If the edges are supported such that D_1, D_2, D_3, or D_4 is restrained, edge forces develop producing bending moments and hoop forces in the shell, which may be conveniently analyzed by the force method, making use of Table 12.16 (Chapter 12) to obtain the flexibility coefficients needed in the analysis. This is explained further by Example 4.5 in Chapter 4.

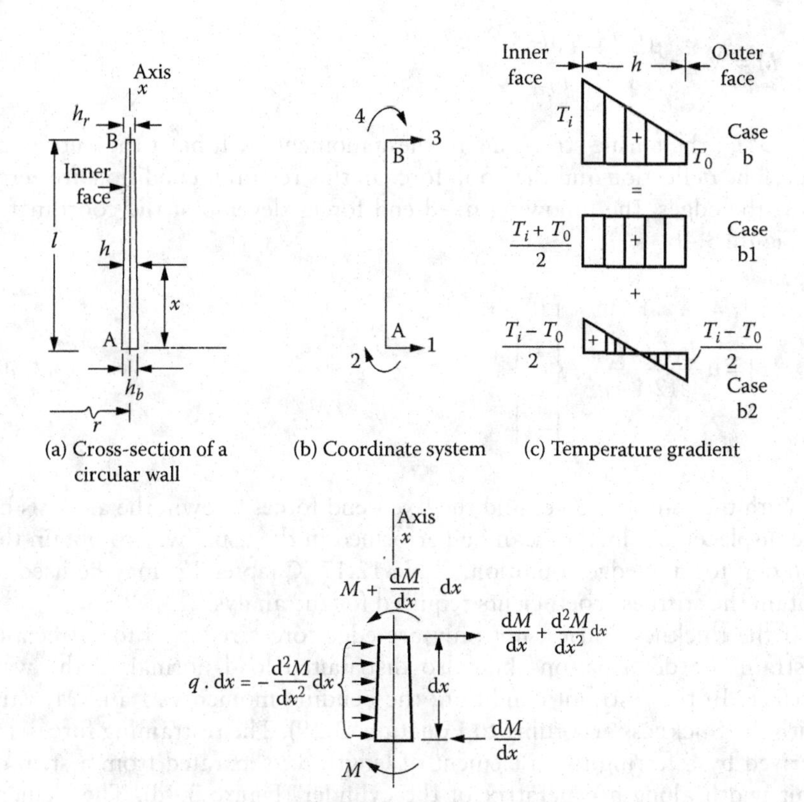

Figure 3.4 **Thermal loading of a circular wall of varying thickness.**

The temperature rise in Case b can be considered as the superposition of (Figure 3.4c):

Case (b1)—A uniform temperature rise $T = (T_i + T_o)/2$, which can be treated in the same way as Case a.

Case (b2)—Temperature gradient $(T_i - T_o)/2$ at the inner face and $-(T_i - T_o)/2$ at the outer face, with zero change in temperature at the middle surface.

For the analysis of Case (b2), the displacement method of analysis is convenient. This is explained here first for a constant-thickness wall. If the edges are totally fixed, bending moments of equal magnitude develop in the longitudinal and circumferential directions. Their value is constant[8]:

$$M = M_\phi = \frac{\mu E(T - T_0)h^2}{12(1 - \nu)}. \tag{3.29}$$

If $T_i > T_o$, the tensile stress due to these moments will be at the outer surface. The deflection and the hoop force in this restraint condition are zero.

At the edges, the following fixed-end forces develop at the coordinates in Figure 3.4b:

$$\{F\} = \mu \frac{E(T_i - T_o)h^2}{12(1 - \nu)} \begin{Bmatrix} 0 \\ 1 \\ 0 \\ -1 \end{Bmatrix}. \tag{3.30}$$

With the internal forces and the fixed-end forces known, the analysis by the displacement method can be performed in the usual way to obtain the solution for any edge condition. Table 12.17 (Chapter 12) may be used to obtain the stiffness coefficients required for the analysis.

If the thickness varies, not only are edge forces required to artificially restrain the deformations but also distributed load normal to the wall surface. In the restraint condition, the bending moments M and M_ϕ vary with the thickness according to Equation (3.29). The restraining forces are derived by referring to an element of length dx separated from a strip of unit width along a generatrix of the cylinder (Figure 3.4d). The element would undergo no deformation due to the change-of-temperature Case (b2) if it were subjected to artificial restraining forces shown in Figure 3.4d. As a result of variation of M, shearing forces must be introduced, and for equilibrium of the forces in the horizontal direction, the load q must be applied. In this restrained condition, the deflection of the element and hence the hoop force are zero at all points. If the whole strip is considered as an assemblage of elements of length dx, the forces on adjacent cross-sections cancel out, leaving the distributed load q and the edge forces $\{F\}$. The intensity of the distributed restraining force must be

$$q = \frac{d^2 M}{dx^2}. \tag{3.31}$$

Substitution of Equation (3.29) into Equation (3.31) gives

$$q = \frac{\mu E(T - T_o)}{12(1 - \nu)} \frac{d^2(h^2)}{dx^2}. \tag{3.32}$$

The restraining forces at the edges are

$$\{F\}=\begin{Bmatrix} -\left(\dfrac{dM}{dx}\right)_{x=0} \\ (M)_{x=0} \\ \left(\dfrac{dM}{dx}\right)_{x=l} \\ -(M)_{x=1} \end{Bmatrix}=\mu\frac{E(T_i-T_o)}{12(1-\nu)}\begin{Bmatrix} -\left(\dfrac{d(h^2)}{dx}\right)_{x=0} \\ (h^2)_{x=0} \\ \left(\dfrac{d(h^2)}{dx}\right)_{x=l} \\ -(h^2)_{x=l} \end{Bmatrix}. \tag{3.33}$$

The internal force due to thermal loading Case (b2) is the sum of the internal forces in the restrained condition (given by Equation 3.29) and the effect of the distributed load of intensity $-q$ and edge forces of magnitude $\{-F\}$ (Equations 3.32 and 3.33).

When the thickness varies as shown in Figure 3.4a, Equations (3.32) and (3.33) become

$$q=\frac{\mu E(T_i-T_o)}{6(1-\nu)}\left(\frac{h_b}{l}\right)^2\left(1-\frac{h_t}{h_b}\right)^2, \tag{3.34}$$

which indicates that q is constant (uniform load), and

$$\{F\}=\frac{\mu E(T_i-T_o)}{12(1-\nu)}h_b^2\begin{Bmatrix} \dfrac{2}{l}\left(1-\dfrac{h_t}{h_b}\right) \\ 1 \\ -\dfrac{2}{l}\left(1-\dfrac{h_t}{h_b}\right) \\ -(h_t/h_b)^2 \end{Bmatrix}. \tag{3.35}$$

It should be noted that, unlike all other loadings considered in this book, the temperature loading Case (b2) (Figure 3.4b) produces circumferential bending moment M_ϕ, which may be of the same importance as the bending moment M in the direction of a generatrix (compare Equation 3.29 with Equation 2.4).

An example of the analysis for thermal loading Case (b) is given in Chapter 4 (see Example 4.4).

Shrinkage of concrete can be treated in the same way as for a decrease in temperature producing the same contraction. It should be noted, however, that creep (which usually occurs at the same time as shrinkage) results in a considerable reduction of stresses as compared with that given by the aforementioned analysis. Some engineers[9] account for this by using a reduced value of E. This is discussed in Chapter 6.

3.9 GENERAL

The finite-difference method constitutes a numerical solution for circular walls or beams on elastic foundations with arbitrary variation in loading, flexural rigidity, foundation modulus, or support conditions.

The accuracy of the finite-difference solution is improved when a larger number of nodes is used. Sufficient accuracy for most practical cases can be achieved when the spacing between the nodes, λ, is chosen as follows: when $\beta l \leq \pi$, $\lambda \leq l/m$; when $\beta l \geq \pi$, $\lambda \leq \pi/(m\beta)$, where β is calculated by Equation (2.13) or Equation (2.17) using average values for EI, k, or h; m is a number between 5 and 10. When working with a computer, use of smaller spacing requires only a little additional effort, which may be justified.

The finite-difference equations presented in this chapter are used for the preparation of the design tables discussed in the following chapter. To ensure a high degree of accuracy, the spacings between the nodes are chosen substantially smaller than the range suggested earlier. A companion Web site for this book provides the computer program CTW (Cylindrical Tank Walls), based on this chapter; see Appendix B.

NOTES

1. See Hetényi, M., 1946, *Beams on Elastic Foundation*, University of Michigan Press, Ann Arbor, pp. 114–19; and also Timoshenko, S. P., and Wojnowsky-Krieger, S., 1959, *Theory of Plates and Shells*, 2nd ed., McGraw-Hill, New York.
2. For more details of the finite-difference method for structural analysis and proof of the equations in this section, see Ghali, A., Neville, A. M., and Brown, T. G., 2009, *Structural Analysis: A Unified and Classical Approach*, 6th ed., Spon Press, London, Chapter 15.
3. The equivalent concentrated loads for straight-line and second-degree parabolic variation of load are given by Ghali et al., 2009. See also Equation (3.20) of the present chapter.
4. For the derivation of the equations in Figure 3.2, see Ghali et al., 2009.
5. For derivation of Equation (3.21), see Ghali et al., 2009.
6. See Ghali et al., 2009.
7. See Steifel, E. L., 1963, *An Introduction to Numerical Mathematics*, Academic Press, New York.
8. See Timoshenko, S., 1956, *Strength of Materials*, 3rd ed., Van Nostrand, New York, Part II, p. 135.
9. See Neville, A. M., Dilger, W., and Brooks, J. J., 1983, *Creep of Plain and Structural Concrete*, Construction Press, Longman Group, London.

Chapter 4

Design tables and examples of their use

4.1 INTRODUCTION

In this chapter, there will be a detailed discussion of tables prepared by the finite-difference method and presented in Chapter 12 of this book, usable for the design of circular walls of constant or variable thickness. This loading and the type of variation in thickness are chosen to represent most practical cases required in the design of containers.

The edges of the wall may be free, hinged, or encastré (totally fixed). To cover all possible combinations of support conditions at the two edges would require a much larger number of design tables than can be included in this book. However, by superposition of the basic cases covered by the tables, the solution can be obtained when the edges are in any of the three mentioned conditions or when they are supported by elastic supports partially restraining the edge deflection or rotation. This is explained by numerical examples.

One value of Poisson's ratio, $\nu = 1/6$, is used for the preparation of the tables, which are intended mainly for concrete walls. For materials with a different value of ν, the use of the tables requires a simple adjustment discussed in Section 4.7.

The spacing λ used in the finite-difference analysis is $\lambda = l/60$, where l is the height of the circular wall. The accuracy of the tables is verified by comparison with closed-form analytical solutions for the cases when these are available.

All tabulated values are dimensionless, thus usable with any system of units.

Beams on elastic foundation can be analyzed using the same tables as discussed in Section 4.8.

4.2 DESCRIPTION OF DESIGN TABLES

A summary of the contents of the design tables in Chapter 12 of this book is given in Table 4.1. The variation of thickness of circular-cylindrical walls

Table 4.1 Summary of the contents of the tables for analysis of circular walls of variable thickness presented in Chapter 12

Group	*Edge conditions*		*Load*	*Values given by the table*	*Table number*
	Bottom	*Top*			
1	Free	Free	Triangular	N	1
			Intensity 0 at top and q at bottom	M, D_2, and D_4	2
			Uniform of	N	3
			intensity q	M, D_2, and D_4	4
			Outward normal line load of	Influence coefficient N	5
			Unit intensity applied at one	Influence coefficient M	6
			Of 11 nodes	Influence coefficients D_2 and D_4	7
			Spaced at one-tenth intervals		
2	Free	Free	$F_1 = 1$	N	8
				M, D_2, and D_4	9
			$F_2 = 1$	N	10
				M, D_2, and D_4	11
			$F_3 = 1$	N	12
				M, D_2, and D_4	13
			$F_4 = 1$	N	14
				M, D_2, and D_4	15
3		Flexibility coefficients		f_{ij}	16
		Stiffness coefficients		S_{ij}	17
4	Encastré	Encastré	Triangular	N, $(V)_{x=0}$, and $(V)_{x=1}$	18
			Intensity 0 at top and q at bottom	M	19
			Uniform of	N, $(V)_{x=0}$, and $(V)_{x=l}$	20
			intensity q	M	21
	Encastré	Free	Triangular	N and $(V)_{x=0}$	22
			Intensity 0 at top and q at bottom	M and D_4	23
			Uniform of	N and $(V)_{x=0}$	24
			intensity q	M and D_4	25
	Hinged	Free	Triangular	N and $(V)_{x=0}$	26
			Intensity 0 at top and q at bottom	M, D_2, and D_4	27
			Uniform of	N and $(V)_{x=0}$	28
			intensity q	M, D_2, and D_4	29

is assumed linear over the whole height. The computer program CTW (Cylindrical Tank Walls), presented in Appendix B and the companion Web site for this book, can analyze walls with any thickness variation and any axisymmetrical load.

The loadings considered are (a) uniformly distributed over the height, (b) linearly varying intensity over the height, and (c) an outward line load of unit intensity applied at any of one-tenth intervals of the height. The tables for loading (c) represent influence coefficients that can be used to analyze circular walls subjected to axisymmetrical loading with arbitrary variation of intensity.

For the tables in Groups 1 and 2 (see Table 4.1), the two edges of the wall are considered free; for Group 4, chosen combinations are considered with the edges free, hinged, or encastré. The tables give the values of the hoop force N and the bending moment M at one-tenth intervals of the wall height. At a supported edge, the value of the shearing force V is given; while at a free or hinged edge, the rotation D_2 or D_4 is listed (Figure 12.1b). The deflection w is not listed because it is directly related to N by Equation (2.1) (see Chapter 2). The tables thus give the forces $\{F\}$ and the displacements $\{D\}$ at the four coordinates shown in Figure 12.1b using the following relations:

$$\{F\} = \begin{Bmatrix} -(V)_{x=0} \\ (M)_{x=0} \\ (V)_{x=l} \\ -(M)_{x=l} \end{Bmatrix}, \tag{4.1}$$

$$\{D\} = \begin{Bmatrix} \frac{r}{E}\left(\frac{N}{h}\right)_{x=0} \\ D_2 \\ \frac{r}{E}\left(\frac{N}{h}\right)_{x=l} \\ D_4 \end{Bmatrix}. \tag{4.2}$$

The tables in Group 2 give the effects of edge force of unit intensity acting at one of the four coordinates in Figure 12.1b. The two tables in Group 3 give elements of the 4 × 4 flexibility and stiffness matrices $[f]$ and $[S]$ corresponding to the same coordinates. Only 10 of the 16 elements of each of the two matrices are listed in the table. The remaining elements are obtained using the property of symmetry of the matrices; thus $f_{ij} = f_{ji}$ and $S_{ij} = S_{ji}$.

Using superposition, the tables in Groups 1 to 3 are sufficient for the analysis of any loading and any edge condition. First, the edges are considered free, then appropriate edge forces are introduced at the coordinates 1 to 4 (Figure 12.1b) to adjust the edge displacement as required. The superposition procedure is explained in detail in Section 4.4.

The tables in Group 4, although they can be derived from the preceding groups, are given because they are most frequently needed, being usable for triangular and uniform loading with specific combinations of edge conditions.

4.3 VARIABLES

The tables are entered with a dimensionless parameter

$$\eta = l^2/(dh_b), \tag{4.3}$$

where l is the wall height, $d = 2r$ is the diameter, and h_b is the thickness at the bottom (Figure 12.1a). Seventeen values of the parameter η chosen between 0.4 and 24 should cover the majority of practical cases.

However, when the parameter $\eta > 24$, the case of a "long" cylinder (see Section 2.9), forces at one edge produce deflections or internal forces that die out before reaching the other end. This makes it possible to use the presented tables for walls having the parameter $\eta > 24$. This is explained in Section 4.6.

The tables are also entered with the ratio of the wall thicknesses at top and bottom h_t/h_b. The tabulated values are for h_t/h_b = 1.00, 0.75, 0.50, 0.25, and 0.05. The last value, which is not practicable, serves only for interpolation. When the wall thickness is constant, h_t/h_b = 1.00, the tables in Chapter 12 may be used.

4.4 ARBITRARY EDGE CONDITIONS

The tables in Groups 1 to 3 will be used here to find the hoop force N, the bending moment M, and the edge forces when the top or bottom edge of the wall is free, hinged, totally fixed, or elastically restrained. The loading may be of the three types considered in the tables in Group 1 (see Table 4.1).

The force method of analysis is applied, choosing the wall with both edges free as a released structure (see Section 1.5). This involves removal of redundant forces $\{F\}$ from the actual structure at a number of coordinates at the two edges. The external load is then applied on the released wall and the inconsistencies in the edge displacements $\{D\}$ are determined, using tables in Group 1. The internal forces at various positions are also calculated using the same tables. The first table in Group 3 is used to find the flexibility matrix of the released structure; the flexibility of an elastic support should be accounted for. The values of the redundant forces $\{F\}$ necessary to eliminate the inconsistencies in the displacements are determined by solving the equations: $[f]\{F\} = -\{D\}$. The tables in Group 2 are then used to find the internal forces due to the calculated redundant forces. Superposition of these on the internal forces of the released structure gives the internal forces in the actual structure.

The numerical examples in the following section demonstrate the use of the tables with the force method of analysis (Examples 4.1 and 4.2). In Example 4.3, the moment distribution method is employed, using the tables to find the end-rotational stiffness and fixed-end moments required for the analysis.

4.5 EXAMPLES

EXAMPLE 4.1
Circular Wall of Varying Thickness, on Bearing Pads, with Trapezoidal Normal Load

Using the tables in Chapter 12, find the variation of the hoop force N and the bending moment M along the height of a circular concrete wall of linearly varying thickness subjected to circumferential prestressing that produces axisymmetrical inward trapezoidal normal load of intensity $0.3q$ per unit area at top and $1.3q$ at bottom (Figure 4.1a). The top edge is hinged and the bottom edge supported on bearing pads providing an elastic support of stiffness $K = 0.2 \times 10^{-3}E$, where E is the modulus of elasticity of the wall material. The wall dimensions are height = l; diameter = $2r = 4l$; $h_b = l/20$; $h_t = h_b/2$. Poisson's ratio is 1/6.

The dimensionless parameter $\eta = l^2/(dh_b) = 5.00$ and the ratio $h_t/h_b = 0.5$.

A released structure is shown in Figure 4.1b. Coordinate 1 in the figure represents the relative radial translation of the bottom edge of the wall and the top surface of the bearing pad; coordinate 2 represents the outward deflection at the top edge.

The loading is treated as the sum of (a) triangular load of intensity zero at top and q at bottom, and (b) uniform load of intensity $0.3q$. The displacements at coordinates 1 and 2 due to these two loadings are determined by using the coefficients in Tables 12.1 (for loading a) and 12.3 (for loading b) and Equation (4.2). Thus,

$$D_1 = \frac{r}{E}\left(\frac{N}{h}\right)_{x=0} = -(1.0176 + 0.3 \times 0.9820)\frac{qr^2}{Eh_b} = -104.98\frac{ql}{E},$$

$$D_2 = \frac{r}{E}\left(\frac{N}{h}\right)_{x=l} = -(0.0246 + 0.3 \times 0.9758)\frac{qr^2}{(Eh_b/2)} = -50.77\frac{ql}{E}.$$

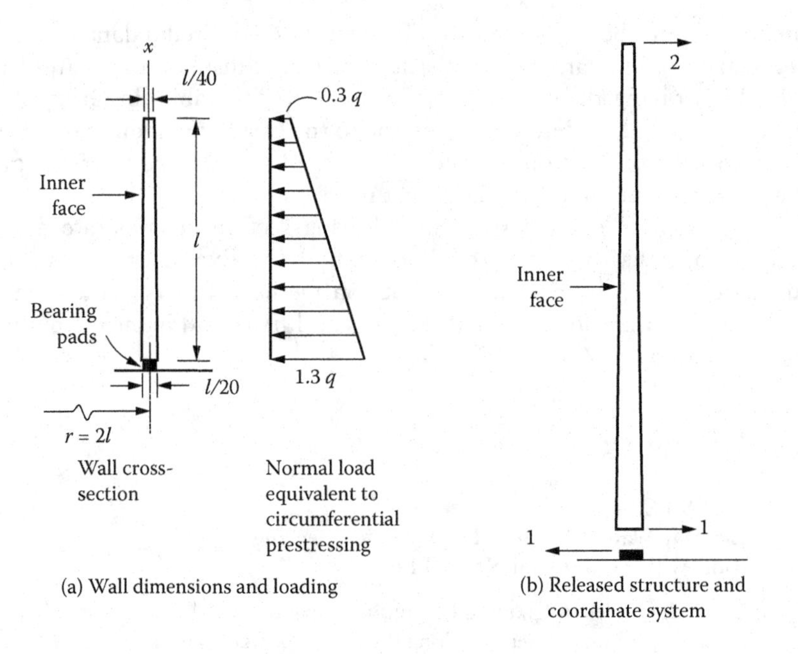

Figure 4.1 Circular wall with variable thickness analyzed using the tables in Chapter 12 (Example 4.1).

Table 4.2 Force method: Hoop force, N_s and bending moment, M_s in a released structure

	x/l						
	0	*0.2*	*0.4*	*0.6*	*0.8*	*1.0*	*Multiplier*
N_s due to load (a)	1.0176	0.7988	0.5951	0.3946	0.1971	0.0246	$-qr$
N_s due to load (b)	0.2946	0.3003	0.3015	0.3016	0.3009	0.2927	$-qr$
Total N_s	1.3122	1.0991	0.8966	0.6962	0.4980	0.3173	$-qr$
M_s due to load (a)	0	200	403	420	228	0	$\frac{-ql^2}{10^6}$
M_s due to load (b)	0	–59	–121	–126	–69	0	$\frac{-ql^2}{10^6}$
Total M_s	0	141	282	294	159	0	$\frac{-ql^2}{10^6}$

The variations of the hoop force N_s and the bending moment M_s in the released structure are obtained by the use of Tables 12.1 to 12.4. (They are elements of $\{A_s\}$; see Chapter 1, Section 1.5.) They are shown in Table 4.2.

The flexibility matrix of the released structure corresponding to the coordinates in Figure 4.1b is (using Table 12.16)

Table 4.3 Superposition Equation (1.5) to determine the required hoop force and bending moment

	x/l						Multiplier
	0	*0.2*	*0.4*	*0.6*	*0.8*	*1.0*	
$N_{u1}F_1$	0.1694	0.0401	–0.0083	–0.0101	–0.0026	0.0021	qr
$N_{u2}F_2$	0.0073	–0.0047	–0.0168	–0.0165	0.0635	0.3153	
N_s	–1.3122	–1.0991	–0.8966	–0.6962	–0.4980	–0.3173	
Sum = N	–1.1355	–0637	–0.9217	–0.7228	–0.4371	0	
$M_{u1}M_1$	0	–1228	–567	–87	15	0	$\frac{ql^2}{10^6}$
$M_{u2}M_2$	0	65	–61	–825	–1999	0	
M_s	0	–141	–282	–294	–159	0	
Sum = M	0	–1304	–910	–1206	–2143	0	

$$[f]=\begin{bmatrix} 0.0932\dfrac{l^3}{Eh_b^3}+\dfrac{5000}{E} & 0.0023\dfrac{l^3}{Eh_b^3} \\ 0.0023\dfrac{l^3}{Eh_b^3} & 0.1970\dfrac{l^3}{Eh_b^3} \end{bmatrix}=\frac{1}{E}\begin{bmatrix} 5745.60 & 18.40 \\ 18.40 & 1576.00 \end{bmatrix}.$$

The value $1/K = 5000/E$ is included in element f_{11} to account for the flexibility of the bearing pad.

The redundant forces are given by solving the equation $[f]\{F\} = -\{D\}$. Thus,

$$\{F\} = E\begin{bmatrix} 5745.60 & 18.40 \\ 18.40 & 1576.00 \end{bmatrix}^{-1}\frac{ql}{E}\begin{Bmatrix} 104.98 \\ 50.77 \end{Bmatrix}=\begin{Bmatrix} 18.169 \\ 32.002 \end{Bmatrix}\frac{ql}{1000}.$$

The values of the hoop force and the bending moment due to unit values of the redundants, N_{ui} and M_{ui} with $i = 1, 2$, are obtained from Chapter 12, Tables 12.8, 12.9, 12.12, and 12.13. (These are elements of $[A_u]$; see Chapter 1, Section 1.5.) Superposition of the effect of F_1 and F_2 on N_s and M_s gives the required values of the hoop force and bending moment in the actual structure as in Table 4.3 (see Chapter 1, Equation 1.5).

EXAMPLE 4.2
Circular Wall, with Circumferential Prestressing Tendons

Figure 4.2 is a cross-section showing the dimensions of a circular prestressed concrete wall and the positions of circumferential prestressing tendons (post-tensioned). In practice, it is advantageous to choose the

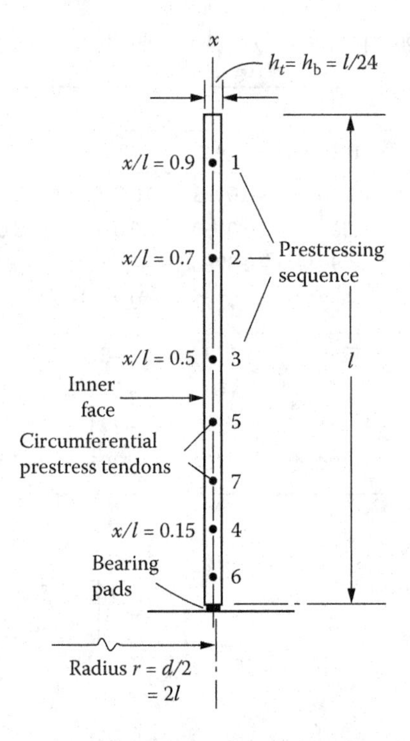

Figure 4.2 Prestressed concrete circular wall analyzed in Example 4.2.

sequence of application of the prestress in the tendons in such a way that no excessive bending moments occur at any stage. This example is to show how to use the tables in Part II to calculate the bending moment at any stage. For a chosen sequence of prestressing indicated in Figure 4.2, it is required to find the bending moment variation in the wall after the application of a force $P = 1.1 \times 10^{-6}El^2$ in each of tendons 1, 2, 3, and 4, where E is the modulus of elasticity of the wall material and l the wall height. The top edge of the wall is free and the bottom supported on bearing pads forming elastic support of stiffness $K = 0.3 \times 10^{-3}E$. Poisson's ratio is $\nu = 1/6$.

The dimensionless parameter $\eta = l^2/(dh_b) = 6.0$. A circumferential force in a tendon produces a line load acting normal to the wall surface of intensity (force/length):

$$Q = -P/r = -1.1 \times 10^{-6} El^2/(2l) = -0.55 \times 10^{-6} El.$$

The minus sign indicates inward load.

The analysis is done by the force method as for Example 2.2 (see Chapter 2), except that the tables in Chapter 12 will be used here. The released structure is a wall with free edges and the released force is represented by one coordinate as in Figure 2.7e (see Chapter 2).

Table 4.4 Bending moments due to prestressing

	x/l						Multiplier
	0	0.2	0.4	0.6	0.8	1.0	
M_s due to Q_1	0	−1.4	−9.4	−22.8	−13.0	0	$-0.55\times10^{-9}El^2$
M_s due to Q_2	0	−5.6	−10.4	16.6	20.4	0	
M_s due to Q_3	0	−4.2	18.3	18.3	−4.2	0	
M_s due to Q_4	0	17.8	−15.3	−10.4	−2.5	0	
Total M_s	0	6.6	−16.8	1.7	0.7	0	

The forces $Q_1, \ldots, 4$ applied at $x = 0.9l, 0.7l, 0.5l$, and $0.15l$, respectively, produce the bending moment M_s in the released structure shown in Table 4.4 (use Table 12.6). (The values of M_s are elements of $\{A_s\}$; see Chapter 1, Section 1.5.)

From Table 12.5, the hoop force at the bottom edge of the released structure due to the Q-forces is

$$(N_s)_{x=0} = -0.55\times10^{-6}El(-0.0193 \quad -0.3943 \quad -0.6186 \quad +3.7132)(r/l)$$

$$= -2.9490\times10^{-6}El.$$

Using Equation (4.2), the displacement of the released structure at coordinate 1 is

$$D_1 = \frac{r}{E}\left(\frac{N}{h}\right)_{x=0} = \frac{2l}{E(l/24)}(-2.9490\times10^{-6}El) = -141.5542\times10^{-6}l.$$

The flexibility of the released structure (Table 12.16) is

$$f_{11} = 0.0628\frac{l^3}{Eh_b^3} + \frac{1}{0.3\times10^{-3}E} = \frac{4201.48}{E}$$

and the redundant force is

$$F_1 = -f_{11}^{-1}D_1 = 33.69\times10^{-9}El.$$

The bending moment due to F_1 is shown in Table 4.5 (Table 12.9; these are elements of $[A_u]$, see Chapter 1, Section 1.5.)

Table 4.5 Bending moment in wall due to F_1

x/l	0	0.2	0.4	0.6	0.8	1.0	Multiplier
$M_{u1}F_1$	0	−0.0701	−0.0350	−0.0066	+0.0008	0	lF_1

Table 4.6 Superposition of M_s and the bending moment due to F_1

x/l	0	0.2	0.4	0.6	0.8	1.0	*Multiplier*
M_s	0	−3.63	9.24	−0.94	−0.39	0	$10^{-9}El^2$
$M_{u1}F_1$	0	−2.36	−1.18	−0.22	+0.03	0	
Sum = M	0	−5.99	8.06	−1.16	−0.36	0	

The bending moment in the actual wall (see Chapter 1, Equation 1.5),

$$\{M\} = \{M_s\} + \{M_{u1}\}F_1$$

is calculated in Table 4.6.

Bending moment local peak values occur at the locations of the prestress tendons; corresponding values of x/l are not included in this example for the sake of simplicity in presentation.

EXAMPLE 4.3
Reinforced Tank, on Rigid Foundation

The reinforced concrete tank in Figure 4.3a is supported on an immovable horizontal surface (e.g., solid rock or thick plain concrete footing). The wall is monolithic with the base and it is required to find the value of the continuity moment at the wall-to-base joint and the shear at the bottom edge of the wall when the tank is filled with liquid of specific weight γ per unit volume. The specific weight of concrete for the base is 2.5γ. Poisson's ratio is $v = 1/6$.

The continuity moment M_r at the outer edge B of the circular plate will cause an annular part of the base to bend as shown in Figure 4.3c, whereas the inside circular part remains flat.

The bending of the base is a nonlinear problem. The width $r_B(1 - \psi)$ of the annular deflected part AB varies with the value of the radial moment at the outer edge. Thus, the deformation of the plate due to an applied radial moment at the outer edge is not directly proportional to the value M_r. A trial-and-error procedure will be applied here. First the value ψ defining the inner radius of the annular deflected part is assumed, and then it is corrected to satisfy the following conditions: at a circle of radius $r = \psi r_B$, the downward deflection w, the moment M_r and the slope dw/dr are zero. The deflected part of the base can be considered as if totally fixed at A, but loaded in such a manner that the radial moment at the fixation is zero.

The loads on the annular plate are (Figure 4.3b) (a) a uniformly distributed load representing the self-weight of the plate and the weight of the liquid above it, which causes negative M_r at A, and (b) a radial moment at the connection of the plate with the wall, which produces positive M_r at A. A correct assumption of the value ψ should satisfy the condition that the sum of M_r caused by loads (a) and (b) is zero.

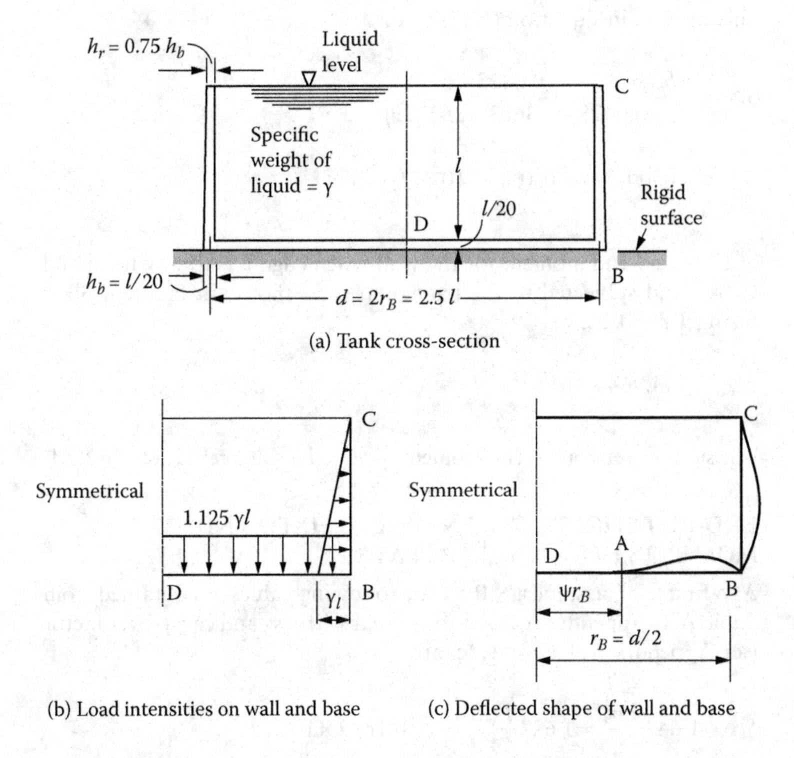

Figure 4.3 A reinforced concrete tank on rigid foundation analyzed in Example 4.3.

This can be quickly checked, using the moment distribution method (Chapter 1, Section 1.7), and the assumption can be modified until the right value of ψ is reached. Usually, no more than three trials should be necessary as shown next.

Gravity load per unit area of the base is equal to the self-weight of the base plus the liquid is $2.5\gamma(7/20) + \gamma l = 1.125\gamma l$.

END-ROTATIONAL STIFFNESS AND FIXED-END MOMENT FOR WALL

The parameter $\eta = l^2/(dh_b) = 8.0$ and $h_t/h_b = 0.75$. The "adjusted" end-rotational stiffness $S_{BC}^{②}$ for a wall having the top edge C free is given by Equation (1.32) (see Figure 1.7):

$$S_{BC}^{②} = \frac{f_{11}}{f_{11}f_{22} - f_{12}^2}.$$

The numerical values of the coefficients f_{ij} are taken from Table 12.16:

$$f_{11} = 0.0428l^3/Eh_b^3;\ f_{22} = 2.3683l/Eh_b^3;\ f_{12} = -0.2259l^2/Eh_b^3.$$

Substitution in Equation (1.32) gives

$$S_{BC}^{②} = \frac{Eh_b^3}{l} \frac{0.0428}{0.0428 \times 2.3683 - (0.2259)^2}$$

$$= 0.8500 \frac{Eh_b^3}{l} = 0.1062 \times 10^{-3} El^2.$$

The fixed-end moment for the wall when edge B is totally fixed and C free and subjected to the triangular load shown in Figure 4.3b is (using Table 12.23):

$$M_{BC} = -14.964 \times 10^{-3} \gamma l^3.$$

The sign convention for end moments is stated in Chapter 1, Section 1.7.1.

END-ROTATIONAL STIFFNESS AND FIXED-END MOMENTS FOR ANNULAR PLATE

As a first trial, assume $\psi = 0.8$. The following values are obtained from Table A.1 (Appendix A). End-rotational stiffness and carry-over factor (see Appendix A, Figure A.2c) are

$$S_{BA} = 1.642 \frac{Eh_b^3}{r_B} = 1.642 \frac{E(l/20)^3}{1.25l} = 0.1642 \times 10^{-3} El^2$$

$$C_{BA} = S_{24}/S_{44} = 0.76696/(0.8 \times 1.6416) = 0.58.$$

Fixed-end moments, from Table A.2 (Appendix A), are

$$M_{BA} = 3.198 \times 10^{-3} q r_B^2 = 3.198 \times 10^{-3} (1.125 \gamma l)(1.25l)^2$$

$$= 5.622 \times 10^{-3} \gamma l^3$$

and

$$M_{AB} = -3.497 \times 10^{-3} q r_B^2 = -3.497 \times 10^{-3} (1.125 \gamma l)(1.25l)^2$$

$$= -6.147 \times 10^{-3} \gamma l^3.$$

MOMENT DISTRIBUTION

The moment distribution procedure is performed in Table 4.7a. The radial moment obtained at A = $M_{AB} = -2.842 \times 10^{-3} \gamma l^3$, which should be zero if the assumption of ψ was correct.

Table 4.7 Moment distribution: Solution of Example 4.3

Assumed value	*Member end*	*AB*	*BA*	*BC*	*Multiplier*
(a) $\psi = 0.8$	Distribution and carry-over factors	$C_{BA} = 0.58$	0.61	0.39	$10^{-3}\gamma l^3$
	Fixed-end moments	−6.147	5.622	−14.964	
	Moment distribution and carry-over	+3.305←	+5.699	+3.643	
	Final end moments	−2.842	11.321	−11.321	
(b) $\psi = 0.9$	Distribution and carry-over factors	$C_{BA} = 0.54$	0.76	0.24	$10^{-3}\gamma l^3$
	Fixed-end moments	−1.497	1.435	−14.964	
	Moment distribution and carry-over	+5.552←	10.282	3.247	
	Final end moments	+4.055	11.717	−11.717	
(c) $\psi = 0.841$	Distribution and carry-over factors	$C_{BA} = 0.56$	0.67	0.33	$10^{-3}\gamma l^3$
	Fixed-end moments	−3.903	3.632	−14.964	
	Moment distribution and carry-over	+4.252←	7.592	3.740	
	Final end moments	0.349	11.224	−11.224	

For the second trial, $\psi = 0.9$, and similar calculations give $M_{AB} = +4.055 \times 10^{-3}\ \gamma l^3$ (Table 4.7b).

A reasonable value of ψ to be assumed for the next trial is obtained by linear interpolation. Thus,

$$\psi = 0.8 + (0.9 - 0.8)\frac{2.842}{2.842 + 4.055} = 0.841.$$

The moment distribution for the third trial with $\psi = 0.841$ is done in Table 4.7c. The corresponding value of the radial moment is $M_{AB} = 0.349 \times 10^{-3}\gamma l^3$, which is small compared to previous values, and no more trials need to be considered. The required continuity moment at the wall-to-base joint is $M_{BA} = -M_{BC} = 11.224 \times 10^{-3}\gamma l^3$.

SHEAR AT BOTTOM EDGE OF WALL

When the bottom edge of the wall is totally fixed (with the top edge free), the shear at the bottom edge is $V_B = 0.1710\gamma l^2$ (from Chapter 12, Table 12.22) and the end moment is $M_{BC} = -14.964 \times 10^{-3}\gamma l^3$. The change in this end moment caused by the rotation at B is $3.740 \times 10^{-3}\ \gamma l^3$;

thus the rotation at B is $3.740\times10^{-3}\gamma l^3/S_{BC}^{②} = 35.22\ (\gamma l/E)$ radian. The rotation results in a change in shear at the bottom edge of the wall given by Equation (1.30) (see Figure 1.7):

$$\Delta V_B = -F_1 = \frac{f_{12}}{f_{11}f_{22} - f_{12}^2}\left(35.22\frac{\gamma l}{E}\right).$$

Substitution of the values of the flexibility coefficients previously calculated, gives

$$\Delta V_B = \frac{-0.2259}{0.0428\times2.3683-(0.2259)^2}\frac{Eh_b^3}{l^2}\left(35.22\frac{\gamma l}{E}\right)$$

$$= -158.1\frac{\gamma h_b^3}{l} = -0.0198\gamma l^2.$$

Thus, the final shear at the bottom edge of the wall is

$$V_B + \Delta V_B = (0.1710 - 0.0198)\gamma l^2 = 0.1512\gamma l^2\ .$$

The variation of the hoop force and the bending moment over the height, if required, can now be determined by superposition of the values for a wall with free edges subjected to triangular loading (Tables 12.1 and 12.2), and to two known values of shear and moment at the bottom edge (Tables 12.8 to 12.11).

The finite-element program Shells of Revolution (SOR) (Appendix B) can produce an alternative analysis for the continuous structure ABC (Figure 4.3c). The computer program SOR and an input file for the structure (with $\psi = 0.825$) are available from the companion Web site for this book.

EXAMPLE 4.4
Circular Wall, with Variable Thickness, Subjected to Temperature Rise Varying Linearly through the Thickness

A circular wall (Figure 4.4a) has the bottom edge encastré and the top edge free and the following dimensions: height = l; diameter, $d = 2r = 4l$; thickness at bottom edge $h_b = l/20$ and at top edge $h_t = 0.5h_b$ with linear variation between the two edges. Poisson's ratio is 1/6; and the coefficient of thermal expansion is μ per degree. Find the variation of M, M_φ, and N due to a rise of temperature that varies linearly across the thickness between T_i degrees at the inner face and $T_0 = 0.5T_i$ at the outer face.

The temperature effect can be treated as the sum of (a) a uniform rise of temperature = $(T_i + T_o)/2 = 0.75T_i$ degrees, and (b) a temperature change varying linearly between $0.25T_i$ at the inner face and $-0.25T_i$

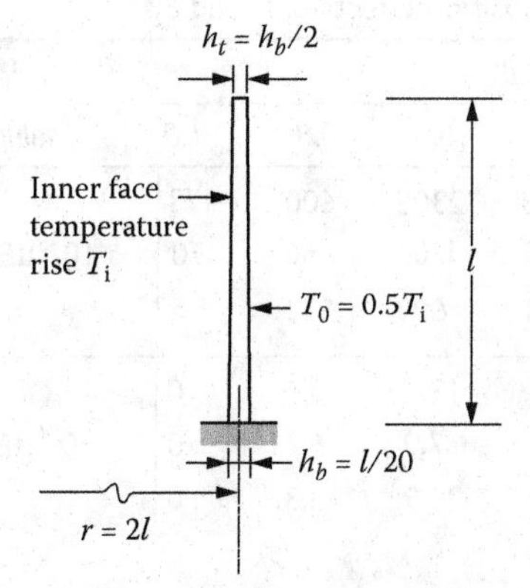

(a) Circular wall cross-section: Temperature rise varies linearly between inner and outer face

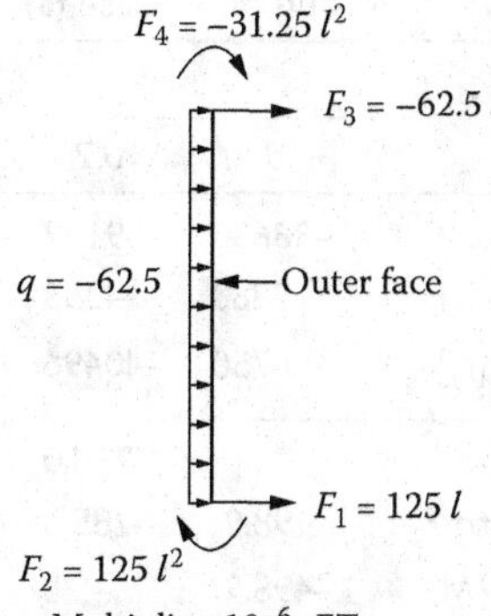

(b) Forces restraining the deflections at all points corresponding to a temperature change varying linearly between 0.25 T_i at inner face and −0.25 T_i, at outer face (Case b, step 1)

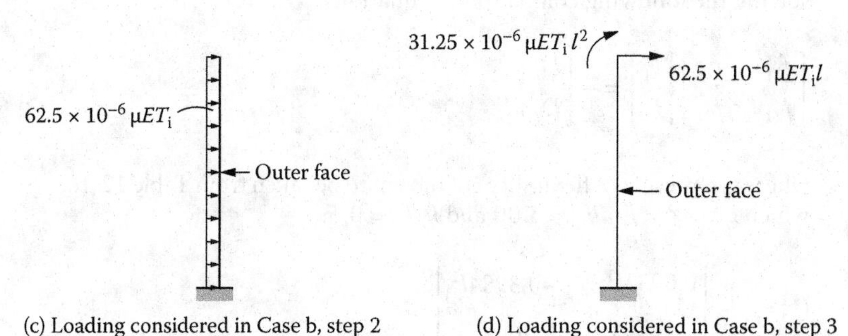

(c) Loading considered in Case b, step 2

(d) Loading considered in Case b, step 3

Figure 4.4 **Analysis of a circular wall of linearly varying thickness for the effect of a temperature rise (Example 4.4).**

at the outer face, with zero temperature change at the middle surface of the wall (see Figure 3.4c). Each of these two cases is treated separately.

CASE (a)

If the two edges of the wall are free, the edge forces are zero and no stresses develop, then the displacements of this released structure at the bottom edge are (see Equation 3.28)

$$\{D\} = \begin{Bmatrix} \mu r(0.75T_i) \\ 0 \end{Bmatrix} = \begin{Bmatrix} 1.5\mu T_i l \\ 0 \end{Bmatrix}. \tag{4.4}$$

The positive directions of $\{D\}$ are as shown in Figure 12.1b. Since the wall is encastré at the bottom edge, the displacements $\{D\}$ should

Table 4.8 N, M, and M_ϕ in Case (a). Superposition effects of F_1 and F_2

	x/l						
	0	*0.2*	*0.4*	*0.6*	*0.8*	*1.0*	*Multiplier*
N due to F_1	−38630	−9143	1898	2305	600	−473	$10^{-6}\mu ET_i r$
N due to F_2	19880	−1353	−4129	−1702	−66	470	$10^{-6}\mu ET_i r$
Sum = $\{N\}_a$	−18750	−10496	−2231	603	534	−3	$10^{-6}\mu ET_i r$
M due to F_1	0	280.0	129.4	19.9	−3.4	0	$10^{-6}\mu ET_i l^2$
M due to F_2	−498.9	−285.3	−64.5	7.0	6.7	0	$10^{-6}\mu ET_i l^2$
Sum = $\{M\}_a$	−498.9	−5.3	64.9	26.9	3.3	0	$10^{-6}\mu ET_i l^2$
$\{M_\phi\} = v\{M\}_a$	−83.2	−0.9	10.8	4.5	0.6	0	$10^{-6}\mu ET_i l^2$

be brought back to zero by redundant forces F_1 and F_2, calculated by solving the following compatibility equations:

$$\begin{bmatrix} f_{11} & f_{12} \\ f_{21} & f_{22} \end{bmatrix} \begin{Bmatrix} F_1 \\ F_2 \end{Bmatrix} = -\begin{Bmatrix} D_1 \\ D_2 \end{Bmatrix}, \tag{4.5}$$

where f_{ij} represents a flexibility coefficient to be taken from Table 12.16, which gives $\eta = l^2(dh_b) = 5.00$ and $h_t/h_b = 0.5$:

$$[f] = \frac{1}{Eh_b^3} \begin{bmatrix} 0.0932l^3 & -0.3984l^2 \\ -0.3984l^2 & 3.3096l \end{bmatrix}. \tag{4.6}$$

Substitution of Equations (4.4) and (4.6) into Equation (4.5) and solving gives

$$F = Eh_b^3 \begin{Bmatrix} -33.1552/l^2 \\ -3.9911/l \end{Bmatrix} \mu T_i = \begin{Bmatrix} -4144.4l \\ -498.9l^2 \end{Bmatrix} 10^{-6}\mu ET_i.$$

Using Tables 12.8 to 12.11 the variations of N and M due to these two edge forces are calculated as in Table 4.8.

In this case, the bending moment in the circumferential direction M_ϕ is equal to Poisson's ratio times the bending moment M in the direction of a generatrix (see Chapter 2, Equation 2.4). The variations of the internal forces in case (a) are shown in Figure 4.5a.

CASE (b)

This is analyzed in three steps as follows:

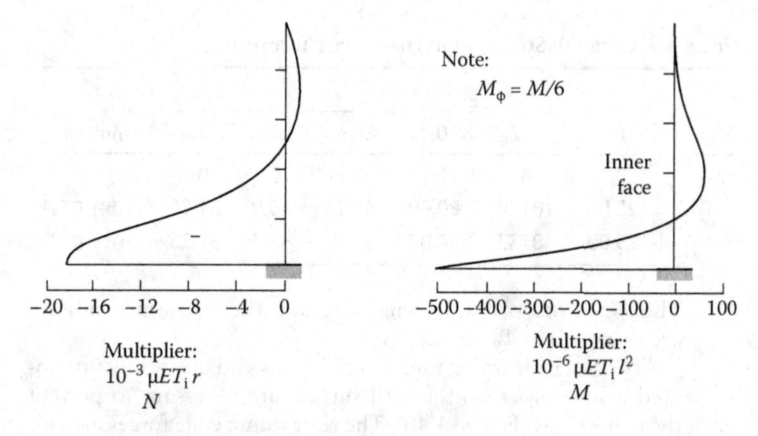

(a) Case a: Uniform temperature rise = 0.75 T_i

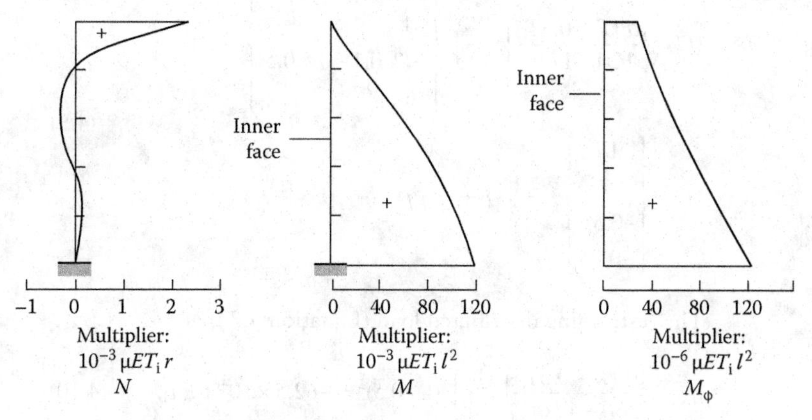

(b) Case b: Temperature change 0.25 T_i rise at inner face and 0.25 T_i drop at outer face

Figure 4.5 Variation of internal forces in a circular wall caused by temperature change (Example 4.4).

1. If the deflections are artificially restrained at all points, the thermal loading in Case (b2) produces bending moments in the direction of a generatrix and in the tangential direction (see Chapter 3, Equation 3.29):

$$M = M_\phi = \frac{\mu E(T_i - 0.5T_i)}{12(1-1/6)} h^2 = 0.05\mu T_i E h^2. \tag{4.7}$$

Since the deflection is prevented, the hoop force is

$$N = 0. \tag{4.8}$$

Table 4.9 Case (b): Step 1—Internal forces in restraint condition

	x/l						
	0	*0.2*	*0.4*	*0.6*	*0.8*	*1.0*	*Multiplier*
$\{N\}_1$	0	0	0	0	0	0	
$\{M\}_1$	125.00	101.25	80.00	61.25	45.0	31.25	$10^{-6}\mu ET_i l^2$
$\{M_\phi\}_1$	125.00	101.25	80.00	61.25	45.0	31.25	$10^{-6}\mu ET_i l^2$

The variations of the internal forces in the restraint condition are as shown in Table 4.9.

Artificial restraining forces at the edges and uniformly distributed loads normal to the wall surface are necessary to prevent the deflections (Figure 4.4b). The restraining edge forces are calculated by substitution in Equation (3.35):

$$\{F\} = \frac{\mu E(T_i - 0.5T_i)}{12(1-1/6)}\left(\frac{l}{20}\right)^2 \begin{Bmatrix} (2/l)(1-0.5) \\ 1 \\ -(2/l)(1-0.5)0.5 \\ -(0.5)^2 \end{Bmatrix} = \begin{Bmatrix} 1/l \\ 1 \\ -0.5/l \\ -0.25 \end{Bmatrix} 125\times 10^{-6}\mu ET_i l^2. \tag{4.9}$$

The restraining distributed load (Equation 3.34) is

$$q = -\frac{\mu E(T_i - 0.5T_i)}{6(1-1/6)}\left(\frac{1}{20}\right)^2 (1-0.5)^2 = -62.5\times 10^{-6}\mu ET_i. \tag{4.10}$$

It is to be noted that the forces $\{F\}$ and the distributed load q represent a set of self-equilibrated forces.

To remove the artificial restraints, a uniform load $-q$ and forces $-F_3$ and $-F_4$ must be applied on a wall with the bottom edge encastré and the top edge free. The effect of these is analyzed in steps 2 and 3.

2. The variations of the internal forces due to uniform load $-q$ (Figure 4.4c) are obtained directly from Tables 12.24 and 12.25, and are given in Table 4.10. (Bottom edge encastré, top edge free; $\eta = l^2/(2rh_b) = 5$; $h_t/h_b = 0.5$.)
3. When the two edges are free, forces of magnitudes $62.5 \times 10^{-6}\mu ET_i l$ and $31.25 \times 10^{-6}\mu ET_i l^2$ at coordinates 3 and 4 produce the following displacements at coordinates 1 and 2 (see Figure 12.1b and use Tables 12.12 to 12.15):

Table 4.10 Case (b): Step 2—Internal forces due to uniform load = $(-q)$

	x/l						
	0	*0.2*	*0.4*	*0.6*	*0.8*	*1.0*	*Multiplier*
$\{N\}_2$ due to $-q$	0	25.18	53.91	64.54	64.56	61.10	$10^{-6}\mu ET_i r$
$\{M\}_2$ due to $-q$	−1.88	−0.10	0.19	0.07	−0.00	0	$10^{-6}\mu ET_i l^2$
$\{M_\phi\}_2$ due to $-q$	−0.31	−0.02	0.03	0.01	−0.00	0	$10^{-6}\mu ET_i l^2$

$$\bar{D}_1 = 10^{-6}\mu ET_i\left(62.5l \times 0.2282\frac{r^2}{Eh_b l} + 31.25l^2 \times 2.1818\frac{r^2}{Eh_b l^2}\right) \tag{4.11}$$
$$= 6595.5 \times 10^{-6}\mu T_i l,$$

$$\bar{D}_2 = 10^{-6}\mu ET_i\left[62.5l \times (-1.8859)\frac{r^2}{Eh_b l^2} + 31.25l^2 \times (-8.7564)\frac{r^2}{Eh_b l^3}\right] \tag{4.12}$$
$$= -31320.5 \times 10^{-6}\mu T_i l.$$

These two displacements are brought to zero by the reactions at coordinates 1 and 2 (Figure 12.1b), the values of which are given by Equation (4.5). This gives

$$\{\bar{F}\} = [f]^{-1}\begin{Bmatrix} -6595.5l \\ +31320.5 \end{Bmatrix} 10^{-6}\mu T_i. \tag{4.13}$$

Substituting the inverse of Equation (4.6) into Equation (4.13) gives

$$\{\bar{F}\} = Eh_b^3\begin{Bmatrix} -62449/l^2 \\ 1946/l \end{Bmatrix} 10^{-6}\mu T_i = \begin{Bmatrix} -7806l \\ 0.243l^2 \end{Bmatrix} 10^{-6}\mu ET_i. \tag{4.14}$$

The variations of N and M due to the two forces applied at coordinates 3 and 4 and their reactions $\{\bar{F}\}$ at coordinates 1 and 2 are obtained by superposition using Tables 12.8 to 12.15. The results are shown in Table 4.11.

The internal forces for Case (b) are equal to the sum of the values calculated in steps 1, 2, and 3. The summation is done in Table 4.12. The variations of the internal forces in Case (b) are shown in Figure 4.5b.

Table 4.11 Case (b): Step 3—Internal forces due to $(-F_3)$ and $(-F_4)$ on a wall totally fixed at the bottom

	x/l						
	0	*0.2*	*0.4*	*0.6*	*0.8*	*1.0*	*Multiplier*
N due to force at 1	−72.78	−17.22	3.58	4.34	1.13	− 0.89	$10^{-6}\mu ET_ir$
N due to force at 2	−9.68	0.66	2.01	0.83	0.03	− 0.23	
N due to force at 3	14.25	−9.18	−32.76	−32.23	124.03	615.7	
N due to force at 4	68.18	7.39	−96.51	−286.90	−210.98	1731.57	
Sum = $\{N\}_3$	0	−18.35	−123.68	−313.96	−85.79	2346.15	
M due to force at 1	0	0.53	0.24	0.04	−0.01	0	$10^{-6}\mu ET_il^2$
M due to force at 2	0.24	0.14	0.03	0.00	0.00	0	
M due to force at 3	0	0.13	−0.12	−1.61	−3.90	0	
M due to force at 4	0	0.97	2.09	−0.98	−14.87	−31.25	
Sum = $\{M\}_3$	0.24	1.77	2.24	−2.55	−18.78	−31.25	
$\{M_\phi\} = \nu\{M\}_3$	0.04	0.30	0.37	−0.43	−3.13	−5.21	$10^{-6}\mu ET_il^2$

Table 4.12 Internal forces in Case (b) = Sum of steps 1, 2, and 3

	x/l					*Multiplier*	
	0	*0.2*	*0.4*	*0.6*	*0.8*	*1.0*	
$\{N\}_1$	0	0	0	0	0	0	$10^{-6}\mu ET_ir$
$\{N\}_2$	0	25.18	53.91	64.54	64.56	61.10	
$\{N\}_3$	0	−18.35	−123.68	−313.96	−85.79	2346.15	
Sum = $\{N\}_b$	0	6.83	−69.77	−249.42	−21.23	2407.25	
$\{M\}_1$	125.00	101.25	80.00	61.25	45.00	31.25	$10^{-6}\mu ET_il^2$
$\{M\}_2$	−1.88	−0.10	0.19	0.07	−0.00	0	
$\{M\}_3$	0.24	1.77	2.24	−2.55	−18.78	−31.25	
Sum = $\{M\}_b$	123.36	102.92	82.43	58.77	26.22	0	
$(M_\phi\}_1$	125.00	101.25	80.00	61.25	45.00	31.25	$10^{-6}\mu ET_il^2$
$(M_\phi\}_2$	−0.31	−0.02	0.03	0.01	−0.00	0	
$(M_\phi\}_3$	0.04	0.30	0.37	−0.43	−3.13	−5.21	
Sum = $\{M_\phi\}_b$	124.73	101.53	80.40	60.83	41.87	26.04	

4.6 LONG CYLINDERS

The highest value of the dimensionless parameter $\eta = l^2/(dh_b)$ in the design tables in Chapter 12 is 24. For longer cylinders with a higher value of η, the tables can still be used with negligible error as follows.

In Chapter 2, Section 2.9, it is shown that for cylinders of constant thickness where $\eta = l^2/(dh) > 7.3$, forces applied at one edge have a negligible effect on the other edge. This can be seen to be valid also for variable-thickness walls by examining the values of the flexibility coefficients f_{13}, f_{14}, f_{23}, and f_{24} in Table 12.16, or by examining the corresponding stiffness coefficients in Table 12.17. Advantage is taken of this property in the analysis of long cylinders.

Figure 4.6 is a cross-section of a long cylinder subjected to axisymmetrical loading of varying intensity and supported at the edges with any type of support (not shown in the figure), providing reactions $\{F\}$ at the four coordinates shown. For the analysis, this wall is treated as a wall with free edges subjected to the sum of the following loadings: (a) the external applied load, (b) the reactions at the bottom edge, F_1 and F_2, and (c) the reactions at the top edge, F_3 and F_4 (see Section 4.4).

For loading (b), the forces $F_1 = 1$ and $F_2 = 1$ are considered to act on the lower edge of a cylinder AC of length $l_1 = (24dh_b)$; Tables 12.8 to 12.11 are used to give the displacements at the lower edge and the variation of N and M, which will die out well before reaching C.

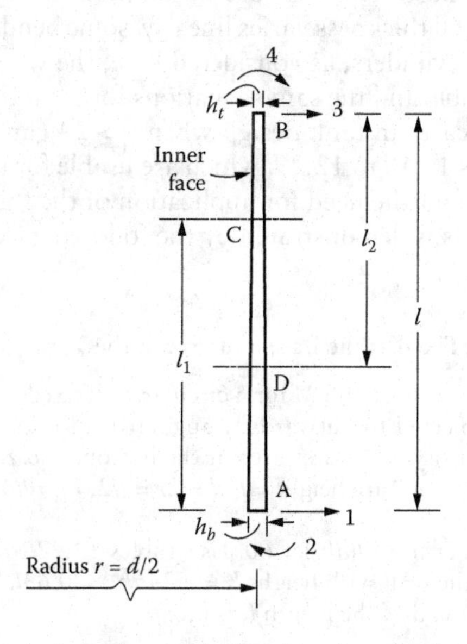

Figure 4.6 Cross-section of circular wall having the value of the parameter $l^2/(dh_b) > 24$.

Similarly, for loading (c), Tables 12.12 to 12.15 are used for a cylinder BD, for which the parameter η is also 24. The appropriate value of h_t/h_b should be used for each of the two cylinders.

The flexibility matrix corresponding to the four coordinates in Figure 4.4 needed in the force method of analysis is to be generated using Table 12.16, considering the two cylinders AC and BD. As mentioned earlier, the coefficients f_{13}, f_{14}, f_{23}, and f_{24} are zero, resulting in uncoupling of the compatibility equations; thus the equations for the redundant forces at each edge can be solved separately.

The two cylinders AC and BD, each with free edges, are again used to analyze for loading (a), using one or more of Tables 12.1 to 12.7. This gives the displacements at the edges, which are to be used in the compatibility equations to find the values of the unknown edge forces $\{F\}$. The variation of the hoop force and moment for loading (a) obtained in this manner will be accurate for the lower part of cylinder AC and for the upper part of BD. As an approximation, the analysis for AC is assumed to be valid from A to a point midway between D and C, while the analysis of the second cylinder is assumed valid for the remainder of the height covered by the second cylinder. For the lengths AC and BD to overlap, the parameter η must be smaller than approximately 96. For walls with a larger value of the parameter, a third cylinder must be used.

In the special case when the wall thickness is constant and the applied load intensity varies linearly and the two edges are free, the hoop force at any point i, $N_i = q_i r$, where q_i is the load intensity at i, and the bending moment, $M_i = 0$. Thus, in such a case no tables need to be used for loading (a). When the wall thickness varies linearly, some bending moment develops. But with long cylinders, as considered here, the values of this bending moment are negligible and the same equations for N_i and M_i can be used.

Many practical cases in tank design when $\eta > 24$ can be solved directly by the use of Tables 12.18 to 12.29, which are usable for uniform and triangular loading, without the need for application of the force method to find the edge forces. This is demonstrated by the following example.

EXAMPLE 4.5
Circular Wall, Fixed at the Base and Free at the Top

Find the variation of N and M for a circular wall fixed at base and free at top and subjected to outwards triangular load normal to the wall surface of intensity q per unit area at the bottom and zero at the top. The wall dimensions are height = l; $d = 2r = l/2$; $h_b = l/30$; $h_t = 0.50h_b$ (Figure 4.6).

The parameter $\eta = l^2/(dh_b) = 60$. Use Tables 12.22 to 12.25 for the lower part of the wall with height $l_1 = \sqrt{(24dh_b)} = 0.63l$. The thickness at C, the upper end of the length l_1 is $0.69h_b$.

Table 4.13 Internal forces read from Tables 12.22 to 12.25 for η = 24 and h_t/h_b = 0.69

	x/l_l						
	0	*0.2*	*0.4*	*0.6*	*0.8*	*1.0*	*Multiplier*
N due to load (a)	0	0.4456	0.3970	0.2514	0.1254	0.0029	qr
N due to load (b)	0	0.3296	0.3817	0.3697	0.3699	0.3693	qr
Total N	0	0.7752	0.7787	0.6211	0.4953	0.3722	qr
M due to load (a)	–3413	647	27	–3	10	0	$10^{-6}ql_l^2$
M due to load (b)	–2253	379	11	–11	–2	0	$10^{-6}ql_l^2$
Total M	–5666	1026	38	–14	8	0	$10^{-6}ql_l^2$
Total M	–2249	407	15	–6	3	0	$10^{-6}ql^2$

The load intensities at the upper and lower ends of the length l_1 are 0.37_q and q, respectively. Consider loading on l_1 as the sum of (a) the triangular load of intensity zero at top and $0.63q$ at bottom and (b) the uniform load of intensity $0.37q$. Tables 12.22 to 12.25 with $\eta = 24$ and $h_t/h_b = 0.69$ give the results shown in Table 4.13.

The aforementioned calculations give the values of N and M at points with $x \le l_1$, where x is the distance from the bottom edge of the wall. For $x > l_1$ $N = qr(l - x)/l$ and $M = 0$.

4.7 POISSON'S RATIO

As mentioned before, the tables in Chapter 12 are intended mainly for use in the design of circular concrete walls. For this reason it is assumed that Poisson's ratio is $v = 1/6$; its effect is thus included in the coefficients of the tables, leaving the entry values and the multipliers for each table in their simplest form. However, the tables can be used for circular walls having a different value for v as explained next.

The tables are to be entered by an adjusted parameter $\bar{\eta}$ to account for the difference in v given by the equation

$$\bar{\eta} = \left[1.01419\sqrt{(1-v^2)}\right]\eta. \tag{4.15}$$

The term between brackets is an adjusting factor; its value is unity when $v = 1/6$. Substituting Equation (4.3) into Equation (4.15):

$$\bar{\eta} = 1.01419\sqrt{(1-v^2)}\left[l^2/(dh_b)\right]. \tag{4.16}$$

The coefficients to be read from the tables are to be employed in the usual way, with the exception of Tables 12.16 and 12.17 of the flexibility and stiffness coefficients.

The flexibility coefficients should be multiplied by a factor $1.028\,57(1-\nu^2)$ and the stiffness coefficients by the inverse of this factor. The factor is, of course, unity when $\nu = 1/6$.

4.8 BEAMS ON ELASTIC FOUNDATIONS

The tables in Chapter 12, which are intended mainly for the design of circular walls, can also be used for beams on elastic foundation when their flexural rigidity EI is constant. This corresponds to the case of a circular wall with $h_t/h_b = 1.0$. When the tables are used for beams, Poisson's ratio does not enter in the calculation.

For the analysis of circular walls of constant thickness h, the tables are entered with the parameter

$$\eta = l^2/(dh), \tag{4.17}$$

where l is the wall height, d the diameter, and h the thickness. For a beam of length l on elastic foundation of modulus k, a characteristic parameter α is calculated by the following equation (also see Chapter 2, Equations 2.13 and 2.18a):

$$\alpha = l\sqrt[4]{\left(\frac{k}{4EI}\right)}. \tag{4.18}$$

A corresponding η value (see Equation 2.19),

$$\eta = (\alpha/1.8481)^2, \tag{4.19}$$

is to be used for entering the tables to give the deflection and bending moment variation along the beam, shear and rotation at the ends, flexibility, and stiffness of the coefficients. Any of these values is calculated as the product of a dimensionless coefficient to be read from the tables in Chapter 12 and a multiplier according to Table 4.14.

Table 4.14 Multipliers of the coefficients in the tables in Chapter 12 when used for beams on elastic foundation

	Action considered and the load causing it	*Multiplier when analysis is for*		*Tables to be used*
		Circular wall of constant thickness h	*Beam on elastic foundation*	
w	Distributed load	$\frac{qr^2}{Eh}$	$\left(\frac{0.02143}{\eta^2}\right)\frac{ql^4}{EI}$	1, 3, 28, 20, 22, 24, 26, 28
	$F_1 = 1, F_3 = 1$ or moving unit load	$\frac{r^2}{Ehl}$	$\left(\frac{0.02143}{\eta^2}\right)\frac{l^3}{EI}$	5, 8, 12
	$F_2 = 1$ or $F_4 = 1$	$\frac{r^2}{Ehl^2}$	$\left(\frac{0.02143}{\eta^2}\right)\frac{l^2}{EI}$	10, 14
M	Distributed load	ql_2	ql_2	2, 4, 19, 21, 23, 25, 27, 29
	$F_1 = 1, F_3 = 1$ or moving unit load	l	l	6, 9, 13
	$F_2 = 1$ or $F_4 = 1$	—	—	11, 15
V	Distributed load	ql	ql	18, 20, 22, 24, 26, 28
D_2 or D_4	Distributed load	$\frac{qr^2}{Ehl}$	$\left(\frac{0.02143}{\eta^2}\right)\frac{ql^3}{EI}$	2, 4, 23, 25, 27, 29
G1	$F_1 = 1, F_3 = 1$ or moving unit load	$\frac{r^2}{Ehl^2}$	$\left(\frac{0.02143}{\eta^2}\right)\frac{l^2}{EI}$	7, 9, 13
	$F_2 = 1$ or $F_4 = 1$	$\frac{r^2}{Ehl^3}$	$\left(\frac{0.02143}{\eta^2}\right)\frac{l}{EI}$	11, 15

continued

Table 4.14 (continued) Multipliers of the coefficients in the tables in Chapter 12 when used for beams on elastic foundation

Action considered and the load causing it	*Multiplier when analysis is for*		*Tables to be used*
	Circular wall of constant thickness h	*Beam on elastic foundation*	
Flexibility coefficients			
f_{11}, f_{33}, f_{13}	$\frac{l^3}{Eh^3}$	$0.08571\frac{l^3}{EI}$	16
		$0.08571\frac{l^2}{EI}$	
		$0.08571\frac{l}{EI}$	
$f_{12}, f_{14}, f_{34}, f_{23}$	$\frac{l^2}{Eh^3}$		
f_{22}, f_{44}, f_{24}	$\frac{l}{Eh^3}$		
Stiffness coefficients			
S_{11}, S_{33}, S_{13}	$\frac{Eh^3}{l^3}$	$11.667\frac{EI}{l^3}$	17
		$11.667\frac{EI}{l^2}$	
		$11.667\frac{EI}{l}$	
$S_{12}, S_{14}, S_{34}, S_{23}$	$\frac{Eh^3}{l^2}$		
S_{22}, S_{44}, S_{24}	$\frac{Eh^3}{l}$		

4.9 EXAMPLE 4.6

EXAMPLE 4.6
Analysis of Beam on Elastic Foundation Using Tables for Circular Cylinders

A horizontal raft foundation is analyzed as a beam on elastic foundation with free ends. The beam has constant flexural rigidity, EI. The length of the beam is l; two downward loads, Q, are applied at distance $x = 0.2l$ and $0.8l$ from a free end. Find the values of the elastic foundation reaction and the bending moment at $x = 0.0l$, $0.2l$, and

Table 4.15 Influence coefficients for (N and M) cylinder, foundation reaction and M for beam

	x/l	0.0 l	0.2 l	0.4 l	0.6 l	0.8 l	1.0 l	Multiplier
$N_{Cylinder}$	Load *Q* at							
	$x = 0.2l$	1.4193	3.0949	1.2067	–0.0081	–0.1220	–0.0013	$Q\,r/l$
	$x = 0.8l$	–0.0013	–0.1220	–0.0081	1.2067	3.0949	1.4193	
	Sum	1.4180	2.9729	1.1986	1.1986	2.9729	1.4180	
$M_{Cylinder}$	Load *Q* at							
	$x = 0.2l$	0.0	0.0426	–0.0077	–0.0060	–0.0009	0.0	$Q\,l$
	$x = 0.8l$	0.0	–0.0009	–0.0060	–0.0077	0.0426	0.0	
	Sum	0.0	0.0417	–0.0137	–0.0137	0.0417	0.0	
Beam reaction	Forces *Q* at 0.2*l* and 0.8*l*	1.4180	2.9729	1.1986	1.1986	2.9729	1.4180	Q/l
Beam moment	Forces *Q* at 0.2*l* and 0.8*l*	0.0	0.0417	-0.0137	-0.0137	0.0417	0.0	$Q\,l$

$0.4l$. Given data: the foundation modulus, $k = 4666\ (EI/l^4)$; this equation combined with Equations (4.18) and (4.19) gives the characteristic parameter, $\eta = 10.0$ for the analogous cylinder. For a unit line load normal to the wall of the cylinder, Tables 12.5 and 12.6 give the influence coefficients listed in Table 4.15; these coefficients are employed to give $N_{cylinder}$ and $M_{cylinder}$ due to the two forces, Q. The same coefficients also give the required values of elastic foundation reaction and the bending moment for the beam. A solution of this example can also be obtained by running the computer program CTW; the program and the input file for this example are given on the Web site for this book (see Appendix B).

4.10 GENERAL

The tables discussed in this chapter and listed in Chapter 12 of this book are intended to simplify the analysis for the design of circular walls. In most practical cases, the analytical or numerical solutions discussed in previous chapters can be substituted by using the tables.

Chapter 5

Finite-element analysis

5.1 INTRODUCTION

The walls, covers, and bases of circular tanks and silos frequently have the shape of axisymmetrical shells of revolution. Any of the structures shown in Figure 1.1 or Figure 5.1a can be idealized as an assemblage of conical shell elements and analyzed by the finite-element method. A typical conical shell element, which is relatively simple but known to give accurate results (Figure 5.1b), will be presented in this chapter. The thickness of the shell within each element is constant but can vary from element to element. For accuracy, small elements should be used, particularly where the stress varies rapidly. (Element length in such a zone may be chosen equal to $\pi/(20\beta)$, where β is defined by Equation 2.13.)

The material of the shell is assumed elastic isotropic; the thickness of the shell is considered small with respect to the radius, such that shear deformation can be ignored. Only the case of axisymmetrical loading is considered here. Because of its frequent practical occurrence, the finite-element analysis of shells of revolution has been treated by many authors,[1] offering refinement to improve accuracy and reduce the number of finite elements required in the idealization. However, the relatively simple element presented in this chapter is adequate for practical design.

The finite-difference method is used in Chapter 3 for analysis of circular-cylindrical shells. Both the finite-element and the finite-difference methods involve the solution of simultaneous equations, and the accuracy of results is increased with an increase in the number of equations. Using the same number of equations, the finite-difference method gives more accuracy. However, the finite-element method applies to axisymmetrical shells of any shape. Thus, the walls, bases, and covers of tanks and silos can be analyzed as continuous structures, without special treatment (see Example 5.4). A computer program, SOR (Shells of Revolution), based on this chapter is available at the companion Web site for this book; see Appendix B.

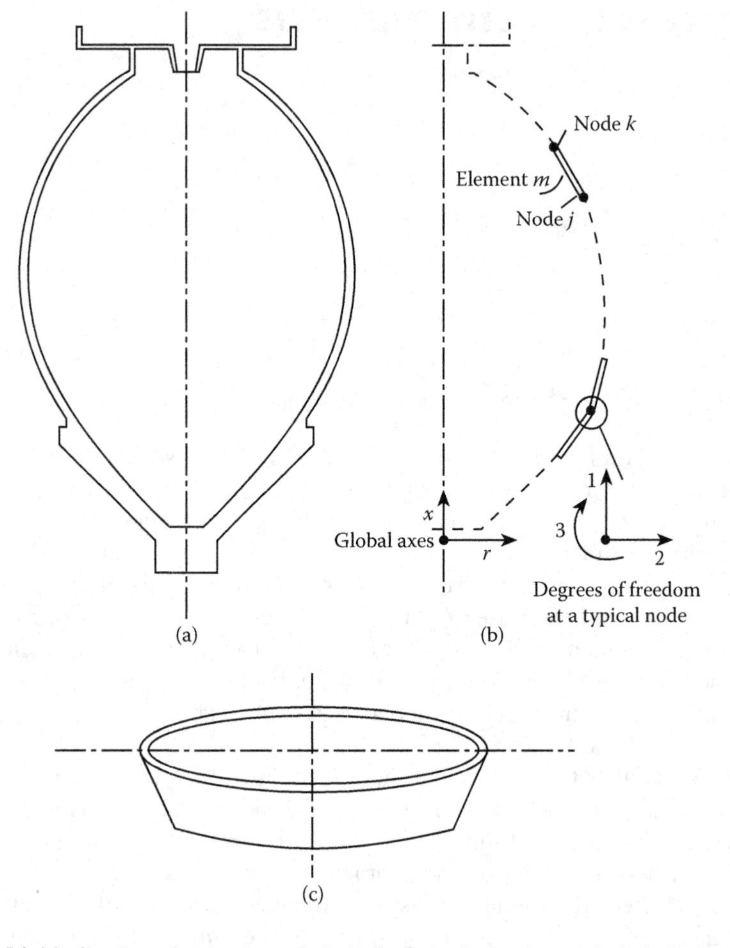

Figure 5.1 Idealization of axisymmetrical shell of revolution by conical shell elements. (a) Vertical section of egg-shaped digestor for sewage water treatment. (b) Finite-element idealization. (c) Typical shell element.

5.2 NODAL DISPLACEMENTS AND NODAL FORCES

Figure 5.1b is a sectional elevation passing through the axis of revolution of an axisymmetrical shell subjected to axisymmetrical loading. The shell is idealized as an assemblage of conical shell elements connected at circular nodal lines. The term "node" is used here to mean a point on the nodal line. The displacements at a typical node are presented by the three coordinates in Figure 5.1b, representing three degrees of freedom per node.

A sectional elevation passing through the axis of revolution of a typical finite element is shown in Figure 5.2a. The element has three degrees of

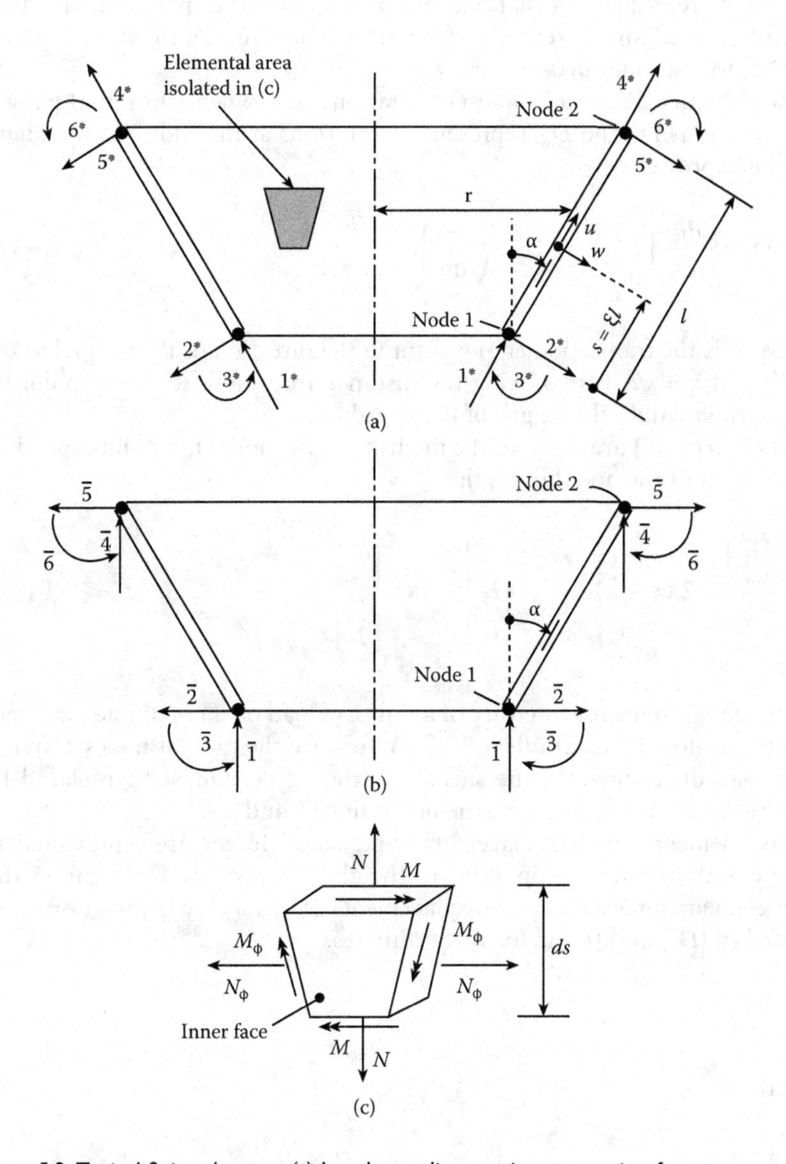

Figure 5.2 Typical finite element. (a) Local coordinates; sign convention for u, w, s, and α. (b) Degrees of freedom in directions of the global axes in Figure 5.1; order of numbering of coordinates at the nodes of a typical element. (c) Pictorial view of an elemental area showing the positive sign convention for N, M, N_ϕ, and M_ϕ.

freedom at each node, represented by six local coordinates. Coordinates 1^* and 4^* represent translations or forces along the meridian line 1–2. Coordinates 2^* and 5^* represent translations or forces in the direction of a normal to the cone surface.

Coordinates 3^* and 6^* represent rotations or moments in radial planes. The rotations D_3^* and D_6^*, representing rotations at the ends of a meridian, can be expressed as

$$D_3^* = \left(\frac{dw}{ds}\right)_{\xi=0}, \qquad D_6^* = \left(\frac{dw}{ds}\right)_{\xi=1}, \tag{5.1}$$

where w is the translation at any point in the direction of the normal to the shell; and $\xi = s/l$, with s being the distance from node 1 to any point on the meridian and l the length of the meridian.

The forces $\{F^*\}$ are equal to the intensity of the nodal forces multiplied by the lengths of the nodal lines; thus

$$\begin{Bmatrix} F_1^* \\ F_2^* \\ F_3^* \end{Bmatrix} = 2\pi r_1 \begin{Bmatrix} q_1^* \\ q_2^* \\ q_3^* \end{Bmatrix}, \qquad \begin{Bmatrix} F_4^* \\ F_5^* \\ F_6^* \end{Bmatrix} = 2\pi r_2 \begin{Bmatrix} q_4^* \\ q_5^* \\ q_6^* \end{Bmatrix}, \tag{5.2}$$

where q represents the intensity of a uniform load on a nodal line (force per length, or (force × length)/length); r_1 and r_2 are the radii at nodes 1 and 2. Any load distributed on the surface of the element must be replaced by static equivalent line loads on the nodal lines 1 and 2.

The element nodal displacements and nodal forces are represented in Figure 5.2b in directions parallel to the global axes x and r (Figure 5.1b). The element forces $\{\bar{F}\}$ and displacements $\{\bar{D}\}$, in global directions, are related to $\{D^*\}$ and $\{F^*\}$ in local coordinates:

$$\{D^*\} = [T]\{\bar{D}\}, \qquad \{\bar{F}\} = [T]^{\mathrm{T}}\{F^*\}, \tag{5.3}$$

where

$$[T] = \begin{bmatrix} [t] & [0] \\ [0] & [t] \end{bmatrix}, \qquad [t] \begin{bmatrix} \cos\alpha & \sin\alpha & 0 \\ -\sin\alpha & \cos\alpha & 0 \\ 0 & 0 & 1 \end{bmatrix}. \tag{5.4}$$

The angle α and its positive sign convention are defined in Figure 5.2a,b; the superscript T indicates matrix transposition.

The finite-element stiffness matrix $[S^*]$, to be derived in Section 5.6, relates the element nodal forces and displacements in the six local coordinates defined in Figure 5.2a:

$$[S^*]\{D^*\} = \{F^*\}. \tag{5.5}$$

5.3 TRANSFORMATION OF THE STIFFNESS MATRIX

The elements of the stiffness matrix $[S^*]$ represent forces due to unit displacements at the local coordinates of an individual element. The stiffness matrices of individual elements are combined to give the structure stiffness matrix. For this purpose, the individual stiffness matrix $[S^*]$ is transformed to $[\bar{S}]$, which corresponds to coordinates in global directions shown in Figure 5.2b:

$$[\bar{S}] = [T]^{\mathrm{T}}[S^*][T], \tag{5.6}$$

where $[\bar{S}]$ is the matrix relating $\{\bar{F}\}$ and $\{\bar{D}\}$,

$$[\bar{S}]\{\bar{D}\} = \{\bar{F}\}. \tag{5.7}$$

The element stiffness matrix $[\bar{S}]$ relates element forces $\{\bar{F}\}$ to element displacements $\{\bar{D}\}$, in global directions. Both $[S^*]$ and $[\bar{S}]$ are symmetrical matrices. To prove Equation (5.6), consider the product $\frac{1}{2}\{\bar{D}\}^{\mathrm{T}}\{\bar{F}\}$, which is equal to the work done by the forces $\{\bar{F}\}$; this is the same as the work done by the forces $\{F^*\}$. Thus,

$$\{\bar{D}\}^{\mathrm{T}}\{\bar{F}\} = \{D^*\}^{\mathrm{T}}\{F^*\}. \tag{5.8}$$

Eliminate $\{\bar{F}\}$ by the use of Equation (5.7); also eliminate $\{F^*\}$ and $\{D^*\}^{\mathrm{T}}$ by the respective use of Equation (5.5) and the first of Equation (5.3) and its transposition. This gives

$$\{\bar{D}\}^{\mathrm{T}}[\bar{S}]\{\bar{D}\} = \{\bar{D}\}^{\mathrm{T}}[T]^{\mathrm{T}}[S^*][T]\{\bar{D}\}. \tag{5.9}$$

Comparison of the two sides of this equation gives Equation (5.6).

5.4 DISPLACEMENT INTERPOLATIONS

At any point on a meridian line, the displacements $\{u, w\}$ are assumed to be related to the nodal displacements by the equation (Figure 5.2b)

$$\begin{bmatrix} u \\ w \end{bmatrix} = \begin{bmatrix} 1-\xi & 0 & 0 & \xi & 0 & 0 \\ 0 & L_1 & L_2 & 0 & L_3 & L_4 \end{bmatrix} \{D^*\}, \tag{5.10}$$

where u is the translation of any point in the direction of the meridian 1–2; w is the transverse deflection in the direction normal to the surface (direction of coordinate 2* or 5* in Figure 5.2a); $\xi = s/l$, with s being the distance between node 1 and the point considered; l is the length of the meridian line 1–2; and L_1 to L_4 are third-degree polynomials describing the deflection w of the meridian line:

$$[L_1 \ L_2 \ L_3 \ L_4] = [1 - 3\xi^2 + 2\xi^3 \quad l\xi(\xi-1)^2 \quad \xi^2(3-2\xi) \quad l\xi^2(\xi-1)]. \tag{5.11}$$

L_1 gives the variation of w over the length l, when $D_2^* = 1$, while the other nodal displacements are zero. Similarly, $(1 - \xi)$ gives the value of u at any point within the element when $D_1^* = 1$, while the other nodal displacements are zero. Likewise, each of the elements of the 2 × 6 matrix in Equation (5.10) can be defined. These are shape functions of ξ interpolating u and w between their nodal values to give the values at any point on the meridian.

5.5 STRESS RESULTANTS

The resultant of stress in meridian and in circumferential directions (Figure 5.2c) can be expressed as

$$\{\sigma\} = [d_e]\{\varepsilon\}, \tag{5.12}$$

where $\{\sigma\}$ and $\{\varepsilon\}$ represent generalized stress and strain vectors defined as

$$\{\sigma\} = \begin{Bmatrix} N \\ N_\phi \\ M \\ M_\phi \end{Bmatrix}, \qquad (\varepsilon) = \begin{Bmatrix} du/ds \\ (w\cos\alpha + u\sin\alpha)/r \\ -d^2w/ds^2 \\ -(\sin\alpha/r)(dw/ds) \end{Bmatrix}, \tag{5.13}$$

where N and M are the normal force and the moment in the meridian direction per unit length; and N_ϕ and M_ϕ are the normal force and the moment in the circumferential (hoop) direction per unit length. The positive sign conversion of the stress resultants is shown in Figure 5.2c. The elasticity matrix relating the generalized stress and generalized strain (Equation 5.12) is

$$[d_e] = \frac{Eh}{1-v^2}\begin{bmatrix} 1 & v & 0 & 0 \\ v & 1 & 0 & 0 \\ 0 & 0 & h^2/12 & vh^2/12 \\ 0 & 0 & vh^2/12 & h^2/12 \end{bmatrix}, \tag{5.14}$$

where h is element thickness, E is modulus of elasticity, and v is Poisson's ratio.

By substitution for u and w from Equation (5.10) in Equation (5.13), the generalized strain can be expressed in terms of the nodal displacements:

$$\{\varepsilon\} = [B]\{D^*\}, \tag{5.15}$$

where

$$[B] = \tag{5.16}$$

$$\begin{bmatrix} -\frac{1}{l} & 0 & 0 & \frac{1}{l} & 0 & 0 \\ \frac{\sin\alpha}{r}(1-\xi) & \frac{\cos\alpha}{r}L_1 & \frac{\cos\alpha}{r}L_2 & \frac{\sin\alpha}{r}\xi & \frac{\cos\alpha}{r}L_3 & \frac{\cos\alpha}{r}L_4 \\ 0 & \frac{1}{l^2}(6-12\xi) & \frac{1}{l}(4-6\xi) & 0 & \frac{1}{l^2(12\xi-6)} & \frac{1}{l}(2-6\xi) \\ 0 & \frac{\sin\alpha}{r}(6\xi-6\xi^2) & \frac{\sin\alpha}{r}(4\xi-3\xi^2-1) & 0 & \frac{\sin\alpha}{rl}(6\xi^2-6\xi) & \frac{\sin\alpha}{r}(2\xi-3\xi^2) \end{bmatrix}$$

Substitution of Equation (5.15) in Equation (5.12) gives the stress resultants:

$$\{\sigma\} = [d_e][B]\{D^*\} + \{\sigma_r\}. \tag{5.17}$$

The vector $\{\sigma_r\}$ represents the stress resultants when the nodal displacements $\{D^*\} = \{0\}$; $\{\sigma_r\}$ is nonzero only when the analysis is for the effect of volume change (e.g., due to temperature, shrinkage, or swelling). Use of Equation (5.17) with $\xi = 0.5$ and $r = (r_1 + r_2)/2$ gives the stress resultants midway between the nodal lines of the element. These are the values commonly included in computer results.

5.6 STIFFNESS MATRIX OF AN INDIVIDUAL ELEMENT

Consider the finite element in Figure 5.2a subjected to forces $\{F^*\}$ producing displacements $\{D^*\}$. When $\{D^*\} = \{0\}$ the initial stress $\{\sigma_r\}$ is considered null. The work done by the forces $\{F^*\}$ is equal to the strain energy. Thus,

$$\tfrac{1}{2}\{F^*\}^T\{D^*\} = \tfrac{1}{2}\int_{\text{area}} \{\sigma\}^{\mathrm{T}}\{\varepsilon\}\mathrm{d}a, \tag{5.18}$$

where the elemental area da is

$$\mathrm{d}a = 2\pi r\ \mathrm{d}s \tag{5.19}$$

with ds = l d < ξ. Use of Equations (5.5), (5.12), (5.15), and (5.19) to eliminate $\{F^*\}$, $\{\sigma\}$, and $\{\varepsilon\}$ gives the element stiffness matrix with respect to its local coordinates (Figure 5.2a):

$$[S^*] = 2\pi l \int_0^1 r[B]^{\mathrm{T}}[d_{\mathrm{e}}][B]\ \mathrm{d}\xi. \tag{5.20}$$

The radius at any point can be expressed as a function of ξ:

$$r = (1-\xi)r_1 + \xi r_2, \tag{5.21}$$

where r_1 and r_2 are the radii at nodes 1 and 2. Gauss numerical integration may be employed to evaluate the integral in Equation (5.20). The computer program employed to solve the examples presented later uses two Gauss sampling points in evaluating the integrals in Equations (5.20) and (5.24). With this choice, the value of the integral is $(g_1 + g_2)/2$, where g_1 and g_2 are the values of the integrand at $\varepsilon = (3-\sqrt{3})/6$ and $(3+\sqrt{3})/6$. Two sampling points give sufficient accuracy for this finite element.

5.7 ANALYSIS FOR THE EFFECT OF TEMPERATURE

Analysis for the effect of temperature variation is frequently required in the design of tanks and silos. Solution of the problem by the finite-element method is discussed below. The element in Figure 5.2a is subjected to an axisymmetrical rise of temperature varying linearly through the thickness between T_{i} and T_{o} at the inner and outer faces, respectively. First, assume that the nodal displacements are artificially prevented by nodal forces $\{F_r^*\}$. Thus, the thermal expansion is restrained, causing development of the restraining generalized stress $\{\sigma_{\mathrm{r}}\}$, which is given by (Equation 5.12)

$$\{\sigma_{\mathrm{r}}\} = -[d_{\mathrm{e}}]\begin{Bmatrix} \mu(T_{\mathrm{o}}-T_{\mathrm{i}})/2 \\ \mu(T_{\mathrm{o}}+T_{\mathrm{i}})/2 \\ \mu(T_{\mathrm{o}}-T_{\mathrm{i}})/\mathrm{h} \\ \mu(T_{\mathrm{o}}-T_{\mathrm{i}})/\mathrm{h} \end{Bmatrix}, \tag{5.22}$$

where μ is the coefficient of thermal expansion (degree^{-1}). The vector on the right-hand side of this equation represents values of the generalized strain that would occur if the elemental part of the shell in Figure 5.2c were allowed to expand freely. In the restraint state, the element is in equilibrium with the nodal forces $\{F_r^*\}$, producing constant stress $\{\sigma_r\}$, given by Equation (5.22). Using virtual work (unit displacement theory[2]), any of the restraining forces F_{ri}^* can be calculated by introducing a virtual displacement $D_i^* = 1$, with all other nodal displacements equal to zero. The force F_{ri}^* is then given by

$$F_{\mathrm{ri}}^* \int_{\mathrm{area}} \{\varepsilon_{\mathrm{u}i}\}^{\mathrm{T}} \{\sigma_{\mathrm{r}}\} \mathrm{d}a, \tag{5.23}$$

where $\{\varepsilon_{ui}\}$ represents the generalized strain when $D_i^* = 1$; this is the same as the ith column of the B matrix (Equation 5.16). Equation (5.23) applied with $i = 1, 2, \ldots, 6$ gives the vector of restraining forces:

$$\{F_{\mathrm{r}}^*\} = 2\pi l \int_0^1 r[B]^{\mathrm{T}} \{\sigma_{\mathrm{r}}\}\, \mathrm{d}\xi. \tag{5.24}$$

The restraining forces $\{F_r^*\}$ at local coordinates (Figure 5.2a) are statical equivalents to nodal forces $\{\bar{F}_r^*\}$ in global directions (Figure 5.2b), whose values are given by Equation (5.3):

$$\{\bar{F}_{\mathrm{r}}\} = [T]^{\mathrm{T}} \{F_{\mathrm{r}}^*\}. \tag{5.25}$$

Equations (5.24) and (5.25) are applied for all elements, and the element restraining forces $\{\bar{F}_r^*\}$ are assembled to give a vector of restraining forces for the structure. These forces are then applied in a reversed direction to eliminate the artificial restraint and produce nodal displacements due to temperature. When the nodal displacements are determined, Equation (5.17) gives the generalized thermal stress in individual elements.

5.8 ASSEMBLAGE OF STIFFNESS MATRICES AND LOAD VECTORS

Let $[\bar{S}_m]$ be the stiffness matrix, with respect to coordinates in global directions, for the mth element, whose nodes are numbered j and k in the idealized structure in Figure 5.1b. Partition $[\bar{S}_m]$ into four 3 × 3 submatrices as follows:

$$[\bar{S}_m] = \begin{bmatrix} [\bar{S}_{11}] & [\bar{S}_{21}] \\ [\bar{S}_{21}] & [\bar{S}_{22}] \end{bmatrix}. \tag{5.26}$$

The idealized structure, with n_j joints, has a stiffness matrix $[S]$ of size $3n_j \times 3n_j$, which can also be partitioned into 3×3 submatrices. The matrix $[S]$ may be expressed as the sum of the stiffness matrices of individual elements arranged in matrices of the same size as $[S]$:

$$[S] = \sum_{m=1}^{n_e} [\overline{\overline{S_m}}], \tag{5.27}$$

where n_e is the number of elements; $[\overline{\overline{S}}]$ is a matrix composed of $n_j \times n_j$ submatrices of size 3×3; and all but four submatrices are null:

$$[\overline{\overline{S_m}}] = \begin{array}{c} \\ j \\ \\ k \\ \\ \end{array} \begin{array}{c} \\ \\ \\ \\ \\ \end{array} \overset{\qquad\qquad j \qquad\qquad\qquad\quad k}{\begin{bmatrix} \cdots & \cdots & \cdots & \cdots & \cdots \\ \cdots & [\bar{S}_{11}] & \cdots & [\bar{S}_{12}] & \cdots \\ \cdots & \cdots & \cdots & \cdots & \cdots \\ \cdots & [\bar{S}_{12}] & \cdots & [\bar{S}_{22}] & \cdots \\ \cdots & \cdots & \cdots & \cdots & \cdots \end{bmatrix}}. \tag{5.28}$$

The nonzero submatrices in Equation (5.28) are the same as the submatrices in Equation (5.26). The submatrix $[\bar{S}_{11}]$ represents forces at node j due to unit displacements at the same node; $[\bar{S}_{21}]$ represents forces at node k due to unit displacements at node j; and so on. Equation (5.28) gives the contribution of the mth element to the structure stiffness matrix, with respect to coordinates in global directions.

If the mth element is subjected to temperature change, its vector of restraining forces in global directions $[\bar{F}_r]$ (Equation 5.25) may be partitioned into two 3×1 submatrices:

$$\{\bar{F}_r\} = \begin{Bmatrix} \{\bar{F}_r\}_j \\ \{\bar{F}_r\}_k \end{Bmatrix}. \tag{5.29}$$

The upper three forces act at node j; the lower three are forces at node k. The vector of restraining forces $\{F_r\}$ for the structure may also be partitioned into 3×1 submatrices, each of which represents forces at an individual joint. The vector $\{F_r\}$ can be expressed as the sum of contributions of individual elements:

$$\{F_r\} = \sum_{m=1}^{n_e} \{\overline{\overline{F_r}}\}_m, \tag{5.30}$$

where $\{\bar{\bar{F}}_r\}_m$ is a vector composed of n_j submatrices of size 3 × 1; and all the submatrices are null except the two associated with nodes j and k:

$$\{\bar{\bar{F}}_r\}_m = \begin{Bmatrix} \cdots \\ \{\bar{F}_{rj}\} \\ \cdots \\ \{\bar{F}_{rk}\} \\ \cdots \end{Bmatrix}_m \begin{matrix} \\ j \\ \\ k \\ \\ \end{matrix} . \tag{5.31}$$

The nonzero subvectors in Equation (5.31) are the same as the subvectors in Equation (5.29).

The displacements $\{D\}_{3nj} \times 1$ at the nodes are determined by solution of the equation

$$[S]\{D\} = \{F\}, \tag{5.32}$$

where $\{F\}$ is a vector of forces in global directions applied at the nodes. When the analysis for the thermal effect is required, the matrix $\{F\} = \{F_a\} = \{F_r\}$, where $\{F_a\}$ is a vector of applied nodal forces and $\{F_r\}$ is a vector of forces restraining thermal expansion. Before the solution of Equation (5.32), [S] must be adjusted[2] to ensure that $\{D\}$ satisfies the prescribed conditions at specified nodes, for example, zero displacement at a support.

5.9 NODAL FORCES

In using conical finite elements for the idealization of an axisymmetrical shell whose meridian is a curve, a physical approximation is involved. This is done by assuming that a continuous curve can be adequately represented by small straight segments. Figure 5.3a shows an elevation of a curved meridian of an axisymmetrical shell subjected to distributed load normal to the shell surface. Figure 5.3b shows the meridian of the idealized structure. The local effect of the distributed load is more accurately represented in the idealized structure by statically equivalent forces at the nodes. In other words, when the meridian is a curve idealized by straight segments, it is simpler and also more accurate to represent the distributed load in Figure 5.3a by statically equivalent concentrated loads at the nodes.

Figure 5.3c,d is half-sectional elevations of an axisymmetrical conical element subjected to uniformly distributed axisymmetrical loads tangential and normal to the surface. Let the intensities (per unit area) of the tangential and the normal loads be q_t and q_n, respectively. The statically equivalent vector of nodal forces for the element is

$$\{F^*\} = \frac{\pi l}{3}\begin{Bmatrix} 2q_{t1}r_1 + q_{t2}r_2 \\ 2q_{n1}r_1 + q_{n2}r_2 \\ 0 \\ q_{t1}r_1 + 2q_{t2}r_2 \\ q_{n1}r_1 + 2q_{n2}r_2 \\ 0 \end{Bmatrix}, \tag{5.33}$$

where the subscripts 1 and 2 refer to nodes 1 and 2, respectively. This equation implies the assumption that the product (qr) varies linearly over the length l of the meridian. Use of Equation (5.33) in all cases, when the meridian of the actual shell is curved or straight, involves approximation, whose effect diminishes as the element size is reduced.

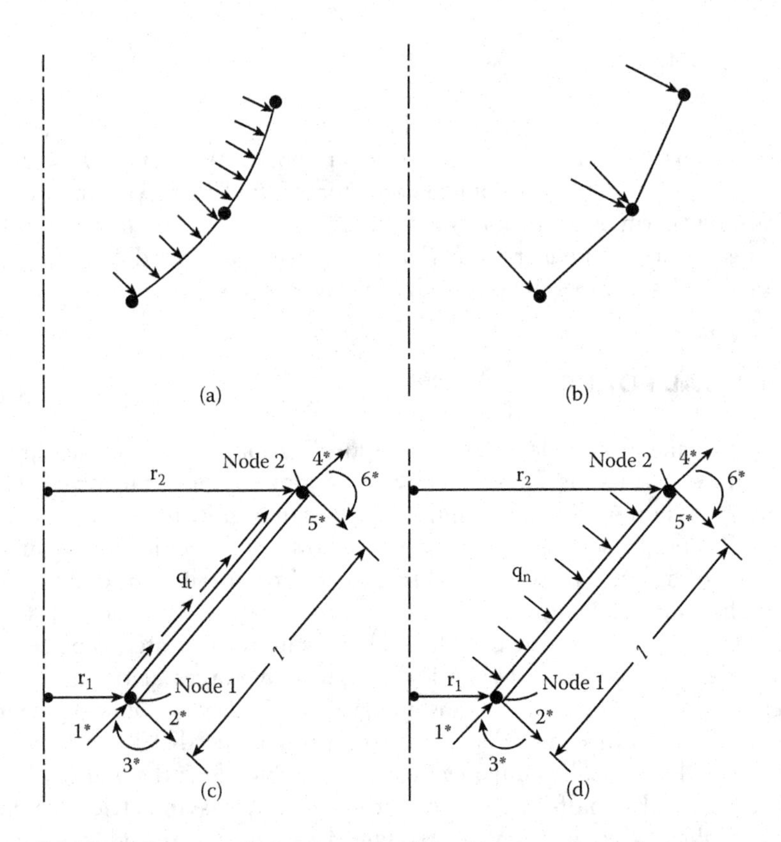

Figure 5.3 Nodal forces equivalent to a distributed load on the surface of an axisymmetrical finite element. (a) Actual load on a shell with a curved meridian. (b) Idealization as conical elements loaded only at the nodes. (c, d) Equivalent nodal forces in local coordinates.

5.10 EXAMPLES

The results of a computer program based on the equations of this chapter are presented next.

EXAMPLE 5.1
Cylindrical Wall, with Uniform Radial Force on Bottom Edge

A cylindrical wall of height l, diameter d = $2.5l$, and thickness $h = 0.05l$ is subjected to uniform inward radial line load of unit intensity at the bottom edge (Figure 5.4a). Poisson's ratio is 1/6.

The variations of displacement w and bending moment M over the wall height are shown in Figure 5.4b,c. To obtain these results, the lower half of the wall height is divided into 10 equal elements. The upper half is divided into six equal elements. For this cylinder, the ratio $l^2/(dh)$ = 8. With this ratio, the wall may be classified as "long" (Chapter 2, Section 2.9); thus, Equation (2.56) can be used to compare the results with the finite-element solution, as shown in Figure 5.4b,c. Figure 5.4d,e shows how the accuracy is affected by the use of a smaller number of elements (five and three, respectively, in the lower and upper halves of the wall).

EXAMPLE 5.2
Cylindrical Wall, Bottom Edge Encastré and Top Edge Free, with Temperature Rise

Consider a circular-cylindrical wall having the same dimensions as in Figure 5.4a, but with the top edge free and bottom edge encastré. The wall is subjected to a rise of temperature, which varies linearly through the wall thickness between T_i at the inner face and $T_o = 0.5T_i$ at the outer face. Poisson's ratio is 1/6.

Figure 5.5a,b shows variations over the wall height of M and M_ϕ, respectively, obtained by finite-element analysis using 40 equal-size elements, compared with the analytical Equations (5.34) and (5.35), which apply for a long cylinder having $l^2/(2rh) > 7.3$:

$$M = -E\mu\left[\frac{r\beta^2 h^3}{12(1-v^2)}(T_o + T_i)Z_3 + \frac{h^2}{12(1-v)}(T_o - T_i)(1-\bar{Z}_1)\right], \qquad (5.34)$$

$$M_\phi = -E\mu\left[\frac{vr\beta^2 h^3}{12(1-v^2)}(T_o + T_i)Z_3 + \frac{h^2}{12(1-v)}(T_o - T_i)(1-v\bar{Z}_1)\right], \qquad (5.35)$$

where

$$\beta = \frac{\left[3(1-v^2)\right]^{1/4}}{\sqrt{rh}}, \qquad (5.36)$$

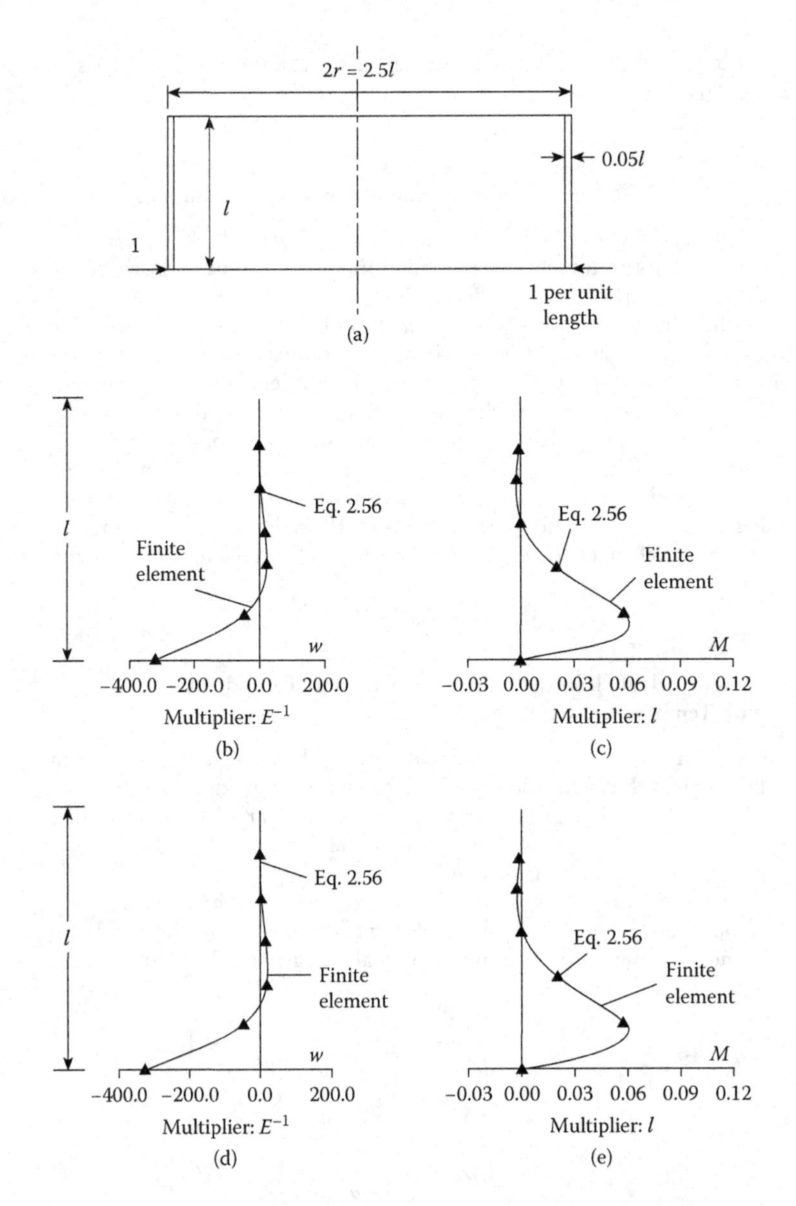

Figure 5.4 Cylindrical wall with free ends subjected to a uniform radial force of unit intensity on the bottom edge (Example 5.1). Comparison of analytical and finite-element solutions. (a) Wall dimensions. (b, c) Radial deflection and meridional moment *M*, respectively, using 16 finite elements. (d, e) Same as (b) and (c), but using eight finite elements.

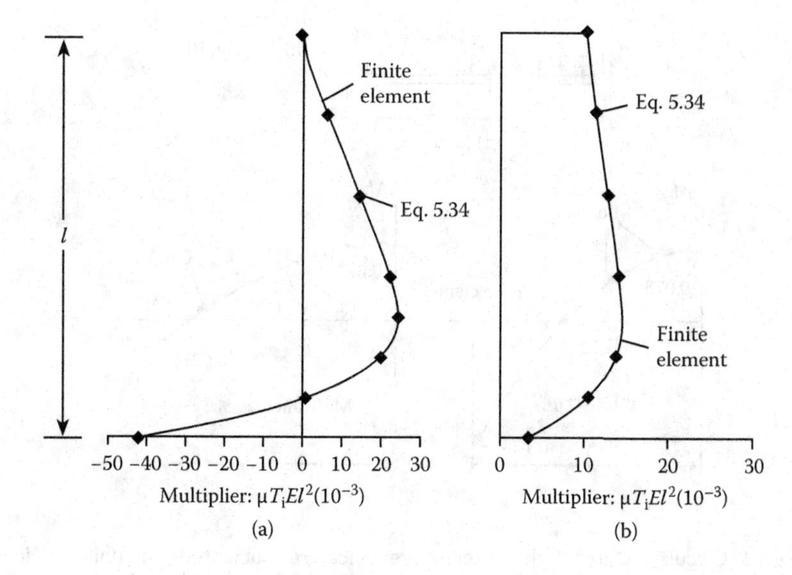

Figure 5.5 The wall in Figure 5.4a with the top edge free and bottom edge encastré (Example 5.2). The effect of temperature rise varying linearly between T_i and $T_o = 0.5T_i$ at the inner and outer surfaces, respectively. (a) Meridional moment M. (b) Circumferential moment Mf.

$$\bar{Z}_1 = e^{-\beta\bar{x}}(\cos\beta\bar{x} + \sin\beta\bar{x}), \tag{5.37}$$

$$Z_3 = e^{-\beta x}(\cos\beta x - \sin\beta x), \tag{5.38}$$

where x is the distance from the bottom edge to any section and $\bar{x} = l - x$. Equations (5.34) and (5.35) are derived in Chapter 7. Equation (5.35) may be considered applicable (with less accuracy) for having $l^2/(2rh) > 5.8$.

EXAMPLE 5.3
Circular Plate, with Outer Edge Encastré, Subjected to Uniform Load

A horizontal circular plate of radius r and thickness $h = r/10$ is shown in Figure 5.6a. With the outer edge encastré, the plate is subjected to a uniform downward load of intensity q per unit area.

The finite-element model, employed to obtain the results in Figure 5.6b,c, has 13 ring-shaped elements: 9 of width $0.1r$ and 4 of width $0.025r$ near the outer edge. Variations of the radial moment M and the circumferential moment M_ϕ are compared with analytical solutions using Equations (A.6) and (A.7) (see Appendix A).

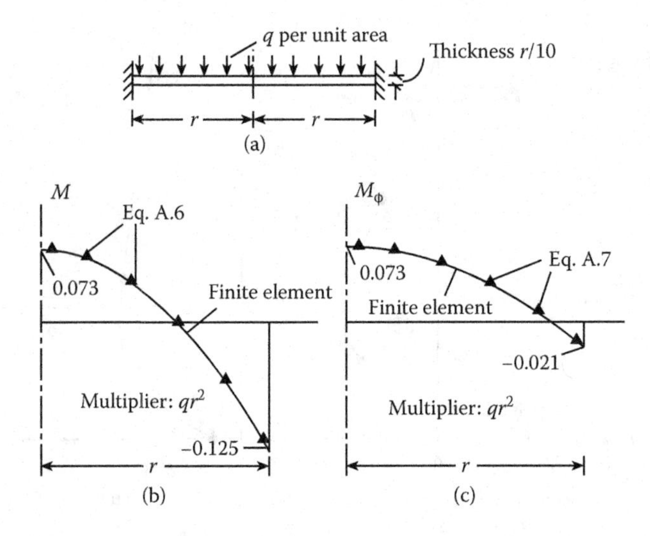

Figure 5.6 Circular plate with outer edge encastré subjected to uniform load (Example 5.3). (a) Plate dimensions. (b) Radial moment M. (c) Circumferential moment M_ϕ.

EXAMPLE 5.4
Circular Cylinder, Monolithic with Spherical Dome

Figure 5.7a is a half sectional elevation of a concrete circular-cylindrical wall monolithic with a concrete spherical dome. The finite-element model is shown in Figure 5.7b. The wall height l is divided into 40 equal-length segments. Poisson's ratio is 1/6 and the specific weight of concrete is γ.

Figure 5.8a,b shows the variations of the meridian bending moment M and the hoop force N_ϕ due to the structure's self-weight. Figure 5.9a,b shows the variations of the same parameters due to three horizontal radial forces of intensities per unit length $p = \{0.25, 0.5, 0.25\}$ on the three nodal lines at A, B, and C (Figure 5.7b). This represents the effect of a hoop prestressing force of magnitude pr on the rim at the top edge of the wall. Comparison of graphs similar to Figures 5.8 and 5.9 can be used in the design for selecting the magnitude of the hoop prestressing force to reduce the peak values of M and N_ϕ to any desired level. The input file and the computer program SOR used for the analysis are available at the companion Web site for this book (see Appendix B).

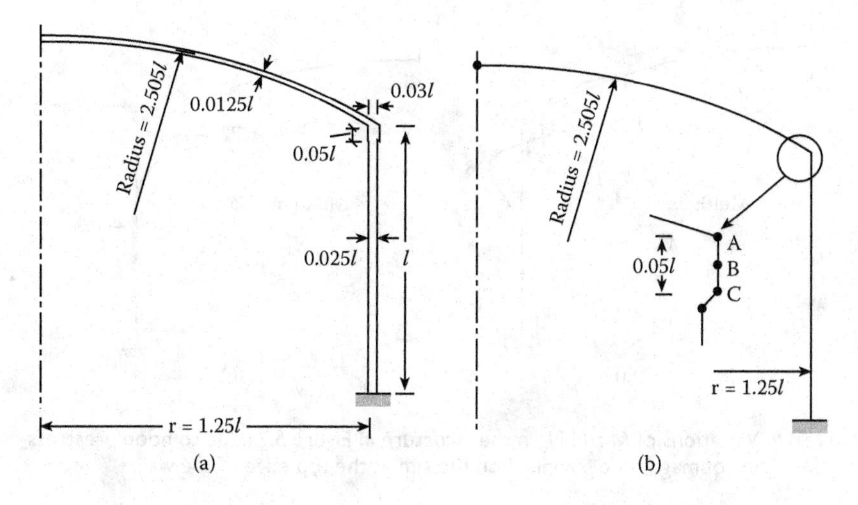

Figure 5.7 Sectional elevation in a circular-cylindrical wall monolithic with a spherical dome (Example 5.4). (a) Structure dimensions. (b) Finite-element idealization, composed of 40 equal-size elements for the cylinder and 60 elements, having equal meridian lengths for the dome.

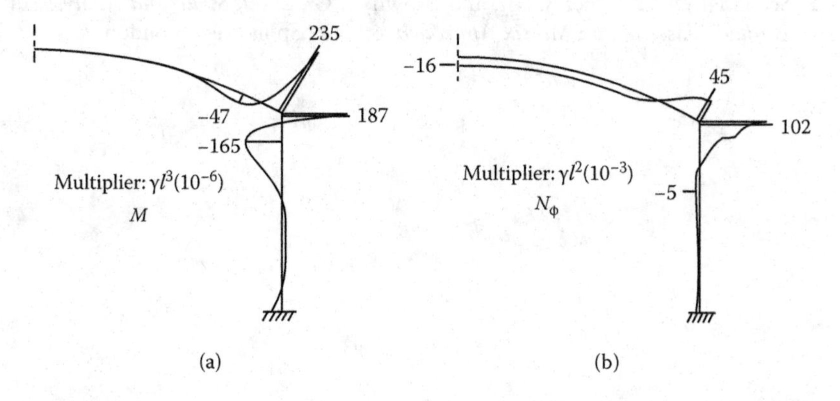

Figure 5.8 Variations of M and N_ϕ in the structure in Figure 5.7a due to the structure's self-weight.

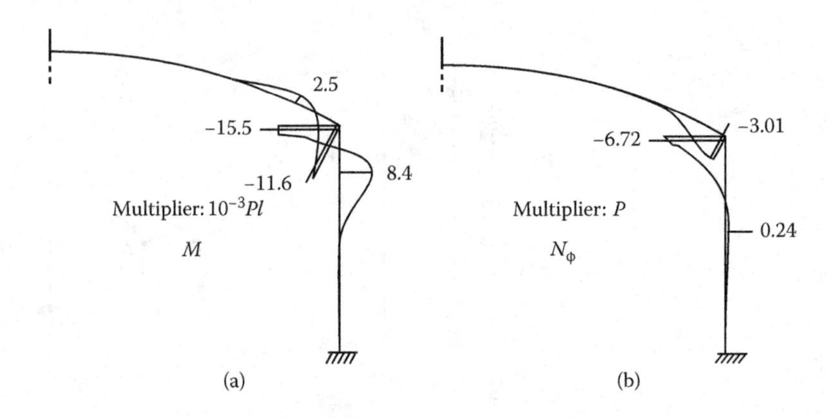

Figure 5.9 Variations of M and N_ϕ in the structure in Figure 5.7a due to hoop prestressing of magnitude p applied on the rim at the top edge of the wall.

NOTES

1. For further reading on the finite-element method and references on its application to axisymmetrical shells, see Zienkiewicz, O. C., and Taylor, R. L., 1989 and 1991, *The Finite Element Method*, 4th ed., McGraw-Hill, New York, vols. 1 and 2, respectively; and Cook, R. D., Makus, D. S., and Plesha, M. E., 1989, *Concepts and Applications of Finite Element Analysis*, 3rd ed., John Wiley, New York.
2. See Ghali, A., Neville, A. M., and Brown, T. G., 2009, *Structural Analysis: A Unified Classical and Matrix Approach*, 6th ed., Spon Press, London.

Chapter 6

Time-dependent effects

6.1 INTRODUCTION

A section of a wall of a tank or a silo composed of concrete with prestressed and non-prestressed reinforcements, subjected to sustained normal force and bending moment, undergoes gradual changes in stress distribution in the three constituents. This is caused by creep and shrinkage of concrete as well as by relaxation of prestressed steel. The same causes can also produce changes in the magnitudes of the normal force and the bending moment at various sections. This chapter presents analyses of these time-dependent effects.

A cylindrical wall subjected to uniformly distributed vertical prestress, at midwall surface, or to linearly varying circumferential prestress, with top and bottom edges free to rotate and translate in the radial direction, is an example of a case where creep, shrinkage, and relaxation gradually change the distribution of stress in the concrete and the reinforcements, without change in stress resultants. However, when the edge displacements are restrained, the internal forces also change with time. A cylindrical wall, having free top edge and bottom edge supported on bearing pads, providing an elastic support, is an example of a frequently occurring case where the magnitude of the reaction at the elastic support and the bending moment due to circumferential prestress undergo significant variations with time. Important time-dependent variations of internal forces can also occur in a tank due to differences in ages of concrete in the base, the wall, and the cover; this is due to differential elasticity moduli, creep coefficients, and magnitudes of shrinkage.

6.2 CREEP AND SHRINKAGE OF CONCRETE

A typical stress–strain graph for concrete is shown in Figure 6.1. In service condition, where the compressive stress rarely exceeds 50% of concrete

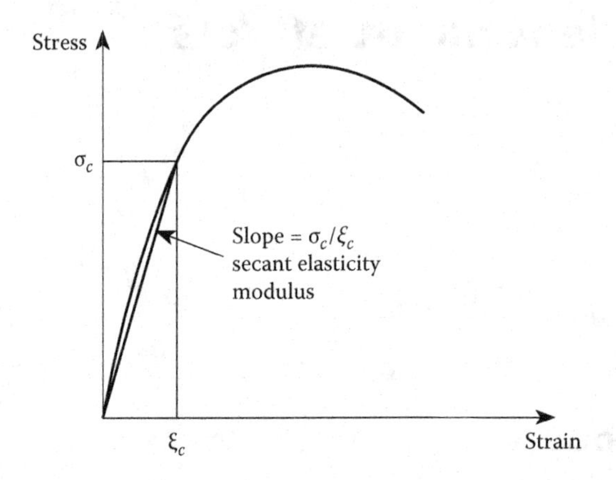

Figure 6.1 Stress–strain relationship for concrete.

strength, it is a common practice to express the immediate strain, occurring within seconds of a stress application, as follows:

$$\varepsilon_c(t_0) = \frac{\sigma_c(t_0)}{E_c(t_0)}, \tag{6.1}$$

where $\sigma_c(t_0)$ is the concrete stress introduced at time t_0; and $E_c(t_0)$ is the modulus of elasticity of concrete at the age t_0. The value of E_c, the secant modulus, defined in Figure 6.1, depends upon the magnitude of the stress, but this dependence is ignored in service condition, which is the condition that this chapter is concerned with.

A stress increment $\sigma_c(t_0)$ introduced at time t_0 and sustained, without change in magnitude, until time t (Figure 6.2a) produces at time t a total strain (instantaneous plus creep) given by (see Figure 6.2b)

$$\varepsilon_c(t) = \frac{\sigma_c(t_0)}{E_c(t_0)}[1 + \phi(t, t_0)], \tag{6.2}$$

where $\phi(t,t_0)$ is creep coefficient, a function of ages of concrete t and t_0, representing the ratio of creep to instantaneous strain. The value of ϕ increases with the decrease in t_0 and the increase in length of the period $(t - t_0)$ during which the stress is sustained.

The dashed line in Figure 6.2a represents a stress increment whose magnitude is gradually increased from zero at t_0 to a final value σ_c at time t. When this case is represented by a number of small stress increments introduced at variable time τ, such that $t_0 < \tau < t$, the total strain at time t

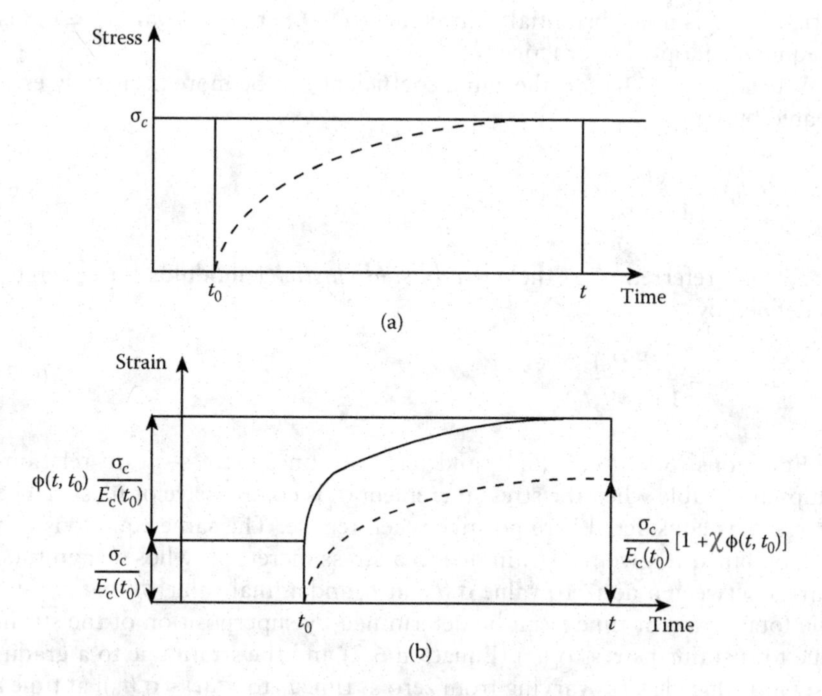

Figure 6.2 Definitions of creep and aging coefficients: (a) stress versus time, (b) strain versus time.

can be expressed as the sum of the effect of individual increments. Using Equation (6.2) for each,

$$\varepsilon_c(t) = \sum_{i=1}^{n} \frac{\Delta\sigma_{ci}}{E_c(\tau_i)}[1+\phi(t,\tau_i)], \tag{6.3}$$

where n is the total number of increments, and $\Delta\sigma_{ci}$ is the ith stress increment introduced at τ_i. In practice this equation is approximated (see Figure 6.2b)

$$\varepsilon_c(t) = \frac{\sigma_c}{E_c(t_0)}[1+\chi\phi(t,t_0)] \tag{6.4}$$

or

$$\varepsilon_c(t) = \frac{\sigma_c}{\bar{E}_c(t,t_0)}, \tag{6.5}$$

where χ, referred to as the *aging coefficient*, is a function of (t,t_0). Variations of E_c and ϕ with time also affect the value of χ.[1] However, in many practical situations χ is not substantially different from 0.8; thus the value $\chi \simeq 0.8$ is frequently adopted in practice.

When $(t - t_0) \geq 1$ year, the aging coefficient can be more accurately estimated by

$$\chi(t,t_0) \simeq \frac{\sqrt{t_0}}{1+\sqrt{t_0}}. \tag{6.6}$$

$\bar{E}_c(t,t_0)$, referred to as the *age-adjusted elasticity* modulus for concrete, is defined by

$$\bar{E}_c(t,t_0) = \frac{E_c(t_0)}{1+\chi\phi(t,t_0)}. \tag{6.7}$$

Equations (6.1), (6.2), (6.4), and (6.5) are linear stress–strain relationships applicable when the stress increment σ_c is compressive or tensile. The stress σ_c is considered here positive when tensile. The same equations can be combined to give the strain due to a stress increment whose magnitude varies between a nonzero value $\sigma_c(t_0)$ at t_0 and a final value $\sigma_c(t)$ at t. Here the total strain at time t can be determined by superposition of the strain due to sustained stress $\sigma_c(t_0)$ (Equation 6.2) and the strain due to a gradually introduced stress varying from zero at time t_0 to $[\sigma_c(t) - \sigma_c(t_0)]$ at time t (Equation 6.4).

Drying of concrete in air causes shrinkage, while concrete in contact with water swells. When free to occur, the volumetric change, the shrinkage or the swelling, develops gradually with time without stress. The symbol $\varepsilon_{cs}(t,t_0)$ will be used for the free (unrestrained) strain due to shrinkage or swelling in the period t_0 to t; ε_{cs} is considered positive when it represents elongation, thus for shrinkage ε_{cs} is a negative value.

In tanks and silos, shrinkage or swelling of concrete is always restrained, thus stresses develop. The restraint can be caused by the presence of reinforcing steel; by the supports at the wall edges; or by the differential values of ε_{cs} for the wall, the base, and the cover. The stress due to shrinkage $\varepsilon_{cs}(t,t_0)$ develops gradually in the period t_0 to t, thus shrinkage is always accompanied by creep. This substantially alleviates shrinkage stresses; analysis of shrinkage stresses will always account for creep.

The strain that develops due to free shrinkage between t_s and later instant t may be expressed by

$$\varepsilon_{cs}(t,t_0) = \varepsilon_{co}\beta_s(t - t_s), \tag{6.8}$$

where ε_{co} is the total shrinkage that occurs after concrete hardening up to time infinity. The value ε_{co} (between -300×10^{-6} and -800×10^{-6}) depends upon the quality of concrete, the ambient humidity, and the thickness of the element. β_s is a function of the length of time $(t - t_s)$, with t_s being the time at which curing stops. The free shrinkage, $\varepsilon_{cs}(t,t_0)$, occurring between any two instants t_0 and t is given by the difference between $\varepsilon_{cs}(t,t_s)$ and $\varepsilon_{cs}(t_0,t_s)$ determined by Equation (6.8).

The concrete parameters required for the time-dependent analyses are $E_c(t_0)$, $\phi(t,t_0)$, $\chi(t,t_0)$, and $\varepsilon_{cs}(t,t_0)$. Codes[2] and technical committee reports[3] give guidance on the values to be used in practice.

6.3 RELAXATION OF PRESTRESSED STEEL

The *intrinsic relaxation* $\Delta\sigma_{pr}$ is defined as the change in stress in a tendon stretched and maintained at a constant length and temperature over a period of time; $\Delta\sigma_{pr}$ is a negative value, representing loss of tension. The intrinsic relaxation magnitude depends upon the quality of steel and the value of the initial stress. For an initial stress of 70% of tensile strength, $\Delta\sigma_{pr}$ that takes place in 1000 h varies between 2.5% and 8.0% of the initial stress. Three times these values are expected after 50 years or more.

The magnitude of the intrinsic relaxation depends heavily on the value of the initial stress. A tendon in a prestressed concrete member does not have a constant length as is the case in a relaxation test to determine the intrinsic relaxation. Owing to creep and shrinkage, the distance between anchorages in the prestressed member is shortened, thus causing faster reduction of tension compared to the reduction in a constant-length test. This has a similar effect on the magnitude of relaxation as if the initial stress were smaller. Thus, the relaxation value to be used in prediction of the loss of prestress in concrete tanks and silos should be smaller than the intrinsic relaxation obtained from a constant-length test. For this purpose, we define the *reduced relaxation* value to be used in analysis of concrete structures:

$$\Delta\bar{\sigma}_{pr} = \chi_r \Delta\sigma_{pr}, \tag{6.9}$$

where $\Delta\sigma_{pr}$ is the intrinsic relaxation (this is the value of relaxation that would occur in a constant-length relaxation test), and χ_r is a dimensionless coefficient, commonly smaller than 1.0. The value of χ_r can be expressed as a function of the initial stress and other parameters depending upon creep and shrinkage.[4] For many practical cases, χ_r is not substantially different from 0.8; the value $\chi \simeq 0.8$ is often adopted.

6.4 BASIC EQUATIONS FOR STRESS AND STRAIN DISTRIBUTIONS IN A HOMOGENEOUS SECTION

Figure 6.3 represents a cross-section subjected to a normal force N at an arbitrary reference point, combined with a bending moment M. This may

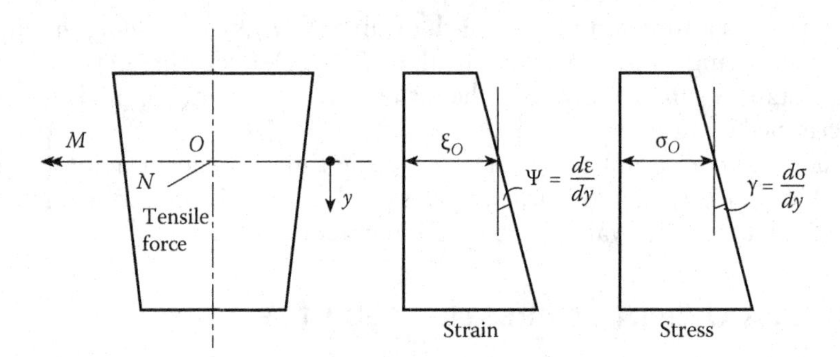

Figure 6.3 Positive sign convention for y, N, M, ε_O, ψ, σ_O, and γ.

represent a horizontal section in a vertical strip representing a wall of a tank subjected to vertical prestressing combined with a bending moment. The distributions of stress σ and strain ε due to N combined with M are shown in Figure 6.3. The common assumption that plane cross-sections remain plane, combined with the assumption that the material is elastic, leads to the straight-line variation of σ and ε:

$$\varepsilon = \varepsilon_O + \psi y, \qquad \sigma = \sigma_O + \gamma y, \tag{6.10}$$

where ε_O and σ_O are values of strain and stress at the reference point O; ψ and γ are equal to $(d\varepsilon/dy)$ and $(d\sigma/dy)$, which are the slopes of the strain and the stress diagrams, respectively; ψ is also equal to the curvature. ε_o and ψ will be referred to as strain parameters, defining the strain distribution. Similarly, σ_O and γ will be defined as stress parameters, defining the stress distribution. The modulus of elasticity E relates the strain and the stress parameters:

$$\sigma_o = E\varepsilon_o, \qquad \gamma = E\psi. \tag{6.11}$$

Now we assume that the section is made of homogeneous material whose elasticity modulus is E.

The forces N and M are equal to the resultants of the stress expressed by Equation (6.10); thus,

$$\begin{aligned} N &= \int \sigma \, dA = \sigma_o \int dA + \gamma \int y \, dA \\ M &= \int \sigma y \, dA = \sigma_o \int y \, dA + \gamma \int y^2 \, dA \end{aligned} \tag{6.12}$$

The integrals $\int dA$, $\int y\,dA$, and $\int y^2\,dA$ are, respectively, equal to the cross-sectional area A, its first moment B, and its second moment I about an axis through the reference point O. Thus, Equation (6.12) may be rewritten:

$$N = \sigma_o A + \gamma B, \qquad M = \sigma_o B + \gamma I . \tag{6.13}$$

Equation (6.13) can be solved for the stress parameters:

$$\sigma_o = \frac{IN - BM}{AI - B^2}, \qquad \gamma = \frac{-BN + AM}{AI - B^2}. \tag{6.14}$$

Substituting these two equations in Equation (6.11) gives the strain parameters:

$$\varepsilon_o = \frac{IN - BM}{E(AI - B^2)}, \qquad \psi = \frac{-BN + AM}{E(AI - B^2)}. \tag{6.15}$$

When the reference point O is at section centroid, $B = 0$. Substituting this equation in Equations (6.14) and (6.15), they will take the more familiar forms:

$$\sigma_o = \frac{N}{A}, \qquad \gamma = \frac{M}{I}, \tag{6.16}$$

$$\varepsilon_o = \frac{N}{EA}, \qquad \psi = \frac{M}{EI}. \tag{6.17}$$

The aforementioned equations will be applied below for cross-sections composed of concrete, non-prestressed steel, and prestressed steel, using for A, B, and I the properties of a transformed section with elasticity modulus equal to that of concrete, E_c. The transformed section is composed of A_c, $(\alpha A)_{ns}$, and $(\alpha A)_{ps}$, where A is the cross-sectional area; the subscripts c, ns, and ps refer to concrete, non-prestressed steel, and prestressed steel; and α is modular ratio

$$\alpha_{\text{ns or ps}} = \frac{E_{\text{ns or ps}}}{E_c}. \tag{6.18}$$

The equations presented in this section will also be used in the following for an *age-adjusted transformed* section whose area properties are $\bar{A}$, $\bar{B}$, and $\bar{I}$. The age-adjusted transformed section has an elasticity modulus $\bar{E}_c$

(defined by Equation 6.7) and an area composed of A_c, $(\bar{\alpha}A)_{ns}$, and $(\bar{\alpha}A)_{ps}$, where

$$\bar{\alpha}_{\text{ns or ps}} = \frac{E_{\text{ns or ps}}}{\bar{E}_c}. \tag{6.19}$$

When the section is cracked, the basic equations presented in this section can be used, with A_c being the area of concrete subjected to compressive stress; the area of concrete in the tension zone is ignored.

Analysis of the time-dependent stresses and strains involves superposition of strain and stress parameters determined for different transformed sections. The superposition is more conveniently done with a noncentroidal reference point O, whose position does not change in the analysis steps.

6.5 TIME-DEPENDENT STRESS AND STRAIN IN A SECTION

Figure 6.4 represents a general concrete section containing prestressed and non-prestressed steel. At time t_0 the section is subjected to a normal force N at reference point O, combined with a bending moment M. The values of N and M include the effect of the prestressing, which is assumed to be introduced at the same instant t_0. The normal force and the moment produced by other loads applied at t_0 are included in the values of N and M. Thus, N and M represent resultants of the normal stress $\sigma(t_0)$ introduced at t_0. The meaning of the symbols N and M will be further explained and demonstrated by Examples 6.1 and 6.2. In tanks or silos, the general cross-section in Figure 6.4 may represent a horizontal section in a vertical strip from a cylindrical wall; it may also represent a cross-section of a member from the cover or the base of a tank.

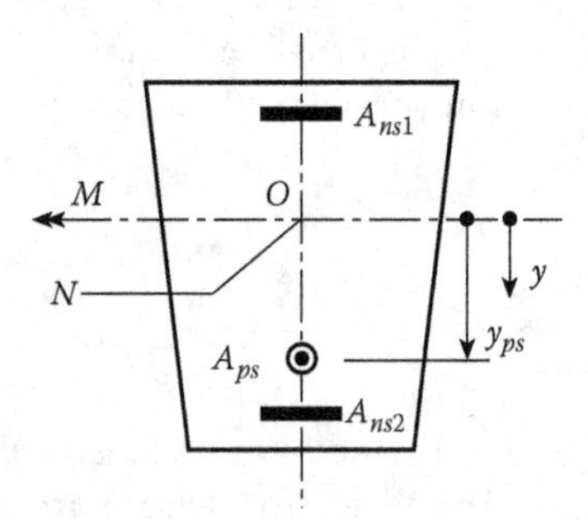

Figure 6.4 Reinforced concrete section considered for analysis of immediate and time-dependent changes of strain and stress.

The following four analysis steps give distributions of strain and stress changes at time t_0 and time t after the occurrence of creep and shrinkage of concrete and relaxation of prestressed steel. Initially, before introducing N and M, the section may have a stress, whose distribution is defined by the parameters $\{\sigma_O, \gamma\}_{initial}$; the values of these parameters are assumed to be known. Thus, when the loading or prestressing is applied in multistages, the four analysis steps presented next can be used to find the immediate and the time-dependent changes in stress at t_0 and in the period $(t - t_0)$ with t_0 and t representing, respectively, an instant at which a load or prestressing is applied, and an instant at which stress or strain distribution is required.

1. Apply Equations (6.14) and (6.15) to determine the instantaneous stress and strain parameter changes: $\sigma_O(t_0)$, $\gamma(t_0)$, $\varepsilon_O(t_0)$, and $\psi(t_0)$. In these calculations, use properties of the transformed section existing at the instant t_0. When post-tensioning is employed, exclude A_{ps} and the cross-sectional area of the duct for each tendon prestressed at t_0. Tendons stressed and ducts grouted at earlier stages should be included.
2. Determine the hypothetical changes in strain parameters that would occur if concrete can deform freely. The hypothetical strain parameter changes are

$$\varepsilon_{O,free} = \phi(t,t_0)\ \varepsilon_O(t_0) + \varepsilon_{cs}, \tag{6.20}$$

$$\psi_{free} = \phi(t,t_0)\psi(t_0). \tag{6.21}$$

3. Calculate the artificial concrete stress that, when gradually introduced on the concrete in the period (t_0 to t), will prevent the occurrence of strain calculated in step 2. The restraining stress at any fiber is (Equation 6.5)

$$\sigma_{restraint} = -\bar{E}_c\{\phi(t,t_0)[\varepsilon_O(t_0) + y\psi(t_0)] + \varepsilon_{cs}\}, \tag{6.22}$$

where $\bar{E}_c = \bar{E}_c(t,t_0)$ is an age-adjusted elasticity modulus of concrete (Equation 6.7); and $\varepsilon_{cs} = \varepsilon_{cs}(t,t_0)$ is the free shrinkage between t_0 and t. The stress parameters that define $\sigma_{restraint}$ are

$$\sigma_{O,restraint} = -\bar{E}_c[\phi(t,t_0)\varepsilon_O(t_0) + \varepsilon_{cs}], \tag{6.23}$$

$$\gamma_{restraint} = -\bar{E}_c\phi(t,t_0)\psi(t_0). \tag{6.24}$$

4. Substitute $\{\sigma_O, \gamma\}_{restraint}$ in Equation (6.13) to determine values of forces restraining creep and shrinkage. Prevent the change in concrete strain due to relaxation of prestressed steel by application of an artificial

restraining force ($A_{ps}\Delta\bar{\sigma}_{pr}$) at the level of prestressed steel. Replace this force by a force of the same magnitude at O plus a couple equal to ($A_{ps}\Delta\bar{\sigma}_{pr}y_{ps}$). Sum up to obtain $\{\Delta N, \Delta M\}_{restraint}$; the restraining force at O; and the couple required to artificially prevent the strain change due to creep, shrinkage, and relaxation. Eliminate the artificial forces by the application of $\{\Delta N, \Delta M\}_{restraint}$ in reversed directions on an age-adjusted transformed section and calculate the corresponding changes in strain and stress parameters by Equations (6.15) and (6.14).

The combined instantaneous plus time-dependent change in strain parameters is the sum of strain parameters determined in steps 1 and 4; the corresponding change in stress parameters is the sum of stress parameters determined in steps 1, 3, and 4.

There are a number of comments that need to be made, as follows:

The flow chart in Figure 6.5 indicates the application sequence of the four steps. If, after step 1, the stress at an extreme fiber exceeds the tensile

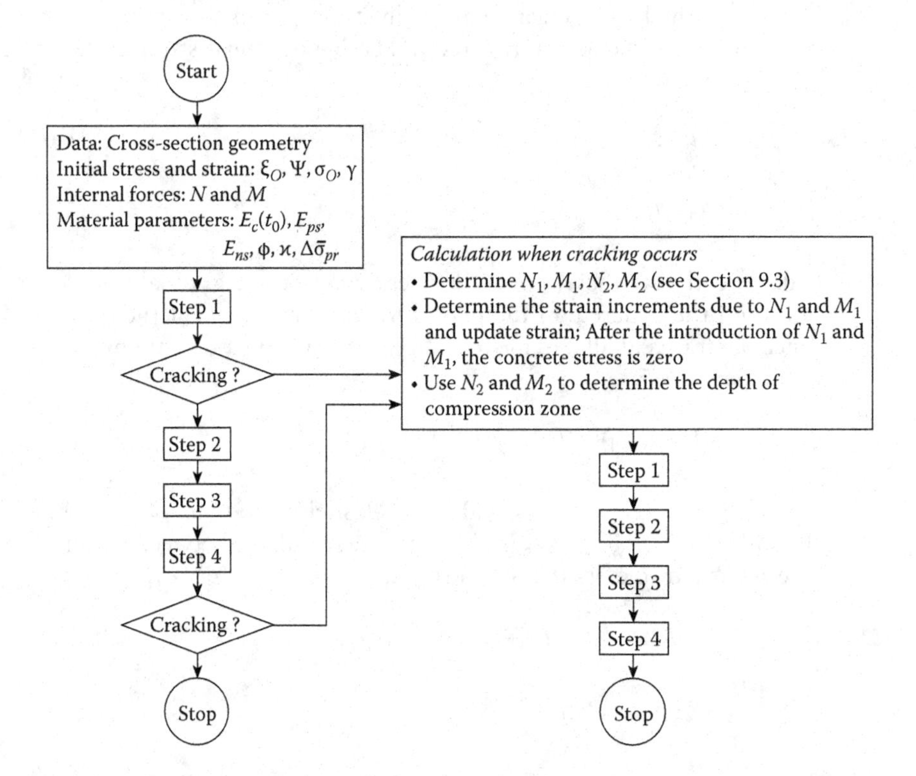

Figure 6.5 Flow chart for the calculation of stress and strain changes in a section due to normal force N and bending moment M introduced at time t_0 and sustained to time t.

strength of concrete, the calculation in step 1 must be repeated, using A, B, and I of the cracked section, in which concrete in tension is ignored. The compression zone depth c must be determined prior to applying steps 2, 3, and 4 to the cracked section in steps 2, 3, and 4. The flow chart also outlines the sequence of analysis steps in a less common case, in which cracking occurs during the period t_0 to t; this will be detected only at the end of step 4. Although the four analysis steps apply to noncracked or cracked sections,[5] the time-dependent analysis examples discussed in this book will ignore cracking.

The four steps give stress and strain at time t, including the effect of loss in prestress force due to creep, shrinkage, and relaxation. Thus, estimation of this prestress loss before the stress analysis is not needed. Only the loss in prestress force due to friction should be included in the calculation of N and M of the input data, when post-tensioned tendons are employed.

The analysis satisfies the requirements of compatibility and equilibrium: the strain changes in any reinforcement layer and adjacent concrete are equal; the time-dependent effects change the partitioning of forces between the concrete and the reinforcements, but do not change the stress resultants.

The same four steps apply to a reinforced concrete section without prestressing, simply by setting $A_{ps} = 0$.

6.5.1 Special case: Section subjected to axial force only

Figure 6.6 represents a section prestressed at time t_0; it is required to determine stress in concrete and reinforcements at time t, after the occurrence of creep and shrinkage of concrete and relaxation of prestressed steel. It is assumed that the centroids of areas of concrete (A_c), the prestressed reinforcement (A_{ps}), and the non-prestressed reinforcement (A_{ns}) coincide. In this special case, ΔP_c, ΔP_{ns}, and ΔP_{ps} are situated at the same centroid; the

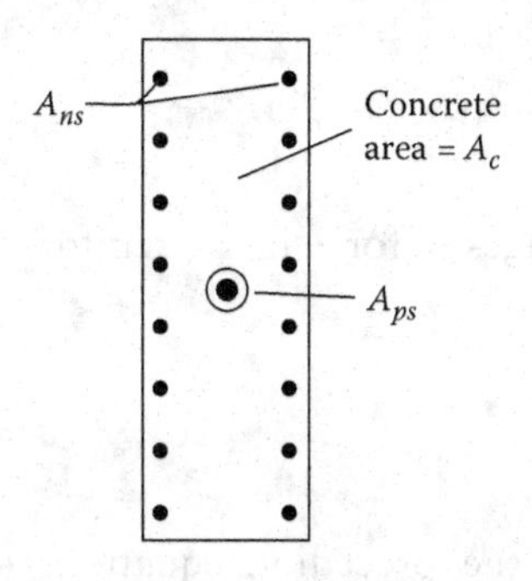

Figure 6.6 A time-dependent prestress loss in a cross-section for which the equations in Section 6.5.1 apply.

symbol ΔP represents change in force in the period t_0 to t; the subscripts c, ns, and ps refer to concrete, non-prestressed reinforcement, and prestressed reinforcement, respectively. The four steps of analysis discussed earlier lead to Equation (6.25), which gives the time-dependent change in resultant force in concrete:

$$\Delta P_c = -\beta^{-1}\left[\phi\sigma_c(t_0)A_s\frac{E_s}{E_c(t_0)} + \varepsilon_{cs}E_sA_s + \Delta\bar{\sigma}_{pr}A_{ps}\right], \tag{6.25}$$

$$\beta = 1 + \frac{A_sE_s}{A_c\bar{E}_c}, \tag{6.26}$$

where $A_s = A_{ns} + A_{ps}$; E_s is the modulus of elasticity of reinforcement (assumed to be the same for the prestressed and the non-prestressed reinforcements); $\phi = \phi(t, t_0)$ and $\varepsilon_{cs} = \varepsilon_{cs}(t, t_0)$; $\bar{E}_c \equiv \bar{E}_c(t,t_0)$ (Equation 6.7); and $\sigma_c(t_0)$ is the stress in concrete at t_0, immediately after prestressing. When post-tensioning is used, $cc(t_0)$ is equal to $(-P/A)$, where P is the absolute value of the prestress force, and A is the area of the transformed section composed of the area of concrete (A_c) excluding the prestress duct plus (αA_{ns}), with $\alpha = E_{ns}/E_c(t_0)$.

The time-dependent change in concrete stress is

$$\sigma_c(t,t_0) = \Delta P_c/A_c. \tag{6.27}$$

This value, commonly positive, represents the loss in compression in concrete. The change in axial strain between t_0 and t is

$$\varepsilon_o(t,t_0) = \varepsilon_{cs} + \phi\frac{\sigma_c(t_0)}{E_c(t_0)} + \frac{\sigma_c(t,t_0)}{\bar{E}_c}. \tag{6.28}$$

The changes in stress in the reinforcements in the same period are

$$\sigma_{ns}(t,t_0) = E_s\varepsilon_O(t,t_0), \tag{6.29}$$

$$\sigma_{ps}(t,t_0) = E_s\varepsilon_O(t,t_0) + \Delta\bar{\sigma}_{pr}. \tag{6.30}$$

The corresponding changes in force in the reinforcements are

$$\Delta P_{ns} = A_{ns}\sigma_{ns}(t,t_0), \tag{6.31}$$

$$\Delta P_{ps} = A_{ps}\sigma_{ps}(t,t_0). \tag{6.32}$$

It can be shown that the preceding equations satisfy the equilibrium requirement

$$\Delta P_c = \Delta P_{ns} + \Delta P_{ps} = 0, \tag{6.33}$$

which means that the resultant force on the section (composed of the three materials) does not change due to creep, shrinkage, and relaxation. Commonly ΔP_{ns} and ΔP_{ps} are negative quantities; the first represents compression picked up by the non-prestressed steel as concrete shortens due to creep and shrinkage; the second quantity represents the time-dependent loss in tension in the prestressed reinforcement. The absolute value $|\Delta P_{ps}|$ is equal to ΔP_c only in the absence of A_{ns}; the difference between these two values is ΔP_{ns}. Thus, ignoring the presence of non-prestressed reinforcement results in underestimation of $\sigma_c(t,t_0)$, which represents the loss in the precompression in concrete. It is this value that the designer should be concerned with to determine whether a section is cracked due to forces applied on the section after the occurrence of prestress loss.

The equations presented in this section apply when, at any instant between t_0 and t, the stress is uniform on the section. This is the case when A_{ps}, A_{ns}, and A_c have the same centroid. This will also be the case when the section in Figure 6.6 represents a vertical section of circular-cylindrical wall subjected to axisymmetrical loads. Because of such loading, or circumferential prestressing, the stress is uniform on all fibers and the equations presented apply, even when the centroids of A_{ps}, A_{ns}, and A_c do not coincide; this will be discussed further in Section 6.6.1.

EXAMPLE 6.1
Time-Dependent Stresses in a Prestressed Section: Effect of Presence of Non-Prestressed Steel

The section in Figure 6.6 is prestressed at time t_0, such that the stress in concrete immediately after prestressing is $\sigma_c(t_0)$ = −5.00 MPa (0.725 ksi). Determine the changes in stress in concrete and in the tension in the prestressed steel due to creep, shrinkage, and relaxation between t_0 and a later time t. Study the effect on the results of varying the non-prestressed reinforcement ratio $\rho_{ns} = A_{ns}/A_c$ from zero to 1.0%. The following data are given: $\rho_{ps} = A_{ps}/A_c = 0.004$; $\phi(t,t_0) = 2$; $\varepsilon_{cs}(t,t_0) = -300 \times 10^{-6}$; $\Delta\bar{\sigma}_{pr} = -50$ MPa (−7.25 ksi); $E_c(t_0)$ = 30 GPa (4350 ksi); E_s = 200 GPa (29000 ksi); $\chi(t,t_0) = 0.8$.

The age-adjusted elasticity modulus of concrete (Equation 6.7):

$$\bar{E}_c(t,t_0) = \frac{30\times10^9}{1+0.8(2.0)} = 11.54 \text{ GPa}, \tag{6.34}$$

$$A_s = A_c(0.004 + \rho_{ns}). \tag{6.35}$$

Table 6.1 Effect of non-prestressed steel ratio on the losses of precompression in concrete and tension in the prestressed reinforcement

ρ_{ns} (%)	0.0	0.2	0.4	0.6	0.80	10
$\sigma_c(t,t_0)$ (MPa)	0.661	0.870	1.066	1.250	1.424	1.588
$\sigma(t,t_0)$ (ksi)	0.096	0.126	0.155	0.181	0.206	0.230
$[\sigma_c(t, t_0)/\sigma_c(t_0)]100$	13.2	17.4	21.3	25.0	28.5	31.8
σ_{ps} (t,t_0)(MPa)	−165	−161	−158	−155	−152	−149
$\sigma_{ps}(t,t_0)$ (ksi)	−23.9	−23.3	−22.9	−22.5	−22.0	−21.6

Substitution in Equations (6.26), (6.25), (6.27), (6.28), (6.30), and (6.32) gives

$$\beta = 1.0693 + 17.33\rho_{ns},$$

$$\Delta P_c = -\beta^{-1}(0.7067 + 126.7\rho_{ns})10^6 A_c,$$

$$\sigma_c(t,t_0) = -\beta^{-1}(0.7067 + 126.7\rho_{ns})10^6, \tag{6.36}$$

$$\varepsilon_O(t,t_0) = -10^{-6}(633) + (11.54\times 10^9)^{-1}\sigma_c(t,t_0) \quad \text{(N, m units)},$$

$$\sigma_{ps}(t,t_0) = 200\times 10^9 \varepsilon_O(t,t_0) - 50\times 10^6 \quad \text{(N, m units)}.$$

Substitution of variable values of ρ_{ns} in the preceding equations gives the results in Table 6.1.

The results indicate that the presence of non-prestressed reinforcement significantly increases the loss in the precompression in concrete (from 13.2% to 31.8%). This is accompanied by a relatively small reduction in the loss in tension in the prestressed reinforcement.

6.6 NORMAL FORCE *N* AND BENDING MOMENT *M* DUE TO PRESTRESSING

Elastic analysis, assuming the material to be homogeneous isotropic, is commonly used to determine the normal force N and the bending moment M at any section of wall, base, or cover of a tank or silo. Here the symbol N means a normal force at the section centroid. The effect of a prestressing tendon on the concrete can be analyzed by application of a system of self-equilibrating forces.

A straight tendon (Figure 6.7a) exerts on the concrete two inward forces at the anchorages. Where a straight tendon changes direction through an angle $\Delta\theta$, it exerts on the concrete a force whose magnitude is $2P\sin(\Delta\theta/2)$, where P is the absolute value of the prestressing force. When $\Delta\theta$ is small,

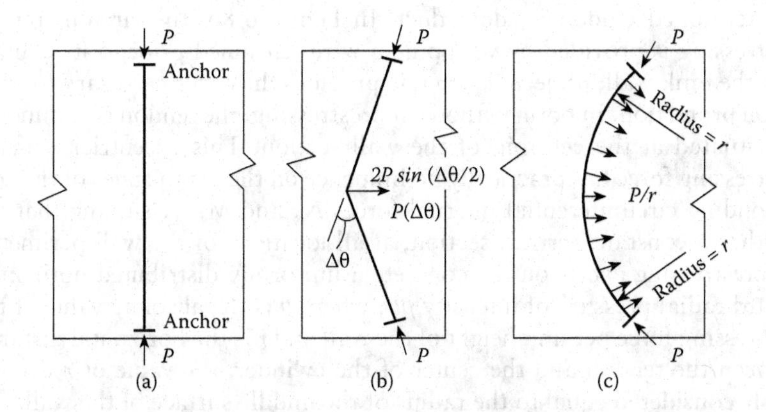

Figure 6.7 Forces exerted on concrete by prestressed tendons: (a) Straight tendon; (b) two straight segments of a tendon; (c) a tendon following an arc of a circle.

the magnitude of this force may be considered equal to $P(\Delta\theta)$. This force acts along the bisector of the angle formed by the two parts of the tendon (Figure 6.7b).

If the change in tendon direction occurs over an arc of a circle of radius r, the tendon exerts on the concrete a uniform radial force of intensity P/r (Figure 6.7c). Owing to friction, which commonly exists between the tendon and the inner wall of its duct, the magnitude P of the prestressing force is variable over the tendon length. Thus, for equilibrium, tangential forces must also exist distributed over the cable length. For simplicity of presentation, the friction forces are ignored and P is considered an average value of the prestressing force over the length considered. The forces in Figure 6.7 represent a system in equilibrium.

6.6.1 Circumferential prestressing

Figure 6.8 represents vertical sections in a circular-cylindrical wall with circumferential prestressing. The prestressing in Figure 6.8a is by

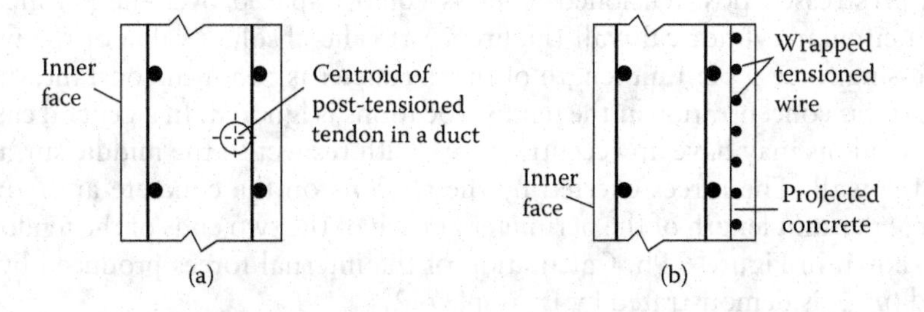

Figure 6.8 Vertical section in a circular-cylindrical prestressed wall. (a) Prestressing by post-tensioned tendon in a duct. (b) Prestressing by tensioned wrapped wire.

post-tensioned tendon inside a duct. In Figure 6.8b, the circumferential prestressing is provided by wrapping a wire tensioned prior to its contact with the tank wall; projected concrete provides the cover necessary for corrosion protection. In both methods of prestressing, the tendon is commonly not situated at the centroid of the wall section. This eccentricity of the prestressing force has practically no influence on the magnitudes of the corresponding circumferential internal forces N_ϕ and M_ϕ. Assuming that the wall has a constant vertical section, at all locations of the wall perimeter, the prestressing exerts on the concrete a uniformly distributed horizontal inward radial pressure of intensity p/r, where p is the absolute value of the prestressing force per unit height of the wall and r is the horizontal distance between the tendon and the center of the cylinder; the value of r is commonly considered equal to the radius of the middle surface of the wall.

Let p represent the absolute value of the circumferential prestressing force per unit height of the wall. In the special case when the ends of the wall at the top and bottom edges are free to translate and rotate in the radial plane, and p varies linearly over the height, the circumferential prestressing produces no stresses in the vertical direction, and the circumferential forces will be

$$N_\phi = -p/r, \qquad M_\phi = 0. \tag{6.37}$$

The hoop force N_ϕ per unit height of the wall has the same shape of variation over the wall height as that of p. In the common case, the thickness of the wall is small compared to r and N_ϕ produces uniform stress through the wall thickness. The equations presented in Section 6.5.1 can be employed to calculate the time-dependent variation of stress in concrete and in reinforcements.

6.6.2 Vertical prestressing of a circular-cylindrical wall

We consider here the normal force N and the moment M produced by vertical prestressed post-tensioned tendons equally spaced over the perimeter of a circular-cylindrical wall (Figure 6.9a). The absolute value of the prestressing force p per unit length of the perimeter is assumed constant; thus the stress concentration at the tendon locations is ignored. In a general case, the tendons may have an eccentricity y_{ps} with respect to the middle surface of the wall. The forces exerted by the tendons on the concrete are p and (py_{ps}) per unit length of the perimeter at each of the two ends of the tendons as shown in Figure 6.9b. Calculation of the internal forces produced by p and (py_{ps}) is demonstrated by Example 6.2.

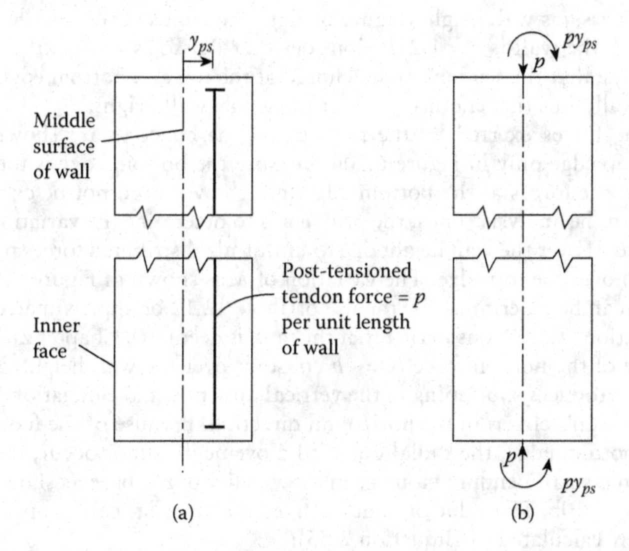

Figure 6.9 Vertical prestressing of a circular-cylindrical wall. (a) Post-tensioning with eccentric tendons. (b) Intensity of uniformly distributed forces exerted by the tendons on the concrete (force/length and force × length/length).

EXAMPLE 6.2
Internal Forces Due to Vertical Prestressing of a Circular Wall

Determine the internal forces N and M in the circular-cylindrical tank wall due to vertical post-tensioning as shown in Figure 6.10a. The top edge of the wall is free and the bottom edge is encastré. Assume that the prestressing tendons have eccentricity $y_{ps} = h/6$, where $h = l/20$ = wall

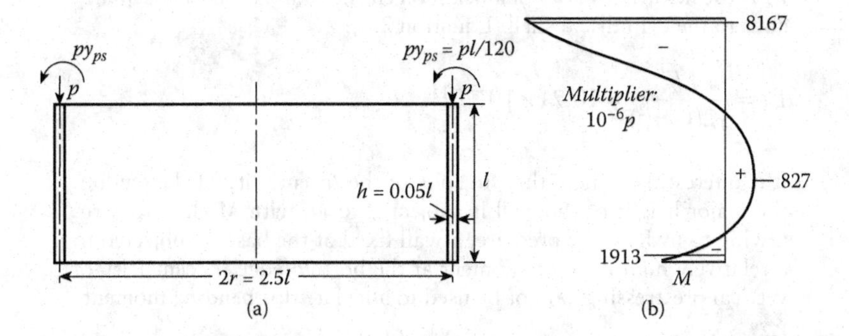

Figure 6.10 Effect of eccentric vertical post-tensioning of a circular-cylindrical wall (Example 6.2). (a) Wall dimensions and forces exerted by tendons on concrete. (b) Bending moment in vertical direction (finite-element result).

thickness; l is wall height (Figure 6.10a). The radius of the middle surface of the wall is $r = 1.25l$. Consider that Poisson's ratio is $\upsilon = 1/6$. Assume that the tendons are anchored at the top and bottom edges of the wall; thus the tendons run over the whole wall height.

The forces exerted by the tendons on the concrete are shown at the top edge only in Figure 6.10a. Because the bottom edge is totally fixed, the forces at the bottom edge (not shown) need not be considered in the analysis. Thus, the problem is to determine the variation of N and M over the wall height due to uniformly distributed forces p and (py_{ps}) over the top edge. The variation of M is shown in Figure 6.10b. This can be determined by the use of Table 12.12 or approximated by Equation (2.56) (considering that the tank is deep) (see Chapter 2). The value of the normal force $N = -p$ constant over the wall height. This force produces shortening in the vertical direction and elongation (due to Poisson's effect) in the horizontal direction. Because of the fixity at the bottom edge, the radial outward movement cannot occur, resulting in a small bending moment in the vicinity of the base as shown in Figure 6.10b. The value of M at the fixed bottom edge can be approximately calculated by (Equation 2.55):

$$M_{\text{bot. edge}} = -2\beta^2 EI\left(\frac{\upsilon pr}{Eh}\right) = -2.4\times 10^{-3} pl.$$

The term $\upsilon pr/(Eh)$ represents the outward radial displacement that would occur if the wall were free to slide on the base; β is a characteristic parameter defined by Equation (2.17)

$$\beta = \frac{\sqrt[4]{3(1-\upsilon^2)}}{\sqrt{rh}} = 5.227/l.$$

EI is the flexural rigidity of a beam on elastic foundation that is analogous to the cylindrical shell (Equation 2.5):

$$EI = \frac{Eh^3}{12(1-\upsilon^2)} = 10.71\times 10^{-6} El^3.$$

It is interesting to note that, in spite of the eccentricity of the tendon, the major height of the wall is subjected to N, with M close to zero. Owing to hydrostatic pressure, a wall fixed at the base is subjected to a relatively high bending moment at the bottom edge. Eccentricity of vertical prestressing cannot be used to alleviate this bending moment.

6.7 TIME-DEPENDENT INTERNAL FORCES

The internal forces caused by creep and shrinkage of concrete, and relaxation of prestressed reinforcement are considered here for an axisymmetrical shell of revolution, assuming that the structure is made of elastic isotropic material. All loads and prestressing are assumed to be axisymmetrical. Analysis of the effect of creep and shrinkage is similar to the procedure used to determine the effect of temperature in Section 5.7. First assume that creep and shrinkage can occur freely in an elemental shell shown in Figure 5.2c and define the free generalized strain vector (Equation 5.13):

$$\{\varepsilon\}_{\text{free}} = \begin{Bmatrix} du/ds \\ (w\cos\alpha + u\sin\alpha)/r \\ -d^2w/ds^2 \\ -(\sin\alpha/r)(dw/ds) \end{Bmatrix}_{\text{free}}, \tag{6.38}$$

where w and u are, respectively, translations in direction normal to the shell surface and along a meridian line (Figure 5.2a); α is the angle defined in Figure 5.2a; and the subscript "free" refers to the values of w and u when they can occur without restraint. The free strain can be expressed as the sum of creep and shrinkage effects:

$$\{\varepsilon\}_{\text{free}} = (\varepsilon)_{\text{creep}} + \{\varepsilon\}_{\text{shrinkage}}, \tag{6.39}$$

where

$$\{\varepsilon\}_{\text{creep}} = \phi(t,t_0)\{\varepsilon\}_{t_0}, \tag{6.40}$$

with $\{\varepsilon\}t_0$ being the vector of generalized instantaneous strain due to loads introduced at t_0 and sustained to a later instant t; and $\phi(t,t_0)$ is the creep coefficient.

Assuming that the value of the free shrinkage $\varepsilon_{cs}(t,t_0)$ in the period t_0 to t is linearly variable through the shell thickness between $(\varepsilon_{cs})_i$ and $(\varepsilon_{cs})_o$ at the inner and outer faces of the shell, respectively, the generalized strain vector due to free shrinkage is

$$\{\varepsilon\}_{\text{shrinkage}} = \begin{Bmatrix} (\varepsilon_0 + \varepsilon_i)/2 \\ (\varepsilon_0 + \varepsilon_i)/2 \\ (\varepsilon_0 - \varepsilon_i)/h \\ (\varepsilon_0 - \varepsilon_i)/h \end{Bmatrix}_{cs}. \tag{6.41}$$

A numerical analysis example of the effect of differential shrinkage between the inner face of a water tank wall, in contact with water, and the outer face, exposed to air, is presented in Example 7.1 (see Chapter 7).

The vector of artificial generalized stress that, when gradually introduced in the period t_0 to t, would prevent the occurrence of the generalized strain $\{\varepsilon\}_{\text{free}}$ is

$$\{\sigma_r\} = -[\bar{d}_e]\{\varepsilon\}_{\text{free}}. \tag{6.42}$$

The subscript r refers to the restraint state; the generalized stress vector is defined by

$$\{\sigma_r\} = \begin{Bmatrix} N \\ N_\phi \\ M \\ M_\phi \end{Bmatrix}_r \tag{6.43}$$

and $[\bar{d}_e]$ is the age-adjusted elasticity matrix related to generalized stress and generalized strain (Equation 5.14):

$$[\bar{d}_e] = \frac{\bar{E}_c h}{1-\nu^2}\begin{bmatrix} 1 & \nu & 0 & 0 \\ \nu & 1 & 0 & 0 \\ 0 & 0 & h^2/12 & \nu h^2/12 \\ 0 & 0 & \nu h^2/12 & h^2/12 \end{bmatrix}, \tag{6.44}$$

where $\bar{E}_c = E_c(t,t_0)$ is the age-adjusted elasticity modulus (Equation 6.7):

$$\bar{E}_c = \frac{E_c(t_0)}{1+\chi\phi(t,t_0)}. \tag{6.45}$$

Equation (6.44) differs from Equation (5.14) in that E is replaced by $\bar{E}_c$ to take into account the effect of creep due to gradually applied stress. The nodal forces $\{F_r^*\}$ that are required at the local coordinates of the element in Figure 5.2a, to maintain equilibrium in the restrained state, are given by Equation (5.24). These forces are transformed to statical equivalent element nodal forces in global directions (Equation 5.25). The element nodal forces are assembled to give a vector of restraining forces for the structure, which are subsequently applied in reversed direction to eliminate the artificial restraint and produce nodal displacements due to creep and shrinkage. When the nodal displacements are determined, the change in the generalized stress vector in individual elements due to creep and shrinkage is given by

$$\{\sigma\} = [\bar{d}_e][B]\{D^*\} + \{\sigma_r\}, \tag{6.46}$$

where $[B]$ is a matrix of functions expressing the generalized strain at any point within the element when the displacement D_i^* is unity at coordinate i (Figure 5.2a). The matrix $[B]$ is given by Equation (5.16).

Relaxation of prestressed reinforcement can be accounted for in the time-dependent analysis by the application of nodal forces equivalent to the relaxed prestress forces on a structure whose elasticity modulus is $\bar{E}_c$. The relaxed prestress force in a tendon is equal to $(A_{ps}\Delta\bar{\sigma}_{pr})$, where A_{ps} is the cross-sectional area of the tendon and $\Delta\bar{\sigma}_{pr}$ is the reduced relaxation value (Equation 6.9).

A computer program performing elastic analysis that can account for the prescribed displacement at any node and the effects of temperature variation can also be used to determine the time-dependent effects as the sum of results from three computer runs:

1. In the first computer run, apply the loads introduced at t_0 on a structure whose modulus of elasticity is $E_c(t_0)$.
2. The second analysis is for a structure whose modulus of elasticity is $\bar{E}_c(t,t_0)$ subjected to prescribed nodal displacements $\phi(t,t_0)\{-D(t_0)\}$, where $\{D(t_0)\}$ is the vector of instantaneous nodal displacements determined in run 1. The results of the second analysis give $\{\sigma_r\}$ and a vector of restraining nodal forces in global directions due to creep. In the input data for this computer run, the prescribed nodal displacement should be entered with as many significant figures as available (preferably six or more); otherwise, the vector of restraining nodal forces can be significantly erroneous. The vector of restraining forces obtained in the second analysis will be the same as the forces on the nodal lines in the first analysis (including the reactions) multiplied by $[-\phi(t,t_0)\bar{E}_c/E_c(t_0)]$. Also $\{\sigma_r\}$ obtained in the second analysis will be the same as the stress resultants $\{\sigma(t_0)\}$ obtained in the first analysis factored by the same multiplier. Thus, by using the results of the first analysis in this way, the second computer run can be omitted.
3. In the third computer run apply the restraining forces determined in run 2 in reversed direction on a structure whose elasticity modulus is $\bar{E}_c(t,t_0)$. In the same run apply nodal forces that are statical equivalents to the time-dependent change in prestress forces. Also enter the free shrinkage values $\varepsilon_{cs}(t,t_0)$ as temperature rise $T = \varepsilon_{cs}(t,t_0)/\mu$, where μ is the coefficient of thermal expansion, and $\varepsilon_{cs}(t,t_0)$ is the value of free shrinkage, commonly a negative quantity.

The sum of the results of computer runs 2 and 3 gives the time-dependent effects of creep, shrinkage, and relaxation. Adding the results of run 1 gives the instantaneous effects combined with the time-dependent effects.

Some comments are needed: If the material of the structure is homogeneous, with constant elasticity modulus $E_c(t_0)$ and constant creep coefficient $\phi(t,t_0)$, creep increases the deformations by a multiplier $\phi(t,t_0)$; but the stresses in the structure remain unchanged. However, creep changes stresses when the structure is composed of parts having different E or ϕ. This is the case of a circular prestressed concrete wall on elastomeric pads; the pads have different material properties from those of the concrete. A structure cast and prestressed in stages is another example where internal forces develop due to creep; the ages of concrete in the parts, the elasticity moduli, and the creep coefficients are different.

EXAMPLE 6.3
Time-Dependent Internal Forces in a Cylindrical Wall on Elastomeric Pads

A circular-cylindrical wall, supported on elastomeric pads, is subjected at time t_0 to post-tensioned circumferential prestress. It is required to find the radial translation and the shearing force at the bottom edge and the bending moment M in the wall at time t_0 and at a later instant t, accounting for the effects of creep and shrinkage of concrete and relaxation of prestressed steel. Assume that the forces on the concrete at time t_0 due to circumferential prestressing are equivalent to linearly varying inward pressure whose intensity p is $p_1 = -7.5$ and $p = -80.0$ kN/m^2 (−160, −1670 lb/ft^2) at the top and bottom edges, respectively; the minus sign indicates inward pressure. The wall dimensions are given in Figure 6.11a. The following data are given: $E_c(t_0) = 30$ GPa (4350 ksi); $\phi(t,t_0) = 2.0$; $\chi(t,t_0) = 0.8$; $\varepsilon_{cs}(t,t_0) = -300 \times 10^{-6}$; Poisson's ratio $\nu = 1/6$.

Assume that the loss of prestress due to creep and shrinkage of concrete and relaxation of prestressed steel during the period t_0 to t is 15% of the value at t_0. The stiffness of the elastomeric pad is $K = 2.0$ MPa (0.29 ksi); this is the radial reaction of the pad at the bottom edge of the wall per unit length of the perimeter per unit radial horizontal translation. Ignore the creep of the pad.

The problem is solved below by hand calculations using the force method[6] of structural analysis and the tables in Chapter 12. The same results, presented in Figure 6.11c,d, are determined by computer (finite-element) analyses, as discussed earlier.

INSTANTANEOUS DEFORMATIONS AND STRESSES

If the bottom edge of the wall is separated from the pad (Figure 6.11a), the wall will be subjected to hoop force pr, without moment, and the radial translation at the bottom edge, immediately after prestressing, will be

$$D_1(t_0) = -p_2 \frac{r^2}{hE_c(t_0)}$$

$$= -80 \times 10^3 \frac{(40)^2}{0.3(30 \times 10^9)} = -14.22 \times 10^{-3} \text{ m (0.5598 inch)}.$$

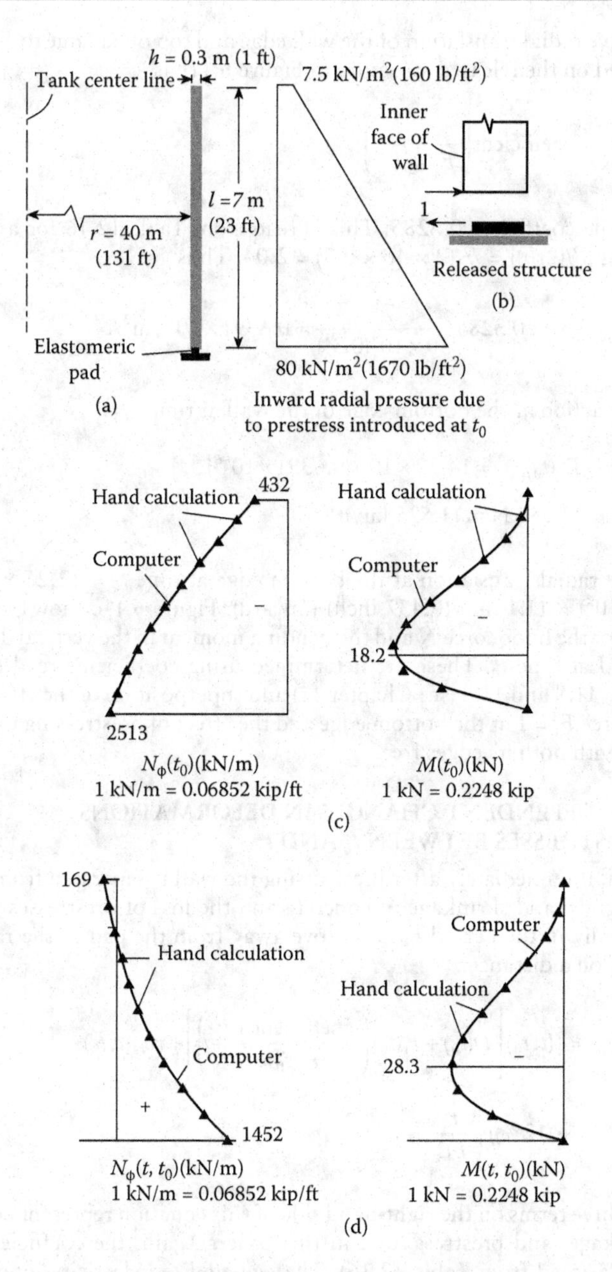

Figure 6.11 Analysis of time-dependent changes in a prestressed concrete tank wall (Example 6.3). (a) Tank dimension and distribution of circumferential prestressing. (b) Released structure and coordinate system. (c) Hoop force and bending moment at t_0. (d) Changes in hoop force and bending moment between time t_0 and t.

Relative radial translation of the wall edge and top of pad due to $F_1 = 1$ applied on the released structure in Figure 6.11b is

$$f_{11} = \frac{1}{K} + \text{coefficient}\left(\frac{l^3}{E_c(t_0)h^3}\right)$$

with the coefficient 0.3285. This is read from Table 12.16 for a wall having $[l^2/(2rh)] = 7^2/(2 \times 40 \times 0.3) = 2.04$. Thus,

$$f_{11} = \frac{1}{2\times10^6} + 0.3285\frac{7^3}{30\times10^9(0.3^3)} = 0.6391\times10^{-6}\ \text{m}^2/\text{N}.$$

The reaction at the bottom edge of the wall at time t_0 is

$$F_1(t_0) = -D_1(t_0)f_{11}^{-1} = 14.22\times10^{-3}(0.6391\times10^{-6})^{-1}$$

$$= 22.25\ \text{kN/m (1.525 kip/ft)}.$$

The radial translation at the bottom edge at time $t_0 = 22.25 \times 10^3/(2 \times 10^6) = 11.1$ mm (0.437 inch) (inward). Figure 6.11c shows variations of the hoop force N and the bending moment in the vertical direction M at time t_0. These are determined using coefficients read from Tables 11.8 and 11.9 (see Chapter 11) and superposition of the effect of the force $F_1 = 1$ at the bottom edge and the effect of prestressing on the wall with both its edges free.

TIME-DEPENDENT CHANGES IN DEFORMATIONS AND STRESSES BETWEEN t_0 AND t

Again, if immediately after prestressing the wall is separated from the pad, creep and shrinkage of concrete and the loss of prestresses will, gradually in the period t_0 to t, move away from the pad in the radial direction a distance:

$$D_1(t,t_0) = \phi(t,t_0)\left[D(t_0) + F_1(t_0)\left(\frac{\text{coefficient}\times l^3}{E_c(t_0)h^3}\right)\right] + r\varepsilon_{cs}(t,t_0)$$

$$+ (12\%)p_2\frac{r^2}{h\bar{E}_c}.$$

The three terms on the right-hand side of this equation represent creep, shrinkage, and prestress loss, in this order. Again, the coefficient is 0.3285 (read from Table 12.16). The age-adjusted elasticity modulus of concrete is used in the last term because the prestress loss develops gradually in the period t_0 to t. The age-adjusted modulus is calculated by Equation (6.7):

$$\bar{E}_c = \frac{30\times10^9}{1+0.82(2.0)} = 11.54\times10^9 \text{ N/m}^2,$$

$$D_1(t,t_0) = 2.0\left[-14.22\times10^{-3} + 22.25\times10^3\left(\frac{0.3285\times7^3}{30\times10^9(0.3^3)}\right)\right]$$

$$+40.0(-300\times10^{-6}) + 0.12(80\times10^3)\frac{40^2}{0.3(11.54\times10^9)}$$

$$= -29.81\times10^{-3}\text{m}.$$

The term in the square brackets is the radial translation at the bottom edge at t_0 (= 11.1 mm). A unit radial force $F_1 = 1$ gradually introduced in the period t_0 to t at the edge of the wall and the top of the pad produces at time t a relative radial displacement:

$$\bar{f}_{11} = \frac{1}{K} + \text{coefficient}\left(\frac{l^3}{\bar{E}_c h^3}\right)$$

with the coefficient 0.3285. Again, $\bar{E}_c$ is used in this equation because the force $F_1 = 1$ is gradually introduced. Thus,

$$\bar{f}_{11} = \frac{1}{2\times10^6} + 0.3285\frac{7^3}{11.54\times10^9(0.3^3)} = 0.8618\times10^{-6}\text{m}^2/\text{N}.$$

The change in reaction at the bottom edge of the wall between t_s and t is

$$\Delta F_1(t,t_0) = -D_1(t,t_0)\bar{f}_{11}^{-1}$$

$$= -(-29.8\times10^{-3})(0.8616\times10^{-6})^{-1}$$

$$= 34.60 \text{ kN/m } (2.371\text{kip/ft}).$$

Change in radial translation at the bottom edge in the period t_0 to t is

$$\Delta F_1(t,t_0)/K = 34.60\times10^3/(2\times10^{16}) = 17.3 \text{ mm (0.681 inch) inward}$$

The changes in the period t_0 to t in $N\phi$ and M are plotted in Figure 6.11d. The values in the figure can be verified using Tables 12.8 and 12.9 and by superposition of the effects of an outward radial force of 34.60 kN/m on the bottom edge and an outward pressure whose intensity equals 12% that of the prestress.

EXAMPLE 6.4
Time-Dependent Internal Forces in a Cylindrical Tank Wall, Monolithic with Base

The tank wall in Figure 6.12a is circumferentially prestressed at time t_0, while its bottom edge is free to rotate and slide on a footing. At a later time t_1, a ring-shaped part of the base is cast to connect the wall to an existing central part of the base, thus making the wall monolithic with the base. It is required to determine the time-dependent changes $N_\phi(t_2,t_1)$ and $M(t_2,t_1)$ in the hoop force and the vertical bending moment in the wall in the period t_1 to t_2, where t_2 is an instant much later than t_1. The circumferential prestressing at time t_0 is equivalent to a radial inward pressure whose intensity is varying linearly as shown in Figure 6.12b; the symbol p in this figure is the pressure intensity at the top edge. The tank dimensions are given in Figure 6.12a in terms of wall thickness h. The following data are given: $E_c(t_0) = 4 \times 10^6 p$; $E_c(\bar{t}) = E_3(3) = 2.4 \times 10^6 p$; Poisson's ratio = 1/6; $\phi(t_1,t_0) = 1.2$; $\phi(t_2,t_0) = 2.3$; $\phi(t_2,t_1) = 1.9$; $\chi(t_2,t_1) = 0.9$; $\varepsilon_{cs}(t_2,t_1) = -300 \times 10^{-6}$; $\varepsilon_{cs}(t_2,\bar{t}) = -375 \times 10^{-6}$; $\phi(t_2,\bar{t}_1) = 2.8$; $\chi(t_2,\bar{t}) = 0.65$; $\bar{t}$ = 3 days (to be defined later).

Assume that the loss of prestress between time t_0 and time t_2 is 15% of its value at t_0; also assume that half the loss occurs in the period t_0 to t_1. Conduct the analysis on the idealized structure in Figure 6.12c. The data are chosen to represent a practical case in which h = 0.3 m (1 ft); p = 7.5 kN/m² (160 lb/ft²); $\{t_0, t_1, t_2\} = \{28, 100, \infty\}$ days; the creep coefficients are approximately in accordance with CEB–FIP Model Code MC-90,[7] for concrete strength 30 MPa (4400 psi) and 50% relative humidity.

Use of the idealized structure in Figure 6.12c implies ignoring the subgrade reaction on BC and neglecting the in-plane strain in the part of the base between B and the center. These assumptions simplify the presentation but may not represent the conditions in practice. While the bottom edge is free to slide, the linearly varying pressure due to prestressing produces no moment but results in linearly varying hoop force:

$$M(t_0) = 0, \qquad N_\phi(t_0) = pr\left(13 - \frac{12x}{l}\right),$$

where x is the distance from B to any point. Creep and shrinkage of concrete and relaxation of prestressed steel in the period t_0 to t reduce the hoop force to 92.5% of its value at t_0; thus,

$$M(t_1) = 0, \qquad N_\phi(t_1) = 0.925\, N_\phi(t_0).$$

After casting and hardening of part BC of the base, the time-dependent translation and rotation at the bottom edge of the wall are restrained, thus causing changes in N_ϕ and M. The age-adjusted elasticity moduli required for the analysis are (Equation 6.45)

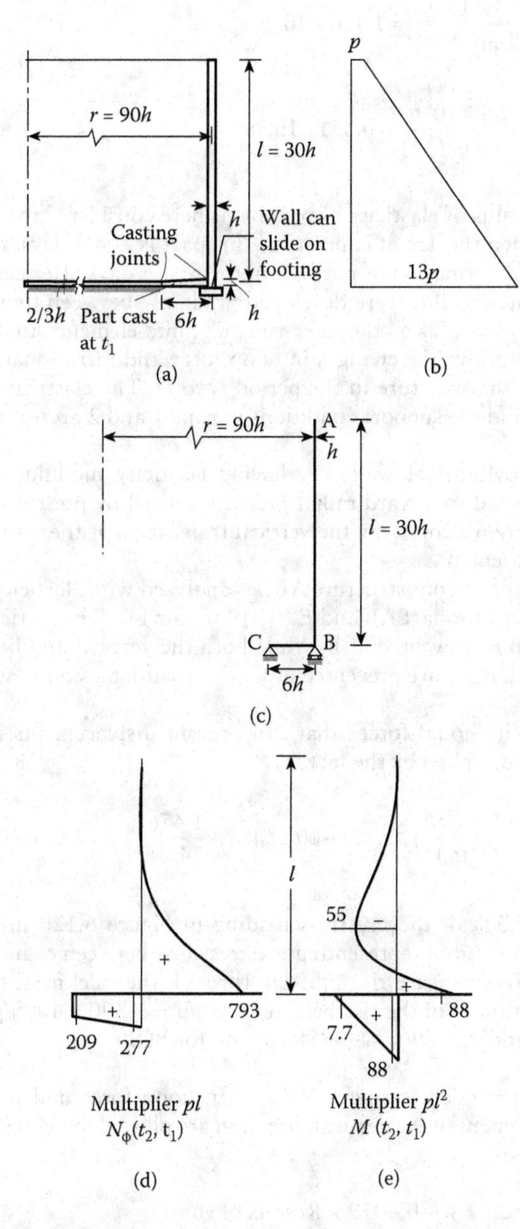

Figure 6.12 Analysis of time-dependent internal forces in a prestressed tank cast in stages (Example 6.4). (a) Tank dimensions. (b) Radial pressure equivalent of circumferential prestressing applied at time t_0. (c) Structure idealization. (d), (e) Changes in hoop force N_ϕ and vertical bending moment M occurring between time t_1 of casting the ring-shaped floor and $t_2 = \infty$.

$$\bar{E}_c(t_2,t_1) = \frac{4\times10^6 p}{1+0.9(1.9)} = 1.476\times10^6 p,$$

$$\bar{E}_c(t_2,\bar{t}) = \frac{2.4\times10^6 p}{1+0.65(2.8)} = 0.851\times10^6 p.$$

The modulus of elasticity of part BC is here considered to affect the analysis when the age of concrete of this part is $\bar{t} = 3$ days; the time-dependent deformation of part AB in the first 3 days after casting BC is here treated as if it were developed gradually between time $(t_1 + 3)$ and time $t_2 = \infty$. Two computer runs of finite-element[8] analyses are performed to give the changes in hoop force and meridional bending moment in the structure in the period t_1 to t_2. The elasticity moduli, the loads, and the support conditions in runs 1 and 2 are:

1. The cylindrical wall AB, having elasticity modulus $E_c(t_0)$, is subjected to inward radial pressure caused by prestressing at t_0 (Figure 6.12b). Only the vertical translation at the bottom edge is prevented.
2. The continuous structure ABC is analyzed with elasticity moduli $\bar{E}_c(t_2,t_1)$ for part AB and $\bar{E}_c(t_2,\bar{t})$ for part BC. The vertical translation is prevented at B, while both the vertical and horizontal translations are prevented at C. The loading is composed of

   - The nodal forces that can prevent displacements in run 1 multiplied by the factor

   $$-\frac{\bar{E}_c(t_2,t_1)}{E_c(t_0)}[\phi(t_2,t_0)-\phi(t_1,t_0)] = -\frac{1.476}{4.0}(2.3-1.2) = -0.4059$$

   - 7.5% of the prestress loading in Figure 6.12b in reversed direction, representing prestress loss between t_1 and t_2
   - Temperature rise uniform through the thickness, the magnitudes of the rise being $\varepsilon_{cs}(t_2,t_1)/\mu = (-300\times10^{-6})/\mu$ for AB and $\varepsilon_{cs}(t_2\bar{t})/\mu = (-375\times10^{-6})/\mu$ for BC.

The time-dependent changes $N_\phi(t_2,t_1)$ in hoop force and $M(t_2,t_1)$ in bending moment in the vertical direction are plotted in Figure 6.12d,e. These are equal to

Results of run 2 + (−0.4059 × Results of run 1).

The finite-element computer program Shells of Revolution (SOR) and its input files for run 1 and run 2 are available from the companion Web site for this book (Appendix B). The loading in the input for run 1 taken equal to the pressure from the prestressing at time t_0 multiplied by the factor (−0.4059). The nodal restraining forces calculated in run 1 are part of the input for run 2.

NOTES

1. Graphs for $\chi(t,t_0)$ are given in Ghali, A., Elbadry, M., and Favre, R., 2012, *Concrete Structures Stresses and Deformations,* 4th ed., Spon Press. The graphs are based on the assumption that the variation of the absolute value of the stress increment has the same shape as the relaxation function for concrete.
2. Fédération Internationale de Béton (fib) CEP-FIP (2012), *Model Code for Concrete Structures, MC2010*, fib bulletin No. 65. fib, Case Postale 88, CH-1015, Lausanne, Switzerland. ACI Committee 209, 2008, *Guide for Modeling and Calculating Shrinkage and Creep in Hardened Concrete,* ACI 209.2R-08, American Concrete Institute, Farmington, MI.
3. See references mentioned in note 2.
4. An equation and graph for χ_r are given in Ghali et al., 2012.
5. Ibid.
6. See Ghali, A., Neville, A. M., and Brown, T. G., 2009, *Structural Analysis: A Unified Classical and Matrix Approach*, 6th ed., Spon Press, London.
7. See note 2.
8. Ghali, A., n.d., Computer Program SOR (Shells of Revolution), available from a companion Web site for this book; see Appendix B.

Chapter 7

Thermal stresses

7.1 INTRODUCTION

Temperature variation produces stresses only when elongation or shortening of the material is restrained. Shrinkage or swelling of concrete has a similar effect. Because shrinkage develops gradually over a long period, the stresses due to shrinkage are commonly alleviated by creep. Also, when the temperature variation develops over a period of time, the thermal stresses can be significantly smaller when the effect of creep is accounted for.[1]

In circular-cylindrical concrete tanks and silos, the restraint can be caused by wall connections with the roof or base and by the presence of reinforcement. When the temperature varies through the wall thickness, a free isolated element would tend to curl in vertical and circumferential directions. But in an axisymmetrical cylindrical wall, vertical radial sections and horizontal sections generally cannot rotate freely. This restraint can produce high vertical and circumferential stresses in the wall, even when displacements are free to occur at the edges.

Code and design recommendations require that the effects of temperature and shrinkage be considered in design for serviceability. This chapter discusses analytical procedures and numerical examples of hand calculations indicating the magnitude of stresses to be expected. The temperature and shrinkage variation is assumed to be axisymmetrical and the shape of variation through the wall thickness is assumed to be of any shape. Computer finite-element analysis of thermal stresses in axisymmetrical shells of any shape (other than cylindrical) is presented in Chapter 5. Both the analytical and the numerical solutions presented in this chapter and in Chapter 5 are linear. Stresses determined by linear analysis should be considered valid only until cracking occurs. The development of cracks due to temperature variation is accompanied by a drop in internal forces. The subsequent increase in the magnitude of temperature variation produces a significantly smaller increase in internal forces than what an elastic analysis would indicate. Cracking reduces the restraint of thermal expansion or contraction, and thus substantially limits the thermal internal forces.[2]

Research has shown that, in mild climates, a temperature differential through the wall thickness of 30°C (54°F), with the outside warmer than the inside, can occur in cylindrical tanks.[3] A reversed gradient, with the inner face 20°C (36°F) higher than the outer face, can also occur. A numerical example in this chapter will show that temperature variation, whose distribution through the wall thickness is linear with 30°C (54°F), between the inner and outer faces, causes tensile stresses at the inner face far exceeding the residual compression of 1.0 to 2.0 MPa (150 to 300 psi) commonly provided by prestressing.

Silos in the cement production industry can have, in operation, a temperature of the contents of 100°C–120°C (210°F–250°F), while the corresponding external temperature is between 0°C (or less) and 40°C or more (32°F–104°F). With such differential temperature, and even with much smaller temperature variations, concrete tanks and silos crack. By allowing cracking to occur at temperature extremes and supplementing prestressing with non-prestressed reinforcement, crack widths can be controlled to a limit of 0.2 or 0.1 mm (0.008 or 0.004 inches).

Temperature gradient due to solar radiation is not axisymmetrical. However, because the rate of temperature variation over the circumference is small, analysis of maximum stresses may be obtained with sufficient accuracy by assuming an axisymmetrical temperature gradient.

In general, distributions of temperature over the thickness of a wall of a tank or a silo is nonlinear and variable with time. Solar radiation can produce nonlinear temperature variation with steep gradients.[4] In this chapter, the analysis of stresses is discussed for the general case of nonlinear temperature distribution.

Analysis for the effect of concrete swelling or shrinkage, ε_{cs}, is the same as that of the effect of variation of temperature of magnitude, ε_{cs}, divided by the coefficient of thermal expansion. The reduction of thermal stresses due to creep, which occurs as shrinkage develops, can be accounted for in the same analysis by the use of the age-adjusted elasticity modulus for concrete (Chapter 6, Equation 6.7). Similarly, the age-adjusted elasticity should also be employed in thermal stress analysis when the temperature variation develops over a period of time. Even when the length of this period is a few hours, creep of concrete can cause significant alleviation of thermal stresses.

Contact of the inner face of a tank with water can cause swelling, while the outer face that is not in contact with water shrinks. Analysis of this volumetric change is treated in the same way as the effects of thermal gradient. Example 7.2 indicates the order of magnitude of stress to be expected from this volumetric change.

The elastic analyses used in the examples in this chapter will indicate that, in the majority of cases, cracking is bound to occur due to temperature and shrinkage. Fortunately, stresses due to temperature and shrinkage are limited in magnitude when cracking occurs. Provision of appropriate amounts of non-prestressed reinforcement controls the width of cracks and satisfactory performance can be expected.

While non-prestressed reinforcement is essential for crack control, its presence restrains the shrinkage and produces tension in concrete. This should not be ignored in evaluating long-term concrete stresses (see Example 6.1). The last section of this chapter will discuss the effects of horizontal cracks in a wall of a cylindrical tank on the distribution of hoop force and bending moment in the vertical direction.

7.2 EFFECTS OF TEMPERATURE VARIATION IN A CYLINDRICAL WALL

Figure 7.1a,b shows distributions of axisymmetrical temperature rise in a circular-cylindrical wall. We consider here the corresponding normal force N and bending moment M in the x direction and the hoop force N_ϕ and bending moment M_ϕ in the circumferential direction. Various combinations of conditions at the top edges are considered. The wall material is assumed elastic isotropic and cracking is ignored.

If thermal expansion is artificially restrained, vertical stress σ and hoop stress σ_ϕ will develop:

$$\sigma_r = \sigma_{\phi r} = -\frac{E\mu}{1-v}T, \tag{7.1}$$

where E is the elasticity modulus; μ is thermal expansion coefficient; v is Poisson's ratio; and $T(= T(y))$ is the temperature rise at any fiber, whose coordinate is y, measured from the middle surface (Figure 7.1a).

The stresses in the restraint state have resultants in the vertical and the circumferential directions:

$$N_r = N_{\phi r} = -\frac{E\mu}{1-v}\int_{-h/2}^{h/2} T \, dy, \tag{7.2}$$

$$M_r = M_{\phi r} = -\frac{E\mu}{1-v}\int_{-h/2}^{h/2} Ty \, dy. \tag{7.3}$$

Forces at coordinates 1 to 6 (Figure 7.1b) must exist at the edges of the cylinder in the restraint state:

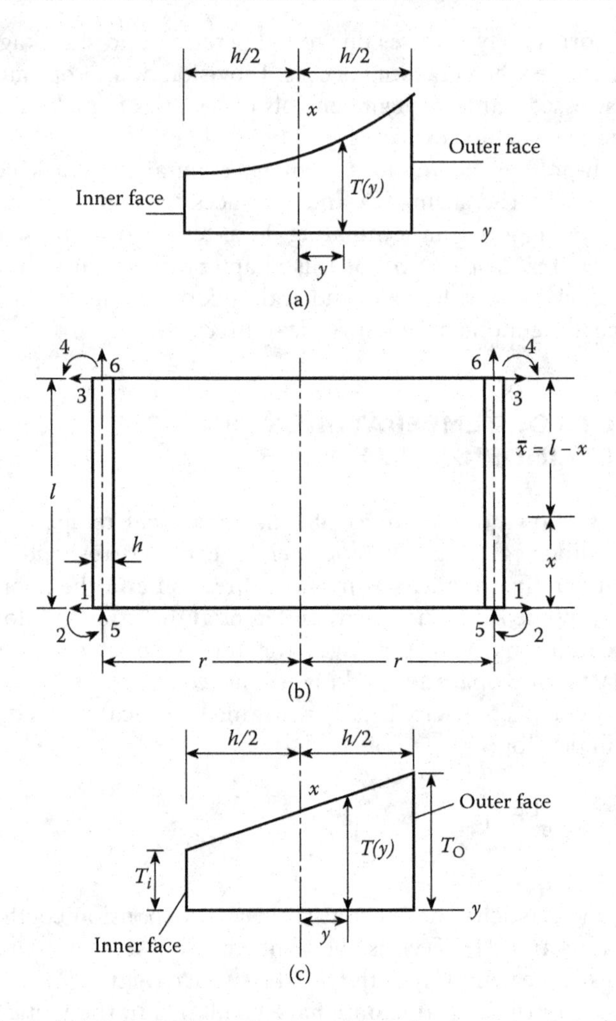

Figure 7.1 Analysis of the effects of temperature rise in a circular-cylindrical wall. (a) Distribution of temperature rise over wall thickness. (b) Coordinate system. (c) Special case of linear temperature distribution.

$$F_{1r} = F_{3r} = 0, \tag{7.4}$$

$$F_{2r} = -F_{4r} = M_r, \tag{7.5}$$

$$F_{5r} = -F_{6r} = -N_r. \tag{7.6}$$

To produce the artificial restraint, these edge forces must be accompanied by radial pressure whose intensity is

$$p = N_{\phi r}/r, \tag{7.7}$$

where r is the radius of the middle surface.

The coordinate system in Figure 7.1b indicates the positive directions of edge forces or edge displacements. N is positive when tensile; positive radial pressure p is outward; and positive M produces tension at the outer surface.

The internal forces in actual conditions are the sum of the values in the restraint state and the internal forces caused by application of the restraining pressure p_r and one or more of the six edge forces $\{F_r\}$ in reversed directions. The effects of p_r and the edge forces can be calculated with the help of the tables in Chapter 12. Alternatively, the equations in Section 7.4 can be used to give the internal forces directly in special cases.

7.3 LINEAR TEMPERATURE VARIATION THROUGH WALL THICKNESS

Consider the effect of a temperature rise that varies linearly through the thickness of a circular-cylindrical wall such that (Figure 7.1c):

$$T = 0.5\left(1 + \frac{2y}{h}\right)T_o + 0.5\left(1 - \frac{2y}{h}\right)T_i, \tag{7.8}$$

where $T = T(y)$ is the temperature rise at any fiber whose coordinate is y, measured outward from the middle surface; and T_o and T_i are values of *temperature rise* at the outer and inner surfaces, respectively. The stress resultants in the restrained state are obtained by substitution of Equation (7.8) in Equations (7.2) and (7.3):

$$N_r = N_{\phi r} = -\frac{E\mu h}{2(1-\upsilon)}(T_o + T_i), \tag{7.9}$$

$$M_r = M_{\phi r} = -\frac{E\mu h^2}{12(1-\upsilon)}(T_o - T_i). \tag{7.10}$$

7.4 THERMAL INTERNAL FORCES IN DEEP TANKS AND SILOS

Presented next are equations for bending moment M in the vertical direction, hoop force N_ϕ, and bending moment M_ϕ in the circumferential direction in a deep circular-cylindrical wall subjected to a rise of temperature varying linearly through the thickness according to Equation (7.8) (Figure 7.1c). A cylindrical wall is classified as "deep" or "long" when the effect of forces applied at one edge dies out before the other end is reached. This is the case when (see Chapter 2, Sections 2.8 and 2.9)

$$l \geq \frac{\pi}{\beta}, \qquad \beta = \frac{[3(1-v^2)]^{1/4}}{\sqrt{rh}} \tag{7.11}$$

or when

$$\frac{l^2}{rh} \geq 5.8, \tag{7.12}$$

where l is cylinder length (wall height), h is thickness, and r is radius of the middle surface. The following equations apply to long (deep) circular-cylindrical walls, with several combinations of edge conditions. In all combinations considered, the top edge of the wall is free to expand vertically; thus the normal force N in the vertical direction is zero. The equations presented in Sections 7.4.1 to 7.4.3 are derived from Equations (7.9) and (7.10), and Equations (2.55) and (2.56) in Chapter 2.

7.4.1 Bottom edge encastré and top edge free

$$N_\phi = -E\mu\left[\frac{h}{2}(T_o + T_i)Z_1 + \frac{1+v}{2r\beta^2}(T_o - T_i)\bar{Z}_3\right], \tag{7.13}$$

$$M_\phi = -E\mu\left[\frac{vr\beta^2h^3}{12(1-v^2)}(T_o + T_i)Z_3 + \frac{h^2}{12(1-v)}(T_o - T_i)(1 - v\bar{Z}_1)\right], \tag{7.14}$$

$$M = -E\mu\left[\frac{r\beta^2h^3}{12(1-v^2)}(T_o + T_i)Z_3 + \frac{h^2}{12(1-v)}(T_o - T_i)(1 - \bar{Z}_1)\right]. \tag{7.15}$$

Graphs of the Z and $\bar{Z}$ functions are shown in Figure 2.6 and equations defining the functions are given at the end of Section 7.4.3.

7.4.2 Bottom edge hinged and top edge free

$$N_\phi = -E\mu\left[\frac{h}{2}(T_o + T_i)Z_4 + \frac{1+v}{2r\beta^2}(T_o - T_i)(Z_3 + \bar{Z}_3 - Z_4)\right], \tag{7.16}$$

$$M_\phi = -E\mu\left\{-\frac{\upsilon r\beta^2 h^3}{12(1-\upsilon^2)}(T_o + T_i)Z_2 + \frac{h^2}{12(1-\upsilon)}(T_o - T_i)[1-\upsilon(Z_1 - Z_2 + \bar{Z}_1)]\right\}, \tag{7.17}$$

$$M = -E\mu\left[\frac{r\beta^2 h^3}{12(1-\upsilon^2)}(T_o + T_i)Z_2 + \frac{h^2}{12(1-\upsilon)}(T_o - T_i)(1 - Z_1 + Z_2 - \bar{Z}_1)\right]. \tag{7.18}$$

7.4.3 Bottom edge free to slide and rotate and top edge free

$$N_\phi = -E\mu\frac{1+\upsilon}{2r\beta^2}(T_o - T_i)(Z_3 + \bar{Z}_3), \tag{7.19}$$

$$M_\phi = -E\mu\frac{h^2}{12(1-\upsilon^2)}(T_o - T_i)(1 - \upsilon Z_1 - \upsilon\bar{Z}_1), \tag{7.20}$$

$$M = -E\mu\frac{h^2}{12(1-\upsilon)}(T_o - T_i)(1 - Z_1 - \bar{Z}_1). \tag{7.21}$$

The variables Z and $\bar{Z}$ used in Equations (7.13) to (7.21) are functions of the dimensionless products βx and $\beta\bar{x}$, respectively, where x is the distance between the bottom edge and any section and $\bar{x} = l - x$ (Figure 7.1b). The Z-functions are defined as follows:

$$Z_1 = e^{-\beta x}(\cos\beta x + \sin\beta x), \qquad Z_2 = e^{-\beta x}\sin\beta x, \tag{7.22}$$

$$Z_3 = e^{-\beta x}(\cos\beta x - \sin\beta x), \qquad Z_4 = e^{-\beta x}\cos\beta x. \tag{7.23}$$

Equations (7.22) and (7.23) also define $\bar{Z}$-functions, replacing x by $\bar{x}$. The values of Z and $\bar{Z}$ are smaller than 1.0, and drop quickly to almost zero as the section moves away from the edges.

EXAMPLE 7.1
Thermal Stresses in a Cylindrical Concrete Wall

Determine the bending moment M in the vertical direction and the circumferential hoop force N_ϕ and the bending moment M_ϕ in the wall of a circular-cylindrical concrete tank subjected to a rise of temperature that varies linearly through the wall thickness between $T_o = 30°C$ and $T_i = 0°C$ (54°F and 0°F). The top edge of the wall is free and the bottom edge is totally fixed. Given data are l = 10 m (33 ft); r = 30 m (98 ft); h = 0.25 m (10 inches); $\mu = 10 \times 10^{-6}$/°C (5.6×10^{-6}/°F); E = 32 GPa (4600 ksi); Poisson's ratio = 1/6.

The results of analysis will be used to examine the possibility of cracking, assuming that the tensile strength of concrete f_{ct} = 2.5 MPa (360 psi) and that the tank is prestressed such that, after losses and with design loads applied, the residual stress in concrete σ_r = −1.5 MPa (−220 psi) in both vertical and circumferential directions. The vulnerability to cracking will also be examined when the bottom edge of the wall is hinged or free to slide.

For this cylindrical wall, the dimensionless parameter $l^2/(rh)$ = 13.3 > 5.8. Thus, the tank can be considered deep and the equations of Section 7.4 apply. The values of M, N_ϕ, and M_ϕ at the bottom edge are calculated next using Equations (7.13) to (7.15) for the case of a wall totally fixed at the bottom edge and free at the top edge:

$$\beta = \frac{\{3[1-(1/6)^2]\}^{1/4}}{\sqrt{30(0.25)}} = 0.4772.$$

At the bottom edge, x = 0, Z_1 = 1.0, Z_3 = 1.0, $\bar{Z}_1 = 0$, $\bar{Z}_3 = 0$. Equations (7.13) to (7.15) give

$$N_\phi = -32\times10^9(10\times10^{-6})\left[\frac{0.25}{2}(30+0)(1.0) + \frac{1+1/6}{2(30)(0.4772)^2}(30-0)(0.0)\right] = -1200\,\text{kN/m},$$

$$M_\phi = -32\times10^9(10\times10^{-6})\left\{\frac{(1/6)(30)(0.4772)^2(0.25)^3}{12[1-(1/6)^2]}(30+0)1.0 + \frac{(0.25)^2}{12[1-(1/6)]}(30-0)[1-(1/6)(0.0)]\right\} = -74.64\text{ kN},$$

$$M = -32\times10^9(10\times10^{-6})\left[\frac{30(0.4772)^2(0.25)^3}{12[1-(1/6)^2]}(30+0)(1.0) + \frac{(0.25)^2}{12[1-(1/6)]}(30-0)(1.0)\right] = -147.8\text{ kN}.$$

Figure 7.2, obtained by finite-element analysis, shows the variation of N_ϕ, M_ϕ, and M over the wall height. The ordinates in the figure can be verified by Equations (7.13) to (7.15). The same figure includes

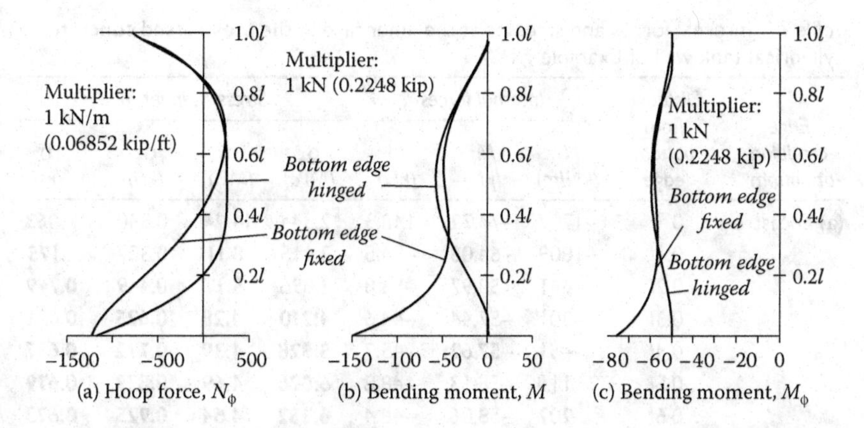

Figure 7.2 Internal forces due to temperature rise varying linearly between $T_i = 0°C$ and $T_o = 30°C$ (0°F and 54°F) (Example 7.1). Cylindrical wall dimensions: $\{l, r, h\}$ = {10, 30, 0.25 m} (33 ft, 98 ft, 10 in). Top edge free and bottom edge encastré or hinged. Positive N is tensile; positive M produces tension at the inner face.

curves for the case when the bottom edge is hinged while the top edge is free. The case when both the top and bottom edges are free to slide is not represented in Figure 7.2; however, in this case the graphs for the top half of the wall are not substantially different from the graphs plotted in the figure; a mirror image of the curves for the top half of the graphs gives the internal forces in the bottom half.

Table 7.1 gives the values of N_ϕ, M_ϕ, M, and the normal stresses σ and σ_ϕ, in vertical and hoop directions at the inner face of a wall. The top edge of the wall is free while the bottom edge is (a) encastré, (b) hinged, and (c) free to slide and rotate. Temperature variation causes cracking when it produces tensile stress at extreme fiber exceeding $(f_{ct} - \sigma_r) = 2.5 - (-1.5) = +4.0$ MPa (580 psi), where σ_r is the residual stress in concrete and f_{ct} is the tensile strength of concrete. The stress values printed in bold in Table 7.1 indicate cracking at the inner face; the cracks due to vertical stress σ and σ_ϕ will be horizontal and vertical, respectively. It can be seen from the table that the thermal loading considered is sufficient to produce cracking in this prestressed tank in all the three specified edge conditions.

If in this example the analysis is repeated for a change in temperature that varies linearly between $T_o = -20°C$ and $T_i = 0°C$ (−36°F and 0°F), the internal forces and the extreme fiber stress due to thermal loading will be the same as in Table 7.1 multiplied by −2/3. Again, the tensile stress will be sufficient to produce horizontal and vertical cracks at the outer face of the wall, with all the specified combinations of edge conditions.

This example indicates that cracking inevitably occurs due to temperature variations in the mild climate considered and that residual prestress of −1.5 MPa (−220 psi) does not prevent cracking. It is

Table 7.1 Internal forces and stresses at the inner face in the prestressed concrete cylindrical tank wall of Example 7.1[a]

Edge condition at bottom[b]	*Distance from bottom edge*	*Internal forces*			*Stress at inner face*[c]			
		N_ϕ *(kN/m)*	M_ϕ *(kN)*	*M* *(kN)*	σ_ϕ *(MPa)*	σ_ϕ *(MPa)*	σ_ϕ *(ksi)*	σ *(ksi)*
(a) Encastré	0	−1207	−74.72	−148.3	2.344	**14.24**	0.340	**2.063**
	0.1*l*	−1009	−64.08	−84.5	2.115	**8.11**	0.307	**1.175**
	0.2*l*	−641	−58.97	−53.8	3.096	**5.17**	0.449	**0.749**
	0.3*l*	−301	−57.44	−44.6	**4.310**	**4.28**	**0.625**	**0.621**
	0.4*l*	−51	−57.62	−45.7	**5.328**	**4.39**	**0.772**	**0.637**
	0.5*l*	111	−58.13	−48.8	**6.026**	**4.69**	**0.873**	**0.679**
	0.6*l*	202	−58.06	−48.4	**6.382**	**4.64**	**0.925**	**0.673**
	0.7*l*	217	−56.90	−41.4	**6.329**	3.97	**0.917**	0.576
	0.8*l*	112	−54.58	−27.5	**5.689**	2.64	**0.824**	0.383
	0.9*l*	−197	−51.74	−10.4	**4.180**	1.00	**0.606**	0.145
	1.0*l*	−810	−50.13	−0.8	1.572	0.08	0.228	0.011
(b) Hinged	0	−1207	−50.08	−0.5	−0.022	0.05	−0.003	0.007
	0.1*l*	−434	−58.84	−53.0	3.915	**5.09**	0.567	**0.738**
	0.2*l*	−7	−62.68	−76.1	**5.990**	**7.30**	**0.868**	**1.059**
	0.3*l*	177	−63.55	−81.3	**6.808**	**7.80**	**0.987**	**1.131**
	0.4*l*	232	−62.93	−77.6	**6.967**	**7.45**	**1.010**	**1.079**
	0.5*l*	239	−61.63	−69.8	**6.871**	**6.70**	**0.996**	**0.971**
	0.6*l*	234	−59.87	−59.2	**6.682**	**5.69**	**0.968**	**0.824**
	0.7*l*	203	−57.55	−45.3	**6.335**	**4.35**	**0.918**	**0.630**
	0.8*l*	84	−54.60	−27.6	**5.579**	2.65	**0.809**	0.384
	0.9*l*	−222	−51.51	−9.1	**4.058**	0.87	**0.588**	0.126
	1.0*l*	−827	−49.87	0.8	1.479	−0.07	0.214	−0.011
(c) Free	0	−827	−50.08	−0.5	1.499	0.05	0.217	0.007
	0.1*l*	−224	−51.82	−10.9	**4.079**	1.05	**0.591**	0.152
	0.2*l*	78	−54.94	−29.6	**5.586**	2.85	**0.810**	0.412
	0.3*l*	190	−57.72	−46.3	**6.299**	**4.45**	**0.913**	**0.644**
	0.4*l*	213	−59.48	−56.9	**6.562**	**5.46**	**0.951**	**0.792**
	0.5*l*	213	−60.08	−60.5	**6.620**	**5.80**	**0.959**	**0.841**
	0.6*l*	213	−59.48	−56.9	**6.562**	**5.46**	**0.951**	**0.792**
	0.7*l*	190	−57.72	−46.3	**6.299**	**4.45**	**0.913**	**0.644**
	0.8*l*	78	−54.94	−29.6	**5.586**	2.85	**0.810**	0.412
	0.9*l*	−224	−51.82	−10.9	**4.079**	1.05	**0.591**	0.152
	1.0*l*	−827	−50.08	−0.5	1.499	0.05	0.217	0.007

[a] Effect of temperature rise varying linearly between T_i = 0°C and T_o = 30°C (0°F and 54°F).

[b] Top edge of wall is free.

[c] Stress values printed in bold are sufficient to cause cracking.

therefore essential for crack control to provide non-prestressed reinforcement in addition to the prestressed tendons. This is discussed in Chapters 9 and 10.

ELASTICITY MODULUS IN ANALYSIS OF THERMAL STRESSES

The magnitude of thermal stresses is proportionate to the value of elasticity modulus E employed in the analysis. The rise or drop of temperature due to climate develops over a period of several hours, during which some creep occurs. Ignoring creep in analysis results in overestimation of thermal stresses. Use of the age-adjusted elasticity modulus of concrete (Equation 6.7) for the value of E is all that is required to account for the creep effect. On the other hand, the value of E for thermal stress analysis in design should be based on mean concrete strength rather than specified concrete strength. Some codes consider that mean concrete strength is greater than specified concrete strength by a designated value (e.g., 8.0 MPa or 1.2 ksi).

7.5 INTERNAL FORCES DUE TO SHRINKAGE OF CONCRETE

Consider the case of a cylindrical concrete tank left empty for some time after construction before first filling. The base, when monolithic with the wall and not subjected to drying, restrains movement of the bottom edge of the wall and thus produces internal forces in the wall. Analysis for this case can be done by assuming that the base undergoes no shrinkage, whereas the wall's free shrinkage is $\varepsilon_{cs}(t,t_0)$, where t_0 and t are the ages of concrete at start and end of the considered period.

When the inner face of the wall is in contact with water it swells, whereas the outer face, which is exposed to wind and sun, shrinks. In analysis, it may be assumed that the free volumetric change varies linearly between ε_{csi} and ε_{cso} at inner and outer wall faces, respectively.

The analysis for effects of shrinkage is the same as for effects of temperature rise, $T = \varepsilon_{cs}/\mu$, where μ is the thermal expansion coefficient, and ε_{cs} is free volumetric change, positive or negative for swelling and shrinkage, respectively. Because of their gradual development with time, shrinkage and swelling are accompanied by creep. Thus, the age-adjusted elasticity modulus should be used in the calculation of the corresponding internal forces. The age-adjusted elasticity modulus is given by

$$\bar{E}_c = \frac{E_c(t_0)}{1+\chi\phi(t,t_0)}, \tag{7.24}$$

where χ and ϕ are functions of t and t_0, with t_0 and t being the start and end of the period in which ε_{cs} occurs; $E_c(t_0)$ is the modulus of elasticity of concrete at age t_0.

EXAMPLE 7.2
Shrinkage Stresses in a Cylindrical Concrete Wall

Determine the circumferential forces N_ϕ and M_ϕ and the bending moment M in vertical direction in a cylindrical prestressed concrete tank wall. The top edge of the wall is free while the bottom edge is encastré. In the period t_0 to t, the wall is subjected to volumetric change whose free value varies linearly between $\varepsilon_{esi} = 70 \times 10^{-6}$ and -260×10^{-6} at the inner and outer wall faces, respectively. Given that the data are $E_c(t_0) = 32$ GPa; $\phi(t, t_0) = 1.3$ and 2.6 at the inner and outer surfaces, respectively; $\chi = 0.72$; $v = 1/6$. The tank has the same dimensions as in Example 7.1, that is: $l = 10$ m, $r = 30$ m, $h = 0.25$ m (33 ft, 98 ft, and 10 inches, respectively). The creep and shrinkage data are selected to represent relative humidities of 100% and 40% at the inner and outer surfaces, respectively; t_0 and t represent concrete ages of 3 months and infinity, respectively.

Table 7.2 gives internal forces and normal stresses at the outer face in vertical and circumferential directions obtained by the use of the equations in Section 7.4.1, with the following equivalent temperature rise at the inner and outer wall faces:

$$T_i = \frac{70\times10^{-6}}{\mu} = 7°\text{C},$$

$$T_o = \frac{-260\times10^{-6}}{\mu} = -26°\text{C},$$

where the coefficient of thermal expansion $\mu = 10 \times 10^{-6}/°\text{C}$. The results in Table 7.2 do not vary with the value selected for μ. For simplicity, a constant value is used for the age-adjusted elasticity modulus:

$$\bar{E}_c = 0.43\bar{E}_c(t_0) = 0.43(32\times10^9) = 13.76 \text{ GPa}.$$

This value is the average of values obtained by Equation (7.24) with $\phi = 1.3$ and 2.6. Figure 7.3 shows the variations of N_ϕ, M_ϕ, and M over the wall height.

Table 7.2 Internal forces and stresses at the outer face in the prestressed concrete cylindrical wall of Example 7.2[a]

Distance from bottom edge	Internal forces			Stress at outer face[b]			
	N_ϕ (kN/m)	M_ϕ (kN)	M (kN)	σ_ϕ (MPa)	σ (MPa)	σ_ϕ (ksi)	σ (ksi)
0	330	32.40	52.52	**4.43**	**5.04**	**0.642**	**0.731**
0.1*l*	276	29.53	35.26	3.94	3.38	0.571	0.491
0.2*l*	174	28.16	27.07	3.40	2.60	0.493	0.377
0.3*l*	77	27.77	24.70	2.98	2.37	0.431	0.344
0.4*l*	2	27.81	24.97	2.68	2.40	0.388	0.347
0.5*l*	–52	27.88	25.38	2.47	2.44	0.358	0.353
0.6*l*	–86	27.67	24.12	2.31	2.32	0.335	0.336
0.7*l*	–93	26.99	20.06	2.22	1.93	0.322	0.279
0.8*l*	–46	25.83	13.06	2.30	1.25	0.333	0.182
0.9*l*	97	24.45	4.81	2.74	0.46	0.397	0.067
1.0*l*	385	23.69	0.00	3.81	0.00	0.553	0.000

[a] Effect of volumetric change varying linearly between 70×10^{-6} and -260×10^{-6} at the inner and outer faces, respectively. Top edge free and bottom edge encastré.

[b] Stress values printed in bold are sufficient to cause cracking.

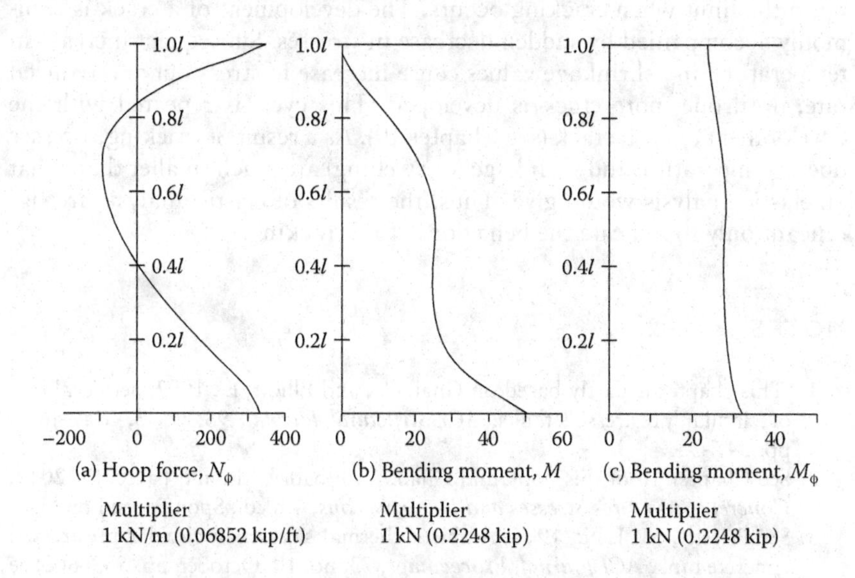

Figure 7.3 Internal forces due to volumetric change varying linearly between $\varepsilon_{csi} = 70 \times 10^{-6}$ and $\varepsilon_{cso} = -260 \times 10^{-6}$ (Example 7.2). Cylindrical wall dimensions $\{l, r, h\} = \{10, 30, 0.25 \text{ m}\}$ (33 ft, 98 ft, 10 in). Top edge free and bottom edge encastré. Positive N is tensile; positive M produces tension at the inner face.

Again, we assume in this example that the residual compression on concrete due to hydrostatic pressure and prestressing, excluding losses, is –1.5 MPa (–220 psi) and that the tensile stress of concrete f_{ct} = 2.5 MPa (360 psi). The stresses due to shrinkage are printed in bold in Table 7.2 when their value exceeds 4.0 MPa (580 psi), the value that produces cracking. While the stresses in this example are smaller than in Example 7.1 (Table 7.1), they are sufficient to produce horizontal and vertical cracks at the outer face of the tank. The risk of cracking at the outer face is increased when the volumetric change of concrete is combined with temperature variation with T_o = –20°C and T_i = 0°C, respectively (36°F and 0°F). Again, this example shows that concrete tanks, although prestressed, are vulnerable to cracking. This cracking can be controlled by the provision of an appropriate amount of non-prestressed reinforcement (see Chapter 10).

7.6 SIGNIFICANCE OF LINEAR ANALYSIS

As mentioned earlier, expansion or contraction of concrete due to temperature and shrinkage (or swelling) produces stresses and internal forces only when restrained. Cracking of a reinforced section, with or without prestressing, partly removes the restraint. Thus, a gradual increase of temperature and shrinkage values produces a proportionate increase in stresses up to the limit when cracking occurs. The development of a crack is commonly accompanied by sudden decrease in stresses. Subsequent increase in temperature and shrinkage values cause increase in stress but at a reduced rate, until one more crack is developed. This cycle is repeated with the development of each crack (see Chapter 10). As a result of cracking, stresses due to temperature and shrinkage (or swelling) are much smaller than what an elastic analysis would give. Thus, the results of elastic analysis are significant only to indicate the behavior before cracking.

NOTES

1. This chapter is partly based on Ghali, A., and Elliott, E., 1992, Serviceability of circular prestressed tanks, *ACI Structural Journal*, 98, no. 3, May–June, pp. 345–55.
2. See Chapter 10 of this book and Ghali, A., Elbadry, M., and Favre, R., 2012, *Concrete Structures Stresses and Deformations*, 4th ed., Spon Press, London.
3. See Priestly, M. J. N., 1976, Ambient thermal stresses in circular prestressed concrete tank, *ACI Journal, Proceedings*, 73, no. 10, October, pp. 553–60. See also Wood, J. H., and Adams, J. R., 1977, Temperature gradients in cylindrical concrete reservoirs, Proceedings of the 6th Australian Conference on the Mechanics of Structures and Materials, Christchurch, No. 2.
4. See Ghali et al., 2012, Chapter 9, "Effects of Temperature."

Chapter 8

Optimum design of prestressing

8.1 INTRODUCTION

Circumferential prestressing of circular-cylindrical concrete tank walls is intended to eliminate the hoop tension due to hydrostatic pressure and maintain residual compressive hoop stress $\sigma_{\phi r}$ (–1.0 to –2.0 MPa or –150 to –300 psi) that is constant over the wall height. This ideal objective can be achieved only when the wall edges at the top and bottom can slide and rotate freely. With these edge conditions, the circumferential prestress that can achieve the ideal objective must produce trapezoidal inward pressure, normal to the wall surface. When the wall has constant thickness, this prestressing alone or combined with triangular hydrostatic pressure over the full height of wall produces zero moments M and M_ϕ in vertical and circumferential directions. The only internal force is normal force N_ϕ in the hoop direction. For this reason, tank walls are often provided with sliding, or partially sliding, support (elastomeric pads) at the bottom edge, while the top edge is left free. The advantage gained is avoiding or significantly reducing the bending moments, at the cost of material, installation, and quality control to produce a nonleaking sliding joint.

A wall monolithic with base and free at the top is easier to build; but the trapezoidal distribution of circumferential prestressing produces a relatively large bending moment at the bottom edge of the wall, with the highest absolute value occurring immediately after prestressing, with the tank empty, before the occurrence of time-dependent prestress loss. This chapter proposes other distributions of circumferential prestress that minimize the absolute values of the bending moment in the vertical direction when the tank is empty and full, and provide residual hoop compression over the major height of the wall.

8.2 TRAPEZOIDAL DISTRIBUTION OF CIRCUMFERENTIAL PRESTRESSING

Figure 8.1 shows a cylindrical wall, of constant thickness, with a free top edge and bottom edge that can rotate and slide freely over a base. The initial intensity p of the pressure, normal to the wall surface, produced by the circumferential prestress is assumed to vary linearly:

$$p_{\text{initial}} = \left[p_{\text{top}} \frac{x}{l} + p_{\text{bot}} \left(1 - \frac{x}{l} \right) \right]_{\text{initial}}, \tag{8.1}$$

$$(p_{\text{top}})_{\text{initial}} = \left(\sigma_{\phi r} \frac{h}{r} \right) \eta_{\text{loss}}, \tag{8.2}$$

$$(p_{\text{bot}})_{\text{initial}} = p_{\text{top}} - (\gamma l) \eta_{\text{loss}}, \tag{8.3}$$

where p is positive when its direction is outward; l, r, and h are height, radius, and thickness of the wall; x is the distance between the bottom edge and any point; $\sigma_{\phi r}$ is specified residual hoop stress (e.g., –1.5 MPa or –220 psi); γ is the specific weight of stored liquid, whose highest level is assumed to be the same as the top of the wall; and η_{loss} is a multiplier equal to the ratio of initial prestress value to the value after time-dependent loss due to creep and shrinkage of concrete and relaxation of prestressed reinforcement. In this ideal condition, the only internal force in the wall is hoop force N_ϕ whose value immediately after prestressing, before filling the tank,

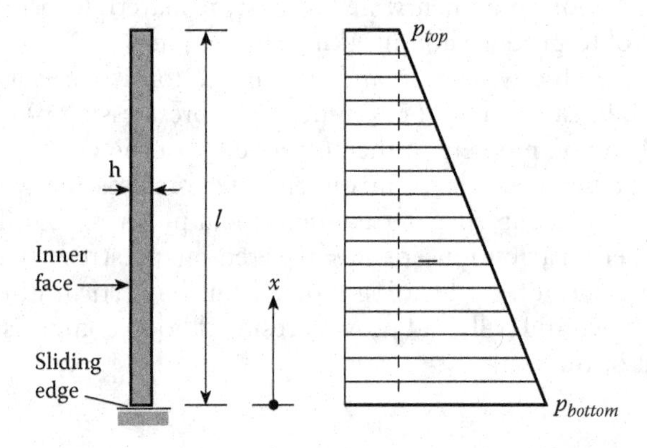

Figure 8.1 Circumferential prestressing on a cylindrical wall free to slide radially. Normal pressure produced by the prestress.

is equal to rp_{intial}. After the occurrence of time-dependent losses, $N_\phi = \sigma_{\phi r} h$ when the tank is full and $N_\phi = (\eta_{loss} - 1) rp_{initial}$ when the tank is empty. Although the distribution of p shown in Figure 8.1 is ideal when the bottom edge is free, the following example will show that the same distribution produces a relatively large bending moment in the vertical direction when the bottom edge is encastré.

EXAMPLE 8.1
Trapezoidal Distribution of Circumferential Prestressing; Cylindrical Wall Encastré at the Bottom Edge

Find N_ϕ and M in a cylindrical wall having the top edge free and the bottom edge encastré due to (a) initial trapezoidal circumferential prestressing (Figure 8.2a, Equations 8.1–8.3) and (b) hydrostatic pressure plus prestressing after the occurrence of time-dependent losses. The tank is assumed to be full to the top of the wall. Given data: $\{l, r, h\}$ = {10, 20, 0.30 m} ({33, 66, 1 ft}); γ = 10.0 kN/m³ (63.6 1b/ft³); $\sigma_{\phi r}$ = −1.5 MPa (−220 psi); η_{loss} = 1.15; Poisson's ratio, v = 1/6.

Initial prestressing produces normal pressure whose intensities at the top and bottom are (Equations 8.2 and 8.3):

$$p_{top} = \left(-1.5 \times 10^6 \frac{0.30}{20}\right) 1.15 = -25.9\,\text{kN/m}^2,$$

$$p_{bot} = -25.9 \times 10^3 - (10.0 \times 10) 1.15 = -140.9\ \text{kN/m}^2.$$

Figure 8.2b,c are plots for N_ϕ and M in the required loading Cases (a) and (b), respectively. These are obtained by computer[1] finite-difference analysis; the plotted values may be verified by use of the tables in Chapter 12.

8.3 IMPROVED DISTRIBUTION OF CIRCUMFERENTIAL PRESTRESSING

The results of Example 8.1 indicate that the initial prestress produces a relatively large bending moment at the bottom edge (Figure 8.2c). This will require a relatively large amount of reinforcement at the outer face of the tank. Or, if vertical prestress is provided, a relatively large prestressing force will be required to eliminate the tensile stress at the outer face. When the tank is filled after the occurrence of prestress loss, the bending moment at the base is reduced but will continue to have the same sign. Because the bottom edge does not slide, the prestressing can only reduce N_ϕ due to hydrostatic pressure but cannot produce residual hoop compression in the vicinity of the base.

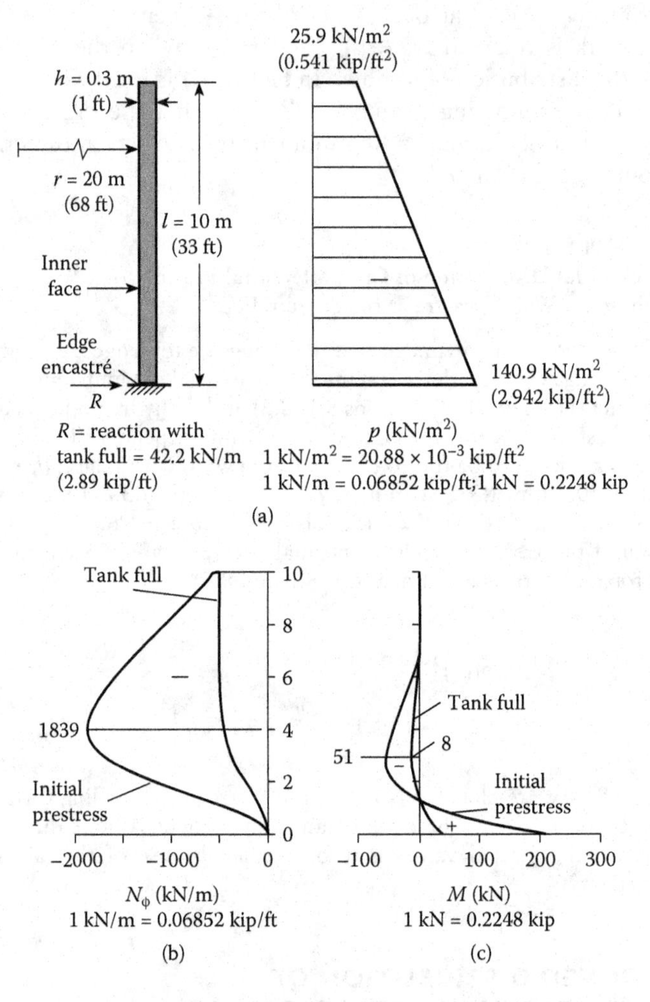

Figure 8.2 Cylindrical tank wall with bottom edge encastré (Example 8.1). (a) Normal pressure produced by circumferential prestressing. (b, c) Hoop force N_ϕ and vertical bending moment M immediately after prestressing and when the tank is full after occurrence of time-dependent loss.

The circumferential prestressing in the vicinity of the bottom edge produces a large bending moment M with a small contribution to the hoop compression N_ϕ. Thus, it can be beneficial to modify the trapezoidal distribution of circumferential prestressing by reducing $|p|$ in the zone close to the base. This can significantly decrease M at a cost of a relatively small decrease in the hoop compression. This will be demonstrated by examples.

EXAMPLE 8.2
Improved Distribution of Circumferential Prestressing

The tank of Example 8.1 is analyzed with two additional distributions of circumferential prestressing as shown in Figures 8.3a and 8.4a. These have the same trapezoidal distribution of Example 8.1, but truncated in two different ways, to eliminate part of the prestress at the bottom.

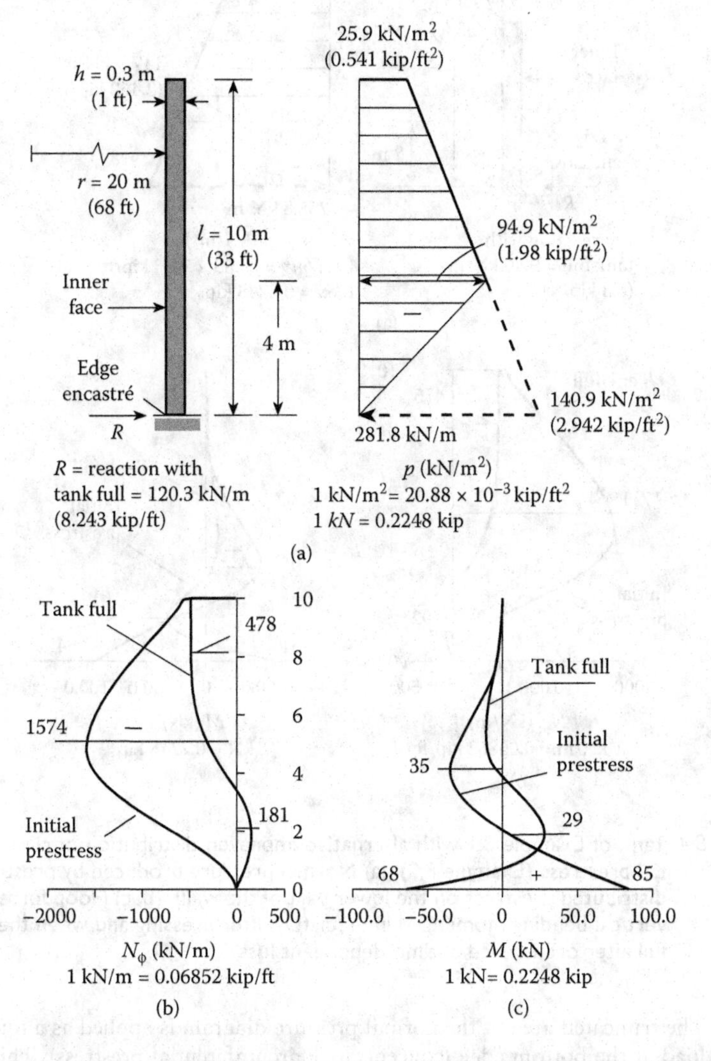

Figure 8.3 Tank of Example 8.1 with improved distribution of circumferential prestress (Example 8.2). (a) Normal pressure produced by the prestress. (b, c) Hoop force N_ϕ and vertical bending moment M immediately after prestressing and when the tank is full after occurrence of time-dependent loss.

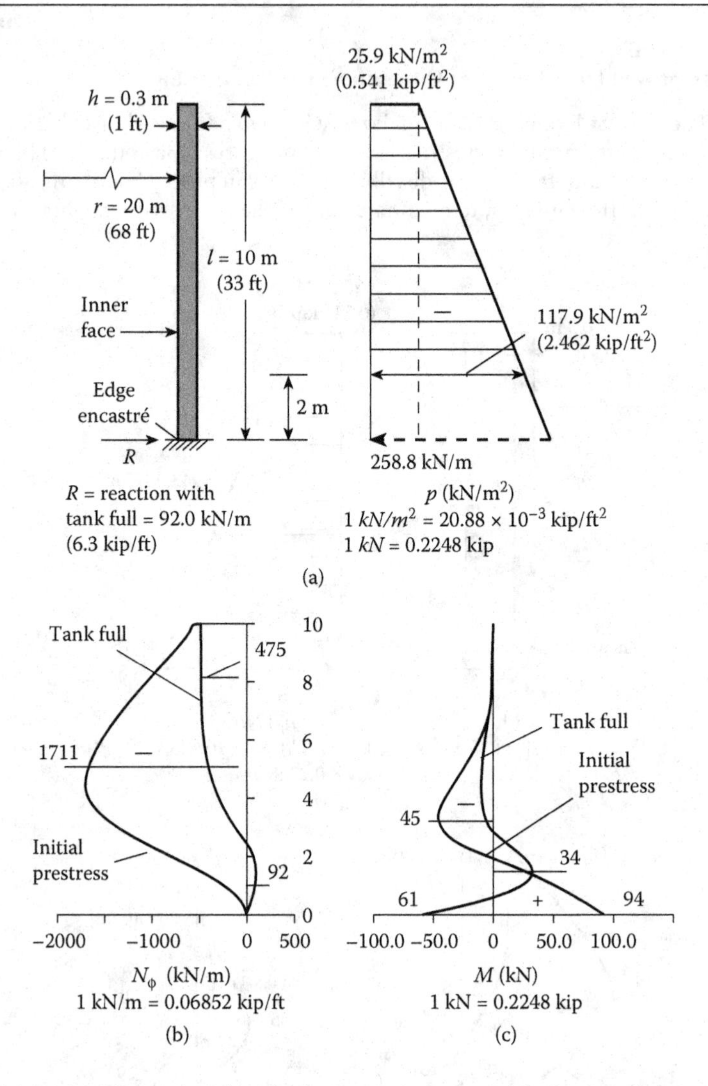

Figure 8.4 Tank of Example 8.1 with alternative improved distribution of circumferential prestress (Example 8.2). (a) Normal pressure produced by prestress (no distributed prestress on the lower part of the wall). (b, c) Hoop force N_ϕ and vertical bending moment M immediately after pressing and when the tank is full after occurrence of time-dependent loss.

The truncated area of the normal pressure diagram is applied as a line load at the bottom edge (concentrated circumferential prestress). This line load produces a reaction at the bottom edge (which compresses the base) but results in no internal forces in the wall.

The N_ϕ and M diagrams in Figure 8.3b,c and Figure 8.4b,c are for the following loading cases: (a) circumferential prestress at time

Table 8.1 Extreme values of bending moment at base and radial normal force in the plane of the base in a cylindrical tank

Circumferential prestress distribution	*Bending moment at base*			*Radial force on edge of base, N (kN/m)*
	M_γ (kN)	$\lvert M_a \rvert$ (kN)	$\lvert M_b \rvert$ (kN)	
Figure 8.2	–143	209	39	–42
Figure 8.3	–143	85	68	–120
Figure 8.4	–143	94	61	–92

Notes: There are three alternative distributions of circumferential prestressing (Figures 8.2, 8.3, and 8.4).

t_0 immediately after prestressing and (b) hydrostatic pressure plus the circumferential prestress at time t after the occurrence of time-dependent losses. The same data as in Example 8.1 apply.

The loading Cases (a) and (b) produce extremes of internal forces. The extreme absolute value of the bending moment in the vertical direction is the larger of

$$|M_a| = |M_p(t_0)|, \quad |M_b| = |M_p(t) + M_\gamma|, \tag{8.4}$$

where M_γ is the moment due to hydrostatic pressure; $M_p(t_0)$ is the moment immediately after prestress; $M_p(t) = M_p(t_0)/\eta_{loss}$ is the moment after time-dependent prestress loss; and η_{loss} is prestress loss multiplier (1.15 in this example). Consider the values of M at the base. The value of $M_\gamma = -143$ kN, while the values of M_p vary with the three circumferential prestress distributions (Table 8.1).

The total amount of circumferential prestress is the same in the three designs considered earlier; however, the extreme absolute value of the bending moment at the base is substantially reduced by truncation of the prestress distribution in the vicinity of the base (Figure 8.3c and Figure 8.4c). A further advantage is the increase in absolute value of the radial compressive force produced on the edge of the base. By trial, the height of the truncated part can be chosen to reduce the extreme absolute value even further. A computer program[2] (using finite differences, Chapter 3) is employed for the analyses in this example.

8.4 DESIGN OBJECTIVES

The purpose of circumferential prestressing of walls of cylindrical concrete tanks is to produce hoop compression that counteracts the hoop tension produced by hydrostatic pressure and to maintain a residual hoop compression when the tank is full. This target can be expressed by the equation

$$\frac{1}{\eta_{loss}} N_{\phi p} + N_{\phi\gamma} = N_{\phi r}, \tag{8.5}$$

where N_ϕ is normal hoop force per unit length; the subscripts p, γ, and r refer, respectively, to initial prestressing, hydrostatic pressure, and specified residual hoop force or stress; and η_{loss} is the prestress loss multiplier, which is equal to the ratio of the initial prestress value to the value after time-dependent loss.

A second purpose of circumferential prestressing is to produce the bending moment M in the vertical direction that partly counteracts the bending moment due to hydrostatic pressure. This may be expressed by a target equation

$$M_p + \eta_{moment} M_\gamma = 0, \tag{8.6}$$

where M is a bending moment in the vertical direction at a specified section (e.g., at base); and η_{moment} is a bending moment multiplier, which can be chosen arbitrarily to achieve a desired target. For example, to make the initial absolute value of M immediately after prestress equal to the moment due to prestress after loss combined with hydrostatic pressure, η_{moment} can be obtained from Equations (8.6) and (8.7):

$$-M_p = \frac{1}{\eta_{loss}} M_p + M\gamma, \tag{8.7}$$

which gives

$$\eta_{moment} = \left(1 + \frac{1}{\eta_{loss}}\right)^{-1}. \tag{8.8}$$

Multiplication of Equation (8.5) by ($1/h$) and Equation (8.6) by ($6/h^2$), respectively, transforms their terms into hoop stress σ_ϕ and vertical stress σ at the outer face of the wall. Thus, the target equations become

$$\frac{1}{\eta_{loss}} \sigma_{\phi p} + \sigma_{\phi\gamma} = \sigma_{\phi r}, \tag{8.9}$$

$$\sigma_p + \eta_{moment} \sigma_\gamma = 0, \tag{8.10}$$

where

$$\sigma_{\phi p} = N_{\phi p}/h, \quad \sigma_{\phi\gamma} = N_{\phi\gamma}/h, \tag{8.11}$$

$$\sigma_p = 6M_p/h^2, \quad \sigma_\gamma = 6M_\gamma/h^2. \tag{8.12}$$

8.5 DESIGN OF CIRCUMFERENTIAL PRESTRESS DISTRIBUTION

The problem is to design a circumferential prestress distribution that closely satisfies target Equations (8.9) or (8.10) applied at n selected sections:

$$\{\sigma_p\}_{n\times 1} = \{\sigma_{target}\}_{n\times 1}, \tag{8.13}$$

where $\{\sigma_p\}$ is a vector of stresses σ_ϕ and/or σ due to initial circumferential prestressing; and $\{\sigma_{target}\}$ is a vector of target stress values (from Equations 8.9 and 8.10):

$$\{\sigma_{target}\} = \begin{Bmatrix} \eta_{loss}\{\sigma_{\phi r} - \sigma_{\phi\gamma}\} \\ -\eta_{moment}\{\sigma_\gamma\} \end{Bmatrix}. \tag{8.14}$$

The circumferential prestressing produces, on the wall, a normal pressure, whose intensity p at any point may be expressed as

$$p = p_{basic} + \Delta p, \tag{8.15}$$

where p_{basic} is the assumed basic distribution such as the trapezoidal distribution given by Equation (8.1) (Figure 8.1); the basic pressure may not produce the target stresses in the vicinity of one or both edges of the wall. Thus, a refinement pressure is added and expressed as

$$Dp = [L_1 \quad L_2 \quad \ldots \quad L_m]\{p\}_{m\times 1}, \tag{8.16}$$

where $L_1, L_2, \ldots, L_m$ are shape functions of x, with x being the distance between the bottom edge and any section; and $\{p\}$ represents unknown multipliers (pressure values). The stresses due to the initial circumferential prestress can also be expressed as

$$\{\sigma_p\} = \{\sigma_{p,basic}\} + [\sigma_u]_{n\times m}\{p\}_{m\times 1}, \tag{8.17}$$

where $\{\sigma_{p,basic}\}$ represents stresses due to p_{basic}; and the second term on the right-hand side represents the effect of Δp. Any column i of $[\sigma_u]$ is composed of n stress values corresponding to normal pressure whose intensity at any point equals L_i. Substitution of Equation (8.16) in Equation (8.13) gives

$$[\sigma_u]_{n\times m}\{p\}_{m\times 1} = \left\{\{\sigma_{target}\} - \{\sigma_{p,basic}\}\right\}_{n\times 1}. \tag{8.18}$$

This represents n simultaneous equations with m unknowns; with $n \geq m$, solution gives the unknown $\{p\}$-values that satisfy or closely satisfy (least-square fit) Equation (8.18). This is done by multiplying both sides of the equation by the transpose matrix $[\sigma_u]^T$ and solving the resulting m equations.

It may be desirable to seek a solution of Equation (8.18) that satisfies more closely one of the simultaneous equations more than the others. This can be achieved by multiplying the corresponding row of $[\sigma_u]$ and $\{\{\sigma_{target}\} - \{\sigma_{pbasic}\}\}$ by a weighting factor > 1.0, then proceeding as discussed.

Once the $\{p\}$-values are known, substitution in Equations (8.15) and (8.14) gives the normal pressure, p. Multiplication by tank radius gives the intensity of circumferential prestressing force.

The presented design procedure may give, on a part of the wall, a positive value of $p(x)$ (outward pressure). The positive part of the p graph may simply be ignored. Alternatively, one can seek a solution that satisfies Equation (8.18) as closely as possible, subject to constraints specified for one or more of the elements of $\{p\}$.

The following equations may be conveniently used as shape functions:

$$L_1 = e^{-\beta x} \sin \beta x, \tag{8.19}$$

$$L_2 = e^{-\beta x} \cos \beta x, \tag{8.20}$$

with

$$\beta = \frac{\sqrt[4]{3(1-v^2)}}{\sqrt{rh}}, \tag{8.21}$$

where r and h are the radius of the middle surface and thickness of the cylindrical wall, respectively; and v is Poisson's ratio. Other shapes may also be useful.[3]

The design for distribution of circumferential prestressing may be performed in steps:

1. Determine the hoop force (N_ϕ) and the bending moment (M) in the vertical direction due to hydrostatic pressure and due to initial basic circumferential prestress (trapezoidal p, Equations 8.1 to 8.3).
2. Select n critical sections at which the vertical stress at the inner face (σ) or the hoop stress (σ_ϕ) will be affected significantly by the refinement pressure (Δp). Use the results of step 1 to generate $\{\sigma_{p,basic}\}$, $\{\sigma_{\phi\gamma}\}$, and $\{\sigma_\gamma\}$. Generate the vector of target stress values at the chosen sections (Equation 8.14).

3. Select m shape functions, $[L]$. Apply normal pressure varying according to each of the shape functions, separately. Calculate the corresponding stresses at each critical section to generate $[\sigma_u]_{n \times m}$.
4. When $m = n$, solve Equation (8.18); or when $n > m$, multiply both sides of Equation (8.18) by $[\sigma_u]^T$ and solve the resulting equation. This gives $\{p\}$.
5. Combine the basic normal pressure with the refinement pressure according to Equations (8.15) and (8.16) and multiply the result by r to give the required distribution of circumferential prestress force.

EXAMPLE 8.3
Design for Distribution of Circumferential Prestressing

Apply the procedure presented in Section 8.5 to design a distribution of circumferential prestressing, for the cylindrical tank wall of Example 8.1, such that

1. At the base, the absolute value of the bending moment $|M|$ in the vertical direction, when the tank is full, after occurrence of time-dependent loss, should be approximately equal to M due to initial prestressing.
2. Over the major part of the wall height (except in the vicinity of the base), the residual hoop stress $\sigma_{\phi r}$ after loss, with a full tank, should be approximately equal to −1.5 MPa (−220 psi).

Given data: $\{l, r, h\} = \{10, 20, 0.3 \text{ m}\}$ (33, 66, 1 ft); $\gamma = 10.0$ kN/m^3 (63.6 lb/ft^3); $\eta_{loss} = 1.15$; Poisson's ratio = 1/6.

To satisfy requirement 1, use the bending moment multiplier (Equation 8.8):

$$\eta_{moment} = \left(1 + \frac{1}{1.15}\right)^{-1} = 0.535.$$

Requirement 2 will be satisfied for the major part of the wall height by using the basic normal pressure, p_{basic}, produced by trapezoidal distribution of circumferential prestress, according to Equations (8.1) to (8.3). Use the shape functions of Equations (8.19) and (8.20) to find the refinement pressure Δp to satisfy requirement 1, without producing excessive tensile stress σ_ϕ, in the vicinity of the base.

We control σ_ϕ, at $x = 0.2l$ and $x = 0.1l$; we also control the vertical stress σ at the outer face of the wall. It is not realistic to require the same value of $\sigma_{\phi r}$ at $x = 0.2l$ and $0.1l$ as for the upper part of the wall. This is so because, with the bottom edge fixed, σ_ϕ tends to zero as the base is approached. We here take $\sigma_{\phi r} = -0.9$ and -0.4 MPa at $x = 0.2l$ and $0.1l$, respectively. Follow the design steps outlined in Section 8.5.

1. Basic normal pressure (Equations 8.1 to 8.3):

$$p_{\text{basic}} = (-140.9 + 11.5x)10^3 \quad (N, \text{ m units}).$$

N_ϕ and M are determined by computer[4] for the effects of p_{basic} and the hydrostatic pressure, separately. The computer results are

$$\begin{Bmatrix} N_{\phi,x=0.2l} = -1172.2 \text{ kN/m} \\ N_{\phi,x=0.1l} = -454.3 \text{ kN/m} \\ M_{x=0} = 209.3 \text{ kN} \end{Bmatrix}_{\text{p,basic}}, \begin{Bmatrix} N_{\phi,x=0.2l} = 779.2 \text{ kN/m} \\ N_{\phi,x=0.1l} = 306.2 \text{ kN/m} \\ M_{x=0} = -142.5 \text{ kN} \end{Bmatrix}_{\gamma}.$$

2. The stress vectors due to p_{basic} and due to hydrostatic pressure are

$$\begin{Bmatrix} \sigma_{\phi,x=0.2l} \\ \sigma_{\phi,x=0.1l} \\ \sigma \end{Bmatrix}_{\text{p,basic}} = 10^3 \begin{Bmatrix} -1172/0.3 \\ -454.3/0.3 \\ 6(209.3)/0.3^2 \end{Bmatrix} = \begin{Bmatrix} -3.907 \\ -1.151 \\ 13.95 \end{Bmatrix} \text{MPa},$$

$$\begin{Bmatrix} \sigma_{\phi,x=0.2l} \\ \sigma_{\phi,x=0.1l} \\ \sigma \end{Bmatrix}_{\gamma} = 10^3 \begin{Bmatrix} 799.2/0.3 \\ 306.2/0.3 \\ 6(-142.5)/(0.3)^2 \end{Bmatrix} = \begin{Bmatrix} 2.597 \\ 1.021 \\ -9.500 \end{Bmatrix} \text{MPa}.$$

The vector of target stresses is (Equation 8.14)

$$\sigma_{\text{target}} = \begin{Bmatrix} \sigma_{\phi,x=0.2l} \\ \sigma_{\phi,x=0.1l} \\ \sigma \end{Bmatrix}_{\text{target}} = 10^6 \begin{Bmatrix} 1.15(-0.9-2.597) \\ 1.15(-0.4-1.021) \\ -0.535(-9.500) \end{Bmatrix} = \begin{Bmatrix} -4.022 \\ -1.634 \\ 5.083 \end{Bmatrix} \text{MPa}.$$

3. The parameter β to be used in the shape functions (Equations 8.19 and 8.20) is

$$\beta = \frac{\sqrt[4]{3[1-(1/6)^2]}}{\sqrt{20}(0.3)} = 0.5335 \text{ m}^{-1}.$$

The same computer program is employed to determine the following internal forces produced by normal pressure whose intensity varies according to the shape functions L_1 and L_2 defined by Equations (8.19) and (8.20):

$$\begin{Bmatrix} N_{\phi,x=0.2l} = 2.232 \\ N_{\phi,x=0.1l} = 0.9431 \\ M_{x=0} = -0.4385 \end{Bmatrix}_{L_1}, \quad \begin{Bmatrix} N_{\phi,x=0.2l} = 0.9912 \\ N_{\phi,x=0.1l} = 0.6567 \\ M_{x=0} = -0.4382 \end{Bmatrix}_{L_2}.$$

The hoop stress σ_ϕ and vertical stress σ corresponding to L_1 form the first column of $[\sigma_u]$; similarly the stresses corresponding to L_2 form the second column of the same matrix:

$$[\sigma_\mu] = \begin{bmatrix} 7.440 & 3.304 \\ 3.146 & 2.189 \\ -29.23 & -29.21 \end{bmatrix}.$$

4. The target stress equations are (Equation 8.18):

$$\begin{bmatrix} 7.440 & 3.304 \\ 3.146 & 2.189 \\ -29.23 & -29.21 \end{bmatrix} \{p\} = 10^6 \begin{Bmatrix} -4.022+3.907 \\ -1.634+1.514 \\ 5.083-13.953 \end{Bmatrix} = \begin{Bmatrix} -0.115 \\ -0.120 \\ -8.870 \end{Bmatrix} 10^6.$$

Multiplication of both sides of the equations by the transposed matrix $[\sigma_u]^T$ and solution of the resulting equations gives $\{p\} = 10^3\{600.9, -297.9\}$. Use of these values with Equation (8.15) will give positive p-values, which are unrealistic. Thus, we set $p_2 = 140.9 \times 10^3$ to give $p = 0$ at $x = 0$. Substitution of $p_2 = 140.9 \times 10^3$ in the preceding target equations gives

$$\begin{bmatrix} 7.440 \\ 3.146 \\ -29.23 \end{bmatrix} p_1 = \begin{Bmatrix} -0.580 \\ -0.428 \\ -4.754 \end{Bmatrix} 10^6.$$

Here, it is considered more important to achieve the target vertical stress σ than the target σ_ϕ. Thus, we multiply the third row of the preceding equation by the arbitrary weighting factor 2.0; then multiply both sides of resulting equation by the transpose of the 3×1 matrix on the left-hand side of the equation, and solve in the same way to obtain $p_1 = 158.0 \times 10^3$.

5. The distribution of normal pressure that the initial circumferential prestress should produce is given by (Equations 8.15 and 8.16):

$$p = 10^3(-140.9 + 11.5x + 158.0L_1 + 140.9L_2)\ \text{N/m}^2.$$

Thus, according to this design, the intensity of the required circumferential prestressing force should have an intensity equal to pr.

Figure 8.5 shows variations of p and the corresponding N_ϕ and M immediately after prestressing. Also, the variations of N_ϕ and M after time-dependent prestress loss with the tank full are shown in Figure 8.5b,c.

The resultant of the pressure p is

$$R_p = \int_0^l p\ dx; \tag{8.22}$$

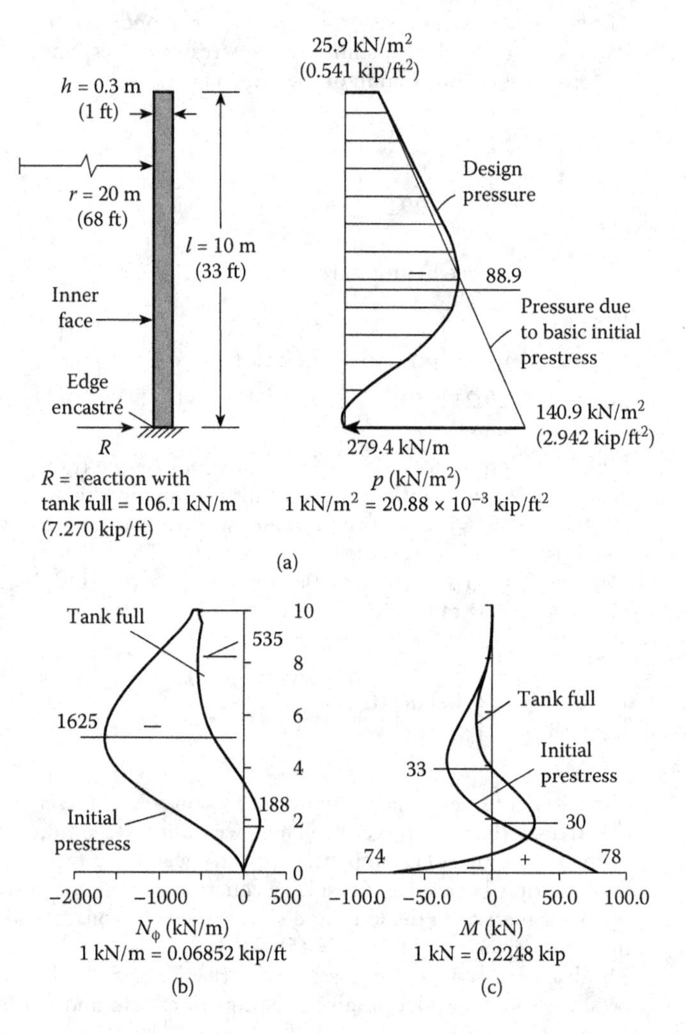

Figure 8.5 Design of circumferential prestress (Example 8.3). (a) Initial normal pressure produced by prestressing (obtained by design). (b, c) Hoop force N_ϕ and vertical bending moment M immediately after prestressing and when the tank is full after occurrence of time-dependent prestress loss.

R_p, the area of the pressure diagram in Figure 8.5a, equals −554.6 kN/m. The difference between the area of the basic pressure diagram and R_p is −279.4 kN/m. Concentrated circumferential prestress equal to $279.4r$ may be applied at the base as shown in Figure 8.5a. When the tank is full, after the occurrence of prestress loss, the tank will be subjected at its base to a radical inward force equal to 106.1 kN/m.

NOTES

1. Ghali, A. (n.d.), Computer program CTW (Cylindrical Tank Walls); see Appendix B of this book and its companion Web site.
2. See note 1.
3. See Ghali, A., and Elliott, E., 1991, Prestressing of circular tanks, *American Concrete Institute Structural Journal*, 88, no. 6, November–December, pp. 721–9.
4. The computer program CTW is employed; see note 1.

Chapter 9

Effects of cracking of concrete

9.1 INTRODUCTION

Cracks occur in reinforced concrete sections, with or without prestressing, when the concrete stress exceeds its tensile strength. After cracking, the internal forces must be resisted by the reinforcement and the uncracked part of the concrete cross-section. The part of the concrete section that continues to be effective in resisting the internal forces is subjected mainly to compression and some tension not exceeding the tensile strength of concrete. This chapter discusses the analysis of strain and stress distributions in cracked sections. It also presents empirical equations that give an estimate of crack widths.

9.2 STRESS AND STRAIN IN A CRACKED SECTION

At a cracked reinforced concrete section, concrete in tension is subjected to a relatively small force, which is commonly ignored. The basic equations in Chapter 6, Section 6.4 apply for a cracked section by substitution of A, B, and I by the cross-sectional area properties A_2, B_2, and I_2 of a transformed section composed of area of concrete in compression A_c plus $(\alpha A)_{ns}$ and $(\alpha A)_{ps}$; here the subscripts ns and ps refer to non-prestressed and prestressed reinforcement, respectively; A_{ns} or A_{ps} is the cross-sectional area of reinforcement; α_{ns} or α_{ps} is the ratio of modulus of elasticity of the reinforcement to that of concrete; the symbol A stands for area; and B and I are its first and second moments about an axis through an arbitrary reference point O. Calculation of A_2, B_2, and I_2 must be preceded by determination of the depth c of the compression zone. Owing to creep and shrinkage of concrete and relaxation of prestressed steel, the depth c changes (commonly increases), and thus A_c changes with time. This change is ignored in the following in the analysis of time-dependent stress and strain, while satisfying equilibrium. This greatly simplifies the analysis but involves negligible error. Analysis of a cracked prestressed section is the same as that of a non-prestressed section, after determining decompression forces, discussed next.

9.3 DECOMPRESSION FORCES

Consider a prestressed section containing prestressed and non-prestressed steel (Figure 9.1a). After the occurrence of creep and shrinkage of concrete and relaxation of prestressed reinforcement, the stress distribution over the section is defined by σ_{Oin} and γ_{in}, which are assumed to be known; where σ_{Oin} is the stress value at reference point O, and γ_{in} (= $d\sigma/dy$) is the slope of a stress diagram. At this stage the section is assumed to be non-cracked. Additional forces N and M are assumed to be applied, causing the section to crack. It is required to determine the stress and strain distribution after cracking.

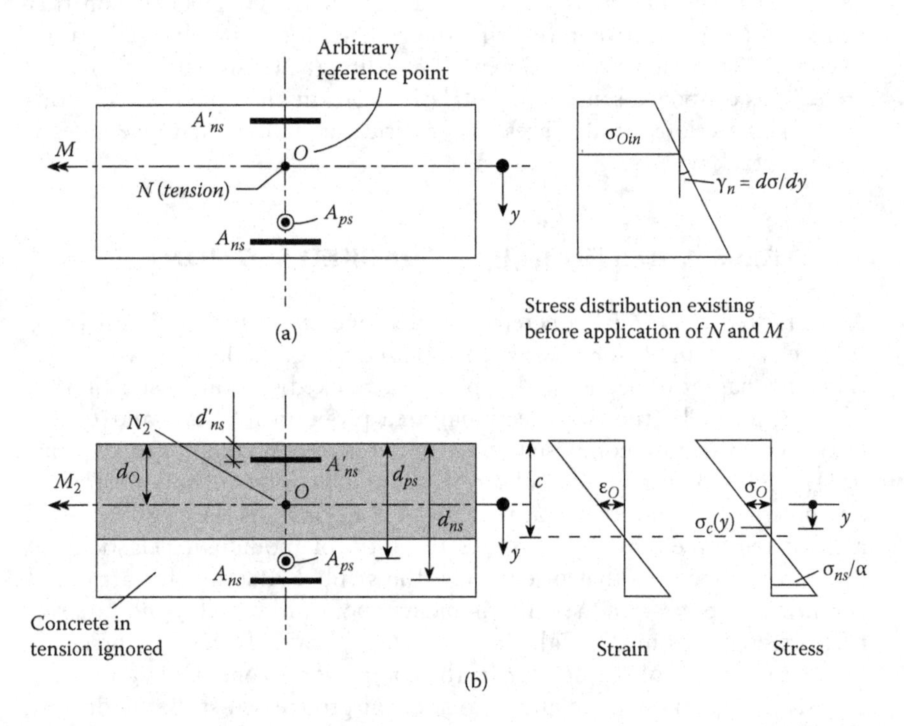

Figure 9.1 Analysis of stress and strain in a cracked prestressed section. (a) Initial stress distribution in a noncracked section; N and M are additional forces producing cracking; positive sign convention for M, N, σ_O, and γ. (b) Additional stress and strain due to N_2 and M_2.

For this purpose, partition N and M as follows:

$$N = N_1 + N_2, \qquad M = M_1 + M_2. \tag{9.1}$$

N_1 combined with M_1 is the part that will bring the stress in concrete to zero; N_2 (= $N - N_1$) combined with M_2 (= $M - M_1$) is the remainder. N_1 and M_1 are called the *decompression forces*. The stress and strain due to N and M can now be determined by superposition of the stress and strain of two analyses: (1) N_1 and M_1 applied on a noncracked transformed section; and (2) N_2 and M_2 applied on a cracked section, for which concrete in tension is ignored. The basic equation in Chapter 6, Section 6.4 applies in both analyses, differing only in that A_1, B_1, and I_1 are to be used in the first analysis and A_2, B_2, and I_2 for the second analysis. The subscripts 1 and 2 refer to the noncracked state and the fully cracked state (ignoring concrete in tension), respectively. The decompression forces are (Chapter 6, Equation 6.13):

$$N_1 = -(\sigma_{\text{Oin}} A_1 + \gamma_{\text{in}} B_1), \qquad M_1 = -(\sigma_{\text{Oin}} B_1 + \gamma_{\text{in}} I_1). \tag{9.2}$$

The parts of N and M to be applied on a fully cracked section are

$$N_2 = N - N_1, \qquad M_2 = M - M_1. \tag{9.3}$$

The positive sign convention for N, M, σ_O, and γ is shown in Figure 9.1a; positive strain and stress represent elongation and tension, respectively.

9.4 EQUILIBRIUM EQUATIONS

Figure 9.1b shows strain and stress distributions on a cracked section subjected to normal force N_2 at reference point O, combined with the bending moment M_2. The assumptions involved in the analyses are linear stress–strain relationship for steel and concrete in compression, with concrete in tension ignored; and the plane cross-section remains plane. The assumptions are expressed in the equations

$$\varepsilon = \varepsilon_O + y\psi, \tag{9.4}$$

$$\sigma_c = E_c \varepsilon_O \left(1 - \frac{y}{y_n}\right) \quad \text{for} \quad y < y_n, \tag{9.5}$$

$$\sigma_c = 0 \quad \text{for} \quad y \geq y_n, \tag{9.6}$$

where ε and σ_c are the strain and stress at any fiber whose coordinate is y (distance, measured downward, from reference point O to any fiber); and y_n is the y coordinate of the neutral axis. The stress in any reinforcement layer is

$$\sigma_{\text{ns or ps}} = E_{\text{ns or ps}}\varepsilon_O\left(1-\frac{y_{\text{ns or ps}}}{y_n}\right). \tag{9.7}$$

The resultants of stress in concrete and reinforcements are equal to N_2 and M_2. This can be expressed in the equations

$$\int \sigma \, \mathrm{d}A = N_2, \quad \int \sigma y \, \mathrm{d}A = M_2, \tag{9.8}$$

$$e = \frac{\int \sigma y \, \mathrm{d}A}{\int \sigma \, \mathrm{d}A}, \tag{9.9}$$

$$e = M_2/N_2, \tag{9.10}$$

where e is eccentricity of N_2, when applied alone as the statical equivalent of N_2 and M_2. When $N_2 = 0$, $e = \infty$, and the neutral axis will be at the centroid of the fully cracked section; thus,

$$y_n = B_2/A_2, \tag{9.11}$$

$$y_n = c - d_O, \tag{9.12}$$

where A_2 is the area of the fully cracked section; B_2 is its first moment about reference point O; d_O is the distance between compressive fiber and reference point; and c is the depth of compression zone.

Equation (9.9) or Equation (9.11) gives c. The solution can be done in trial cycles. In a typical ith cycle, a value c_i is assumed; the right-hand side of the equation is evaluated numerically; the difference represents residual R to be eliminated by Newton's method. In the next trial c is assumed equal to

$$c_{i+1} = c_i + R\left(\frac{\mathrm{d}R}{\mathrm{d}c}\right)^{-1}. \tag{9.13}$$

The derivative $(\mathrm{d}R/\mathrm{d}c)$ may be determined by differentiation or by finite difference.

The depth of compression zone c depends only on e, not on individual values of N_2 and M_2. Thus, in evaluating the right-hand side of Equation (9.9) or (9.11), σ_c can be considered to vary between $\sigma_{extreme}$ at compression fiber and zero at the neutral axis; thus,

$$\sigma_c = -\left(\frac{y_n - y}{c}\right)\sigma_{extreme}. \tag{9.14}$$

Unit stress (or arbitrary value) may be used for $\sigma_{extreme}$.

The trial cycles converge to the correct value of c only when the trial value c_1 in the first cycle is close to the answer. A value close to the answer can be achieved by calculating A_2, B_2, and I_2, using a guessed value of c (e.g., half-depth of section), then using Equation (6.14) to obtain the corresponding values of σ_O and γ, $\gamma_n = -\sigma_O/\gamma$ and $c = y_n + d_O$. The value of c thus obtained is the value used as the first trial value in Newton's iteration.

The aforementioned equations apply to a non-prestressed cracked section with zero initial stress. In this case, there are no decompression forces; thus, $N_1 = 0$ and $M_1 = 0$. Also, the equations apply when cracking is caused during the prestressing. There is also no initial stress in this case, and N_1 and M_1 are zero. The given forces N and M should include the effect of prestressing and other forces that come into action at the same time as the prestressing; A_{ps} and the cross-sectional area of the prestressed duct (if any) should not be included in the calculation of A_2, B_2, and I_2.

9.5 RECTANGULAR SECTION ANALYSIS

When the section analyzed is a rectangle (e.g., strip of unit width of a wall of a tank or a silo), the calculation of the depth of the compression zone[1] can be done by equations presented in this section. The term "top fiber" is used to mean extreme compression fiber (Figure 9.1b). Coordinate y is the distance from reference point O, measured downward to any fiber; coordinate $\bar{y}$ has the same meaning as y, but the distance is measured from the bottom reinforcement layer; e_s is the eccentricity of the resultant of N_2 and M_2, measured downward from the bottom reinforcement; subscripts ps and ns refer to prestressed and non-prestressed reinforcement, respectively; a prime refers to the top reinforcement; d is the distance from the top fiber to reference point O or to any reinforcement layer; subscript O indicates the reference point; b is the section width; and α is the ratio of the modulus of elasticity of reinforcement to that of concrete. The eccentricity e and the coordinate y may be expressed as

$$e = d_{ns} - d_O + e_s, \qquad y = d_{ns} - d_O + \bar{y}. \tag{9.15}$$

Substitution in Equation (9.9) gives

$$\int \sigma \bar{y}\, da - e_s \int \sigma\, da = 0, \tag{9.16}$$

$$\begin{aligned} &b(\tfrac{1}{2}c^2)(d_{ns} - \tfrac{1}{3}c) + (\alpha_{ns} - 1)A'_{ns}(c - d'_{ns})(d_{ns} - d'_{ns}) \\ &-\alpha_{ps}A_{ps}(d_{ps} - c)(d_{ns} - d_{ps}) + e_s\left[b(\tfrac{1}{2}c^2) + (\alpha_{ns} - 1)A'_{ns}(c - d'_{ns})\right. \\ &\left.-\alpha_{ps}A_{ps}(d_{ps} - c) - \alpha_{ns}A_{ns}(d_{ns} - c)\right] = 0. \end{aligned} \tag{9.17}$$

This is a cubic polynomial in c that can be expressed as

$$c^3 + a_1c^2 + a_2c + a_3 = 0, \tag{9.18}$$

where

$$a_1 = -3(d_{ns} + e_s), \tag{9.19}$$

$$\begin{aligned} a_2 = -\frac{6}{b}\big[&A'_{ns}(\alpha_{ns} - 1)(d_{ns} + e_s - d'_{ns}) \\ &+ \alpha_{ps}A_{ps}(d_{ns} + e_s - d_{ps}) + \alpha_{ns}A_{ns}e_s\big], \end{aligned} \tag{9.20}$$

$$\begin{aligned} a_3 = \frac{6}{b}\big[&A'_{ns}d'_{ns}(\alpha_{ns} - 1)(d_{ns} + e_s - d'_{ns}) + \alpha_{ps}A_{ps}d_{ps}(d_{ns} + e_s - d_{ps}) \\ &+ \alpha_{ns}A_{ns}d_{ns}e_s\big]. \end{aligned} \tag{9.21}$$

Closed-form or trial solution of the polynomial cubic Equation (9.18) gives c, when $N_2 \neq 0$. But when $N_2 = 0$, use Equation (9.11) written in the form

$$\begin{aligned} &\tfrac{1}{2}bc^2 + \left[A'_{ns}(\alpha_{ns} - 1) + \alpha_{ps}A_{ps} + \alpha_{ns}A_{ns}\right]c \\ &-\left[A'_{ns}(\alpha_{ns} - 1)d'_{ns} + \alpha_{ps}A_{ps}d_{ps} + \alpha_{ns}A_{ns}d_{ns}\right] = 0 \end{aligned} \tag{9.22}$$

or

$$\bar{a}_1c^2 + \bar{a}_2c + \bar{a}_3 = 0, \tag{9.23}$$

where

$$\bar{a}_1 = b/2, \qquad \bar{a}_2 = A'_{ns}(\alpha_{ns} - 1) + \alpha_{ps}A_{ps}\alpha_{ns}A_{ns}, \tag{9.24}$$

$$\bar{a}_3 = -A'_{ns}(\alpha_{ns} - 1)d'_{ns} - \alpha_{ps}A_{ps}d_{ps} - \alpha_{ns}A_{ns}d'_{ns}. \tag{9.25}$$

Solution of the quadratic Equation (9.23) gives c. When depth c of the compression zone and transformed section properties A_2, B_2, and I_2 are determined (ignoring concrete below the neutral axis), they can be substituted in the equations in Chapter 6, Section 6.4 to give parameters defining strain and stress distributions in the fully cracked section.

9.6 TENSION STIFFENING

At a cracked section, we ignore concrete in tension and use properties of a fully cracked section A_2, B_2, and I_2 to calculate axial strain ε_{O2} and ψ_2. The subscript 2 refers to the fully cracked state. If ε_{O2} and ψ_2 are employed in the calculation of displacements, they will result in overestimation. Because of the bond between concrete and reinforcement, concrete between cracks shares resisting the tension with the reinforcement and contributes to the rigidity of the section. To account for this effect, referred to as tension stiffening, we use mean values ε_{Om} and ψ_m in displacement calculation. The mean values are given by the empirical equations

$$\varepsilon_{Om} = (1 - \zeta)\varepsilon_{O1} + \zeta\varepsilon_{O2}, \tag{9.26}$$

$$\psi_m = (1 - \zeta)\psi_1 + \zeta\psi_2, \tag{9.27}$$

$$\zeta = 1 - \beta\left(\frac{f_{ct}}{\sigma_{1,extreme}}\right)^2, \tag{9.28}$$

where ε_{O1} and ψ_1 are unreal values of strain at reference point O and slope of strain diagram, calculated with the assumption of no cracking; $\sigma_{1,extreme}$ is the extreme fiber tensile stress, calculated with the same assumption; f_{ct} is the tensile strength of concrete; and β is the coefficient that may be taken as 0.5 when deformed steel bars are used and when the loading is repetitive or sustained. The mean values given by Equations (9.27) and (9.28) are interpolated quantities between two extremes: states 1 and 2 that underestimate and overestimate deformations. Accounting for the effect of tension stiffening and the prediction of crack width rely on empirical equations adopted in codes[2] (e.g., Equations 9.27 and 9.28, European codes).

9.7 CRACK WIDTH

The coefficient ζ defined by Equation (9.28) is also used to give mean crack width

$$w_m = s_r \zeta \varepsilon_{s2}, \tag{9.29}$$

where ε_{s2} is the tensile strength in the reinforcement caused by N_2 and M_2 on a fully cracked section (Equation 9.3); and s_r is crack spacing given empirically by

$$s_r = \text{constant} + \kappa_1 \kappa_2 \frac{d_b}{4\rho_r}. \tag{9.30}$$

The "constant" is 50 mm (2 in); d_b is bar diameter; and $\kappa_1 = 0.8$ for high bond bars and 1.6 for plain bars. When cracking is due to the restraint of intrinsic or imposed deformations (e.g., restraint of shrinkage), the coefficient κ_1 is to be replaced by $(0.8\kappa_1)$. κ_2 is a coefficient depending upon the shape of strain diagram; $\kappa_2 = 0.5$ in the case of bending without axial force; $\kappa_2 = 1.0$ in the case of axial tension. In the case of eccentric tension,

$$\kappa_2 = \frac{\varepsilon_1 + \varepsilon_2}{2\varepsilon_1}, \tag{9.31}$$

where ε_1 and ε_2 are, respectively, the greater and the lesser tensile strain values (assessed for a fully cracked section) at the upper and lower boundaries of the effective tension area A_{cef} defined in Figure 9.2. The steel ratio ρ_r is defined as

$$\rho_r = \frac{A_s}{A_{cef}}. \tag{9.32}$$

The effective tension area A_{cef} is generally equal to 2.5 times the distance from the tension face of the section to the centroid of A_s (Figure 9.2); but the height of the effective area should be less than or equal to $(h - c)/3$, where h is the height of the section and c is the depth of the compression zone.

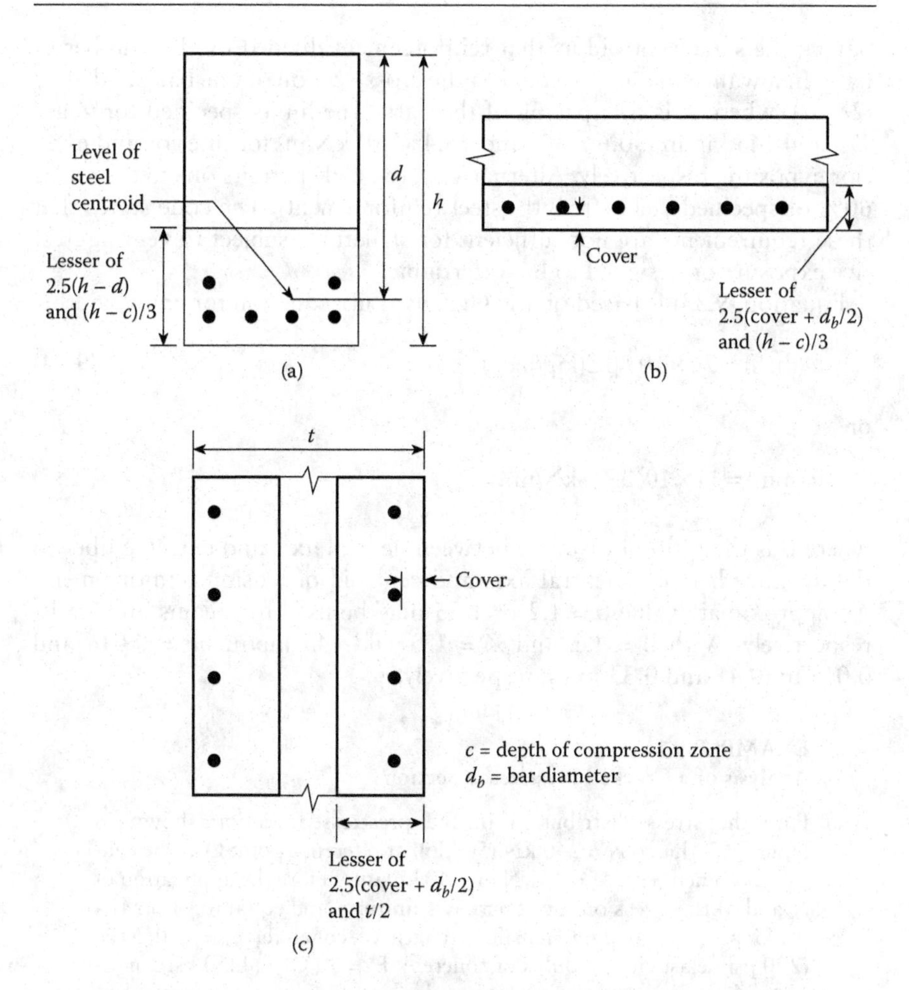

Figure 9.2 Effective area A_{cef} for use in Equation (9.32): (a) beam, (b) slab, (c) member in tension. (From Mode Code 2010; see note 2.)

As a means of controlling width of cracks, the American Concrete Institute's code requires that a parameter Z (force/length) does not exceed the specified limit, where Z is defined by

$$Z = \sigma_s \sqrt[3]{d_c A_c}, \tag{9.33}$$

where σ_s is the stress in reinforcement (steel) at service load, calculated for a fully cracked section; d_c is the cover distance between the extreme tension fiber to the center of the bar located close thereto; and A_e is the effective tension area of concrete surrounding the flexural tension reinforcement and having the same centroid as that reinforcement divided by the number of bars. In a wall, A_e is simply equal to the cross-section of one bar divided by $(2d_c\ s_b)$, where s_b is the spacing of the bars. The limits specified for Z are 175 and 145 kip/in (30.6×10^3 and 25.4×10^3 kN/m) for interior and exterior exposure, respectively. Alternatively, the code permits one to take σ_s as 60% of specified yield strength (steel reinforcement). The code states that these requirements are not sufficient for structures subject to very aggressive exposure or designed to be watertight.

Equation (9.33) is based on the Gergely–Lutz equation for crack width:

$$w(\text{inch}) = 76 \times 10^{-6} \beta Z(\text{kip/inch}) \tag{9.34}$$

or

$$w(\text{mm}) = 11 \times 10^{9} \beta Z(-\text{kN/m}), \tag{9.35}$$

where β is the ratio of distance between neutral axis and extreme fiber to the distance between neutral axis and centroid of tension reinforcement. An approximate value $\beta = 1.2$ or 1.35 may be used for beams and walls, respectively. With $\beta = 1.2$ and $Z = 175$ and 145 kip/in, $w = 0.016$ and 0.013 in (0.41 and 0.33 mm), respectively.

EXAMPLE 9.1
Analysis of a Cracked Prestressed Section

Find the stress distribution in the prestressed section shown in Figure 9.3, due to $N = 240$ kN (54 kip), at reference point O at the center, combined with $M = 50$ kN m (440 kip in). Before the application of N and M the stress on the concrete is uniform and equal to −1.0 MPa (−145 psi). Given data: tensile strength of concrete, $f_{ct} = 2.0$ MPa (290 psi); elasticity modulus of concrete, $E_c = 30$ GPa (4350 ksi); modulus of elasticity of prestressed and non-prestressed reinforcement, $E_s = 200$ GPa (29000 ksi). Determine the crack width by Equation (9.29), assuming $s_{rm} = 200$ mm (15 in).

The properties of the noncracked transformed section are

$$A_1 = 320.4 \times 10^{-3}\,\text{m}^2, \qquad B_1 = 0.5440 \times 10^{-3}\,\text{m}^3,$$

$$I_1 = 2.420 \times 10^{-3}\,\text{m}^4.$$

The decompression forces are (Equation 9.2)

$$N_1 = 1.0 \times 10^6 (320.4 \times 10^{-3}) = 320.4\ \text{kN},$$

$$M_1 = 1.0 \times 10^6 (0.5440 \times 10^{-3}) = 0.8160\ \text{kN m}.$$

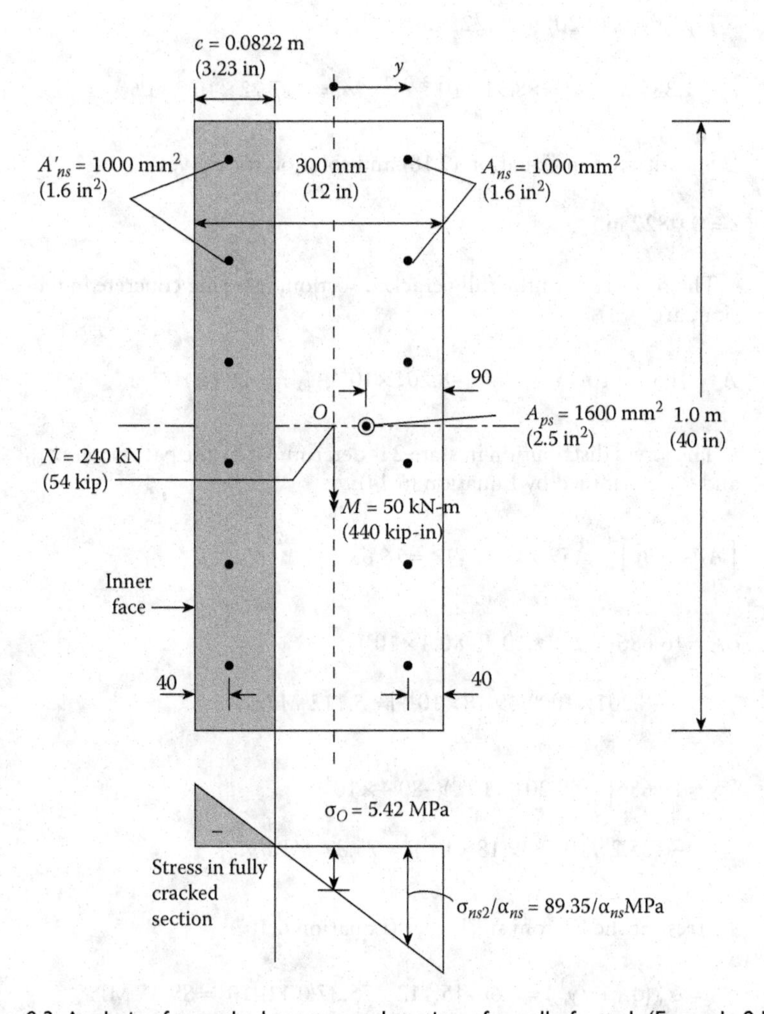

Figure 9.3 **Analysis of a cracked prestressed section of a wall of a tank (Example 9.1).**

Forces to be applied on a fully cracked section are (Equation 9.3)

$$N_2 = 240.0 \times 10^3 - 320.4 = -80.4 \text{ kN},$$

$$M_2 = 50.0 \times 10^3 - 0.8160 \times 10^3 = 49.18 \text{ kN m}.$$

Application of Equations (9.10) and (9.15) gives

$$e_s = e - d_{ns} + d_O = \frac{49.18}{-80.4} - 0.26 + 0.15 = -0.7217 \text{ m}.$$

Equations (9.20) to (9.22) give

$$a_1 = 1.385 \text{ m}, \quad a_2 = 88.91\times10^{-3} \text{ m}^2, \quad a_3 = -17.22\times10^{-3} \text{ m}^3.$$

Substitution in Equation (9.18) and solution for c gives

$$c = 0.0822 \text{ m}.$$

The properties of the fully cracked section, ignoring concrete in tension, are

$$A_2 = 105.2\times10^{-3} \text{ m}^2, \; B_2 = -8.201\times10^{-3}, \; I_2 = 1.209\times10^{-3} \text{ m}^4.$$

The stress distribution in state 2 is determined by the parameters σ_{O2} and γ_2 determined by Equation (6.14):

$$\left[A_2I_2 - B_2^2\right]^{-1} = (59.93\times10^{-6})^{-1} = 16\ 686,$$

$$\sigma_O = 16\ 686\big[1.209\times10^{-3}(-80.4\times10^3)$$
$$-(-8.201\times10^{-3})(49.18\times10^3)\big] = 5.112\ MPa,$$

$$\gamma = 16\ 686\big[-(-8.201\times10^{-3})(-80.4\times10^3)$$
$$-(105.2\times10^{-3})(49.18\times10^3)\big] = 75.37\ MPa/m.$$

Stress at the bottom steel layer (Equation 6.10):

$$\sigma_{ns} = \alpha_{ns}(\sigma_O + \gamma y_{ns}) = 6.667\left[5.112 + 75.37(0.11)\right]10^6 = 89.35 \text{ MPa}.$$

When cracking is ignored, the hypothetical tensile stress at the bottom fiber due to N_2 combined with M_2 is

$$\sigma_{1,\text{extreme}} = \sigma_{O1} + \gamma_1 y_{\text{extreme}}.$$

Application of Equation (6.14) as done earlier using for area properties the values of A_1, B_1, and I_1 gives

$$\sigma_{O1} = -0.2855 \text{ MPa}, \quad \gamma = 20.39 \text{ MPa/m}.$$

At $y_{extreme} = 0.15$ the stress is

$$\sigma_{1,extreme} = -0.2855 + 20.39(0.15) = 2.773 \text{ MPa}.$$

The interpolation coefficient between states 1 and 2 (Equation 9.28, with $\beta = 0.5$) is

$$\zeta = 1 - 0.5\left(\frac{2.0}{2.773}\right)^2 = 0.740.$$

The mean crack width (Equation 9.29) is

$$w_m = 200(0.740)\frac{89.35 \times 10^6}{200 \times 10^9} = 0.07 \text{ mm } (3 \times 10^{-3} \text{ inch}).$$

EFFECT OF REDUCTION OF THE NON-PRESTRESSED STEEL ON CRACK WIDTH

If in this example the non-prestressed steel area at each face of a wall is reduced from 1000 mm² (1.6 in²) to 500 mm², what will be the effect on crack width? The reduction in steel area has a significant effect on the crack spacing s_r and strain ε_{s2} in the bottom reinforcement in the fully cracked states; both parameters increase. Thus w_m, by Equation (9.30), will increase even further. Assuming that s_r increases to 250 mm and repeating the analysis[3] gives $\varepsilon_{s2} = 589 \times 10^{-6}$ and $w_m = 0.11$ mm (4.3×10^{-3} in). When the analysis is repeated once more with the same non-prestressed steel area, the same crack spacing, the same $N = 240$ kN (54 kip), but with M changing sign to become $M = -50$ kN m (−440 kip in), the results are

$$\varepsilon_{s2} = 1370 \times 10^{-6}, \quad w_m = 0.25 \text{ mm } (10 \times 10^{-3} \text{ inch}).$$

9.8 TIME-DEPENDENT STRAIN AND STRESS IN CRACKED SECTIONS

The analysis presented in the previous sections applies to reinforced concrete sections with or without prestressing. Four steps of analysis are presented in Section 6.5 to give the time-dependent changes in stresses and strains due to creep, shrinkage, and relaxation. The flow chart in Figure 6.5 shows how the four steps can be applied for time-dependent analysis of a cracked section. In the chart it is assumed that forces N and M are applied on a section for which the initial stress is known. When these forces produce cracking, the chart indicates additional calculations to be done before proceeding with the same four steps as for a noncracked section. The only difference is that, with cracking, the cross-section properties exclude concrete in tension.

9.9 INFLUENCE OF CRACKING ON INTERNAL FORCES

A cracked zone of a reinforced concrete tank wall is less stiff than the part away from the cracks. This causes redistribution of the internal forces. Values of the internal forces in a cracked zone are generally smaller than the values obtained by linear analysis in which cracking is ignored. This is shown by the example of Figure 9.4. The cylindrical reinforced concrete water tank wall shown is prestressed only in the circumferential direction. The top edge of the wall is free, while the bottom edge is totally fixed (Figure 9.4a). The bending moment M and the hoop force diagrams shown in Figure 9.4b are due to hydrostatic pressure with the tank full to the top. Results of three analyses are discussed next. The data used in the analyses are specific weight of water = 9.81 kN/m³ (62.4 lb/ft³); E_c = 32 GPa (4600 ksi); f_{ct} = 2.5 MPa (360 psi); Poisson's ratio = 1/6; vertical reinforcement ratios, A_s/d = 0.006 and 0.002 at the inner and outer faces,

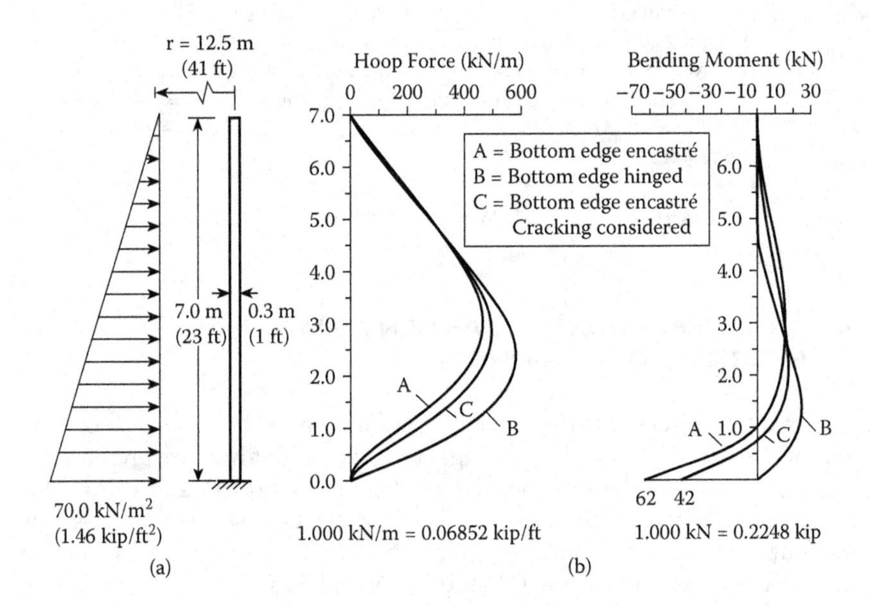

Figure 9.4 Effects of cracking on internal forces in a reinforced concrete cylindrical tank wall free at the top edge and encastré at the bottom. The wall is prestressed in the vertical direction only.

respectively, where A_s is the steel area per unit length of the parameter and d = 0.25 m (10 in); E_s = 200 GPa (29 000 ksi). The wall dimensions are shown in Figure 9.4a.

The curves labeled A in Figure 9.4b result from linear analysis, ignoring cracking. At the inner face at the bottom edge, the moment M = 62 kN (14 kip). It can be verified that this moment produces tension at the inner face of the wall greater than the tensile strength of concrete, indicating that cracking occurs.

A nonlinear analysis,[4] which takes into account the effect of the reduction in stiffness due to cracking, gives the results presented by the curves labeled C. Cracking reduces the rigidity over a short distance (0.3 m, 1 ft) adjacent to the base, causing the bending moment at the edge to drop to M = 42 kN (9.4 kip); this is accompanied by a small increase in the hoop force in the lower part of the wall.

The curve labeled B is the result of linear analysis that is the same as for curves A, but with the bottom edge hinged instead of fixed. As expected, M at the bottom edge is reduced to zero; at the same time the hoop force increases in the lower part of the wall, compared to curve C. This comparison indicates that cracking at the bottom causes M and N_ϕ diagrams to lie between the two extremes of bottom-edge rotation: prevented or allowed to rotate freely.

The bending moment value at the base in this example depends upon the steel reinforcement ratio at the inner face. Nonlinear analysis for the same problem with reinforcement ratio 0.006, 0.010, and 0.014 gives values of M = 42, 48, and 52 kN (9.4, 10.8, and 11.7 kip), respectively.

NOTES

1. Graphs and tables giving the depth of compression zone in reinforced concrete rectangular or T-shaped sections are provided in Ghali, A., Elbadry, M., and Favre, R., 2012, *Concrete Structures Stresses and Deformations*, 4th ed., Spon Press, London.
2. Fédération Internationale de Béton (fib) CEP-FIP, 2012, *Model Code for Concrete Structures, MC2010*, fib bulletin No. 65, fib, Case Postale 88, CH-1015, Lausanne, Switzerland; ACI Committee 209, 2008, *Guide for Modeling and Calculating Shrinkage and Creep in Hardened Concrete*, ACI 209.2R-08, American Concrete Institute, Farmington, MI; ACI Committee 318, 2011, *Building Code Requirements for Structural Concrete*, ACI318-11 and Commentary, American Concrete Institute, Farmington, MI.
3. Analysis can be done by computer program: Ghali, A., n.d., CGS (Cracked General Sections), www.sponpress.com.
4. See Elbadry, M., and Ghali, A., n.d., User's Manual and Computer Program CPF: Cracked Plane Frames in Prestressed Concrete, research report no. CE85-2, Department of Civil Engineering, University of Calgary, Canada.

Chapter 10

Control of cracking in concrete tanks and silos

10.1 INTRODUCTION

Cracking inevitably occurs in reinforced concrete tanks and silos. Temperature variation, shrinkage, or swelling normally produce stresses that are sufficiently high to produce cracking. Provision of prestressing can partly prevent cracking or reduce crack width. Cracking is accompanied by stiffness reduction, causing the internal forces due to temperature, shrinkage, or swelling to significantly drop. Thus, these effects produce higher internal forces if the prestressing level is increased in an attempt to minimize cracking. Therefore, it is beneficial to use partial prestressing, allowing cracking to occur due to these effects, while controlling the width of cracks by provision of an appropriate amount (to be discussed in this chapter) of non-prestressed reinforcement.

The first difficulty in analysis and design to control cracking is prediction of the tensile strength of concrete, f_{ct}. The magnitude of this parameter needs to be known to predict whether a section will crack. In a statically indeterminate cracked structure, as is the case with tanks and silos, the magnitude of the internal forces due to temperature, shrinkage, or swelling is commonly limited, and the limits depend largely on the value of f_{ct}. At a cracked section, the tensile force resisted by concrete immediately before cracking is transferred to the reinforcement, causing an increase in its tensile stress. A minimum amount of reinforcement, depending on the value of f_{ct}, is required to avoid yielding of the reinforcement and excessive widening of the crack. Because f_{ct} cannot be accurately predicted, the upper and lower limits of the probable value are used in design.

Cracking due to sustained load, such as hydrostatic pressure or prestressing, will be referred to as *force-induced cracking*. When the elongation or shortening due to temperature, shrinkage, or swelling is prevented, internal forces develop and cracking can occur. This type of cracking is referred to as *displacement-induced* cracking. The analysis of and the difference between the two types of cracking will be discussed here. The amount of reinforcement required to limit the mean width of cracks to a specified value will also be explored. However, the equations include the crack spacing, which can be predicted only by empirical equations. Furthermore, in different sections of a member, crack widths vary from crack to crack, even when the internal forces are the same. Also, the width of the same crack is variable from point to point. Thus, it is impossible to predict the width of cracks with certainty. Nevertheless, it is easier to predict the strain in the reinforcement in a cracked section. This gives a good indication of the sum of the width of cracks to be expected within a unit length of the member; the amount of reinforcement can be selected to limit their strain (or stress) to a specified limit.

10.2 CAUSES OF CRACKING

Reinforced concrete members crack when the tensile stress at a section due to the normal force N or the bending moment M reaches the tensile strength of the concrete, f_{ct}. The internal forces may be caused by external forces applied on the structure, with their magnitude unchanged because of cracking. This type of cracking, referred to as force-induced cracking, is discussed in Section 10.4.

Cracking caused by restrained expansion or contraction due to temperature variation, shrinkage, or swelling of concrete and settlement of the subgrade, referred to as displacement-induced cracking, is discussed in Section 10.5. This type of cracking is accompanied by a reduction of the values of N and M existing just before cracking. The remainder of this chapter is concerned with the design of reinforcement to control cracking caused by applied forces and by restrained deformations.

In walls of circular storage tanks and silos, force-induced cracking does not occur separately from displacement-induced cracking. However, the two types of cracking are discussed separately, assuming ideal conditions that will help us to understand how cracks develop.

Other causes of cracking not discussed in this chapter include plastic shrinkage, plastic settlement, and corrosion of steel reinforcement. The first two occur a few hours after casting; proper curing and good compaction are among the preventive measures that can be taken.[1] A rusted steel bar occupies a larger volume than the original bar. The increase in volume builds up internal pressure, leading to a surface crack running parallel to the bar.

10.3 TENSILE STRENGTH OF CONCRETE

The value of the tensile strength of concrete, f_{ct}, is required to determine whether a section is cracked and to design the minimum reinforcement required to control cracking. However, f_{ct} is a more variable property of concrete than the compressive strength. The value of f_{ct} in a member varies from section to section; thus, cracking does not occur at the same level of internal forces. Values of f_{ct} determined by experiments can vary within ±30% from the mean tensile strength. Furthermore, because of micro-cracking and surface shrinkage, the value of f_{ct} in a structure can be smaller than the value measured by the testing of cylinders. The tensile strength f_{ct} is about 10% to 15% of the compressive strength. Empirical equations to predict the value of f_{ct} can be written in the form

$$f_{ct} = \alpha_1 (f_c')^{\alpha_2}, \tag{10.1}$$

where f_c' is the specified compressive strength of concrete. European codes[2] take $\alpha_2 = 2/3$ and give lower and upper values for $\alpha_1 = 0.20$ and 0.40, respectively, with f_{ct} and f_c' in MPa ($\alpha_1 = 1.1$ and 2.1, when f_{ct} and f_c' are in psi). North American codes[3] take $\alpha_2 = 1/2$ and $\alpha_1 = 0.625$ when f_{ct} and f_c' are in MPa ($a = 7.5$ when f_{ct} and f_c' are in psi). Because the tensile strength in a structure is smaller than the strength obtained by testing of cylinders, the effective tensile strength may be considered[4] equal to the value given by Equation (10.1) reduced by a multiplier, varying linearly between 0.8 and 0.5 for thickness h varying between 0.3 and 0.8 m (1.0 and 2.7 ft), respectively. The reduction factor is 0.8 or 0.5, when h is smaller than 0.3 m (1 ft) or greater than 0.8 m (2.7 ft), respectively.

10.4 FORCE-INDUCED CRACKING

Figure 10.1 shows a reinforced concrete member subjected to an axial force N of gradually increasing magnitude. The member may represent a developed ring-shaped strip of a circular-cylindrical wall. The first crack, corresponding to axial force N_{r1}, occurs at the weakest section, when the stress in concrete reaches its tensile strength f_{ct1} (the corresponding strain $\cong 0.0001$). At a crack, the stress in concrete is zero and the reinforcement alone carries the entire axial force. On both sides of the crack, the major part of the force in the reinforcement is transmitted by the bond from the reinforcement to the concrete. The transmission develops over a length equal to the crack spacing S_r. Further increase of the axial force to N_{r2} produces a crack at the second weakest section when the stress in concrete reaches a value f_{ct2}, slightly greater than f_{ct1}. The distance between the two cracks cannot be smaller than S_r. Further increase of the value of N produces successive cracks at axial force values $N_{r1}, N_{r2}, \ldots, N_{rn}$. The maximum number, n, of cracks that can occur is l/S_r, where l is the length of the member. Subsequent increase of N causes widening of the existing cracks but does not produce new ones. When the number of cracks is equal to n, a stabilized crack condition is reached.

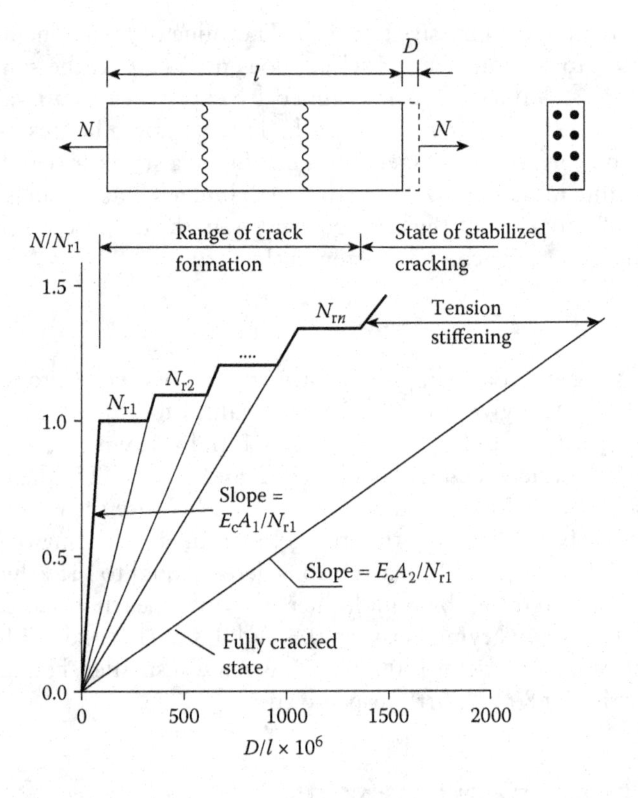

Figure 10.1 A reinforced concrete member subjected to axial force N of increasing magnitude.

Figure 10.1 shows a graph of the axial force N versus nominal strain $\varepsilon_{nominal}$ defined as equal to D/l, with D being the elongation of the member. The graph can be plotted using the following equations:

$$\frac{D_i}{l} = \frac{N_{ri}}{E_c A_1}\left[l - (i-1)s_r\right]\frac{1}{l} + \frac{N_{ri}}{E_c A_m}(i-1)\frac{s_r}{l} \quad \text{with} \quad 1 \le i \le n, \tag{10.2}$$

$$\frac{D_i}{l} = \frac{N_{ri}}{E_c A_1}(l - is_r)\frac{1}{l} + \frac{N_{ri}}{E_c A_m} i \frac{s_r}{l} \quad \text{with} \quad 1 \le i \le n. \tag{10.3}$$

Here, i is the crack number ($0 \le i \le n$); E_c is the modulus of elasticity of concrete; A_1 is the transformed area of noncracked section (state 1), composed of area of concrete A_c plus αA_s, where A_s is the cross-sectional area of reinforcement and $\alpha = E_s/E_c$, with E_s being the modulus of elasticity of the reinforcement; and A_m is the mean cross-sectional area given by

$$A_m = \frac{A_1 A_2}{\zeta A_1 + (1-\zeta) A_2}, \tag{10.4}$$

$$\zeta = 1 - \beta\left(\frac{f_{ct}}{\sigma_c}\right)^2, \tag{10.5}$$

$$\sigma_c = N / A_1, \tag{10.6}$$

where $A_2 = \alpha A_s$, the transformed area of the section in the fully cracked state 2; ζ is a coefficient for interpolation between states 1 and 2, to account for tension stiffening (Chapter 9, Equation 9.29); and σ_c is the hypothetical concrete stress value calculated by ignoring cracking. When N has a value between N_{r1} and N_{rm}, it is assumed that σ_c and f_{ct} are equal; thus, $\zeta = 1 - \beta$; β is a coefficient depending upon the quality of the bond between concrete and reinforcement, load repetition, and load duration; a value $\beta = 0.5$ may be used for most cases.

At load level N_{ri}, the ith crack occurs, with member elongation equal to D_i; the elongation increases to a value $\bar{D}_i$ before further increase of N. In derivation of Equation (10.3) the member is considered to be composed of a noncracked part of length $(l - iS_r)$ and flexibility equal to $(E_c A_1)^{-1}$ and a cracked part, whose length is (iS_r) and flexibility equal to $(E_c A_1)^{-1}$. For any value N_{ri}, there are two values of D_i/l given by Equations (10.2) and (10.3); see graph in Figure 10.1.

The tensile strength of concrete at which successive cracks occur may be assumed to vary as[5]

$$f_{cti} = f_{ct1}\left[1 + \eta(\varepsilon_{i,\ \text{nominal}} - \varepsilon_{\text{nominal}})\right], \tag{10.7}$$

where the subscripts 1 and i refer to the first and ith crack, respectively; η is a dimensionless constant; and $\varepsilon_{\text{nominal}}$ is the nominal strain value defined by

$$\varepsilon_{\text{nominal}} = D/l. \tag{10.8}$$

The ith crack occurs when σ_c (by Equation 10.6) reaches f_{cti} and the axial force value is

$$N_{ri} = f_{cti} A_1. \tag{10.9}$$

The first crack occurs when the normal force and nominal strain values are

$$N_{r1} = f_{ct1}/A_1, \qquad \varepsilon_{\text{nominal}} = f_{ct1}/E_c. \tag{10.10}$$

For any i value, Equations (10.2), (10.7), (10.9), and (10.10) can be solved for N_{ri} and (D_i/l). Substitution of the calculated value of N_{ri} in Equation (10.3) gives $(\bar{D}_i/l)$. The solution can be written in the form

$$N_{ri} = N_{r1}\left(\frac{1-\eta D_1/l}{1-\kappa_i \eta D_1/l}\right) \tag{10.11}$$

where

$$\kappa_i = 1 + \frac{i-1}{(l/s_r)}\left(\frac{A_1}{A_m} - 1\right), \tag{10.12}$$

$$D_i/l = \kappa_i N_{ri}/(E_c A_1), \tag{10.13}$$

$$\bar{D}_i/l = \kappa_{i+1} N_{ri}/(E_c A_1). \tag{10.14}$$

Substitution of $i = 1, 2, \ldots, n$ gives all the points needed to plot the graph in Figure 10.1. Equation (9.29) can be employed here to predict mean crack width, corresponding to any value of N:

$$w_m = s_r \zeta \varepsilon_{s2}. \tag{10.15}$$

In this application, ε_{s2} is the strain in the reinforcement caused by N, with concrete in tension ignored (state 2):

$$\varepsilon_{s2} = \frac{N}{E_c A_2}. \tag{10.16}$$

Equations (10.15) and (10.16) indicate that the graph in Figure 10.1 can represent w_m versus (D/l) when the ordinates are multiplied by the constant $N_{r1} s_r \zeta/(E_c A_2)$.

10.4.1 Effect of prestressing

Consider the case when the member in Figure 10.1 is prestressed. Assume that the stress in concrete, immediately before application of the axial force, is $\sigma_{prestress}$ (a negative value). After application of an axial force N, the stress in concrete becomes

$$\sigma_c = \frac{N}{A_1} + \sigma_{prestress}. \tag{10.17}$$

The ith crack will occur when σ_c reaches the value f_{cti} and the axial force value is

$$N_{ri} = A_1(f_{cti} - \sigma_{prestress}). \tag{10.18}$$

The values of N and $\varepsilon_{nominal}$ when the first crack occurs are

$$N_{r1} = A_1(f_{ct1} - \sigma_{prestress}), \quad \varepsilon_{1,\,nominal} = D_1/l = (f_{ct1} - \sigma_{prestress})/E_c. \tag{10.19}$$

The strain in the reinforcement in state 2, ε_{s2}, for use in prediction of mean crack width (Equation 10.15) is

$$\varepsilon_{s2} = \frac{N + A_1\sigma_{prestress}}{E_c A_2}. \tag{10.20}$$

The product $(-A_1\sigma_{prestress})$ represents the decompression force (see Chapter 9, Section 9.3). The equations of Section 10.4 apply for the prestressed member when Equations (10.17), (10.18), (10.19), and (10.20) replace Equations (10.6), (10.9), (10.10), and (10.16), respectively. Also, the symbol A_s will represent the sum of cross-sectional areas of the non-prestressed and the prestressed reinforcements. With these changes, the equations can be used to plot $[(N + A_1\sigma_{prestress})/N_{r1}]$ versus D/l; the resulting graph will be similar to the graph in Figure 10.1. But, when prestressing is provided, the value of N will be larger; for any given level of N_{ri}, cracking may either be avoided or crack width reduced when prestressing is provided.

EXAMPLE 10.1
Hoop Force versus Elongation for a Circular Reinforced Concrete Ring

Figure 10.2a represents a cross-section and elevation of a developed ring, of unit height, isolated from a circular-cylindrical concrete wall. Plot a graph for the hoop force N versus the elongation D. Also find the mean crack width at the start and at the end of the state of crack formation. The developed length of the ring is $l = 8.0$ m (26 ft). Assume a mean crack spacing, $s_r = 0.4$ m (16 in); $\beta = 0.5$ (for use in Equation 10.5). The dimensions of the cross-section are 1.00×0.25 m^2 (40×10 in^2); $A_s = 2400$ mm^2 (3.7 in^2); $E_s = 200$ GPa; $E_c = 30$ GPa; and $f_{ct1} = 2.0$ MPa, $f_{ctn} = 2.6$ MPa, where f_{ct1} and f_{ctn} are the tensile strength of concrete at the first and last cracks (just before reaching stabilized cracking).

The maximum number of cracks is $n = l/s_r = 8.0/0.4 = 20$. The transformed cross-sectional areas in the noncracked and the fully cracked states are

$$A_1 = 0.2636\,\text{m}^2, \quad A_2 = 16.00 \times 10^{-3}\ \text{m}^2.$$

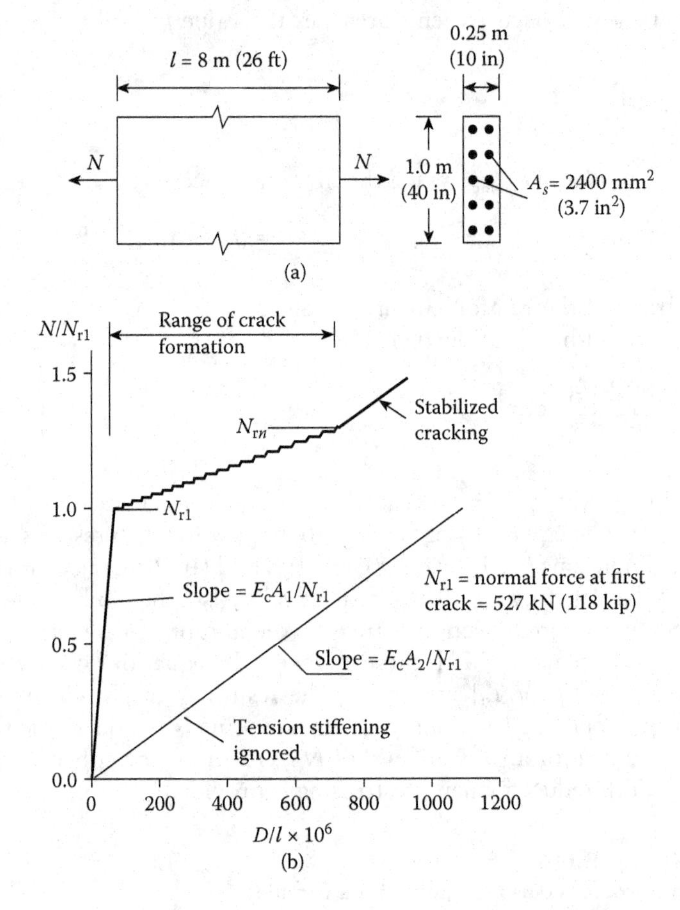

Figure 10.2 A circular ring subjected to hoop force N of increasing magnitude (Example 10.1). (a) Cross-section and developed elevation. (b) (N/N_{r1l}) versus (D/l).

The mean cross-sectional area during the crack development stage is (Equation 10.4, with $\zeta = 0.5$)

$$A_m = \frac{0.26361(16.00\times10^{-3})}{0.5(0.2636)+(1-0.5)(16.00\times10^{-3})} = 30.17\times10^{-3}\ \text{m}^2.$$

The first and last cracks occur when the values of N are (Equation 10.9)

$$N_{r1} = 2.0\times10^6(0.2636) = 527\ \text{kN},$$

$$N_{r1} = N_{r20}\times2.6\times10^6(0.2636) = 685\ \text{kN}.$$

Substitution of $i = 1$ and $i = n = 20$ in Equations (10.12) and (10.13) gives

$$\kappa_1 = 1.00, \qquad \varepsilon_{1,\text{nominal}} = \frac{D}{l} = \frac{1.00(527 \times 10^3)}{30 \times 10^9 (0.2636)} = 67 \times 10^{-6},$$

$$\kappa_n = \kappa_{20} = 8.35, \qquad \varepsilon_{20,\text{nominal}} = \frac{D_{20}}{l} = \frac{8.35(685 \times 10^3)}{30 \times 10^9 (0.2636)} = 723 \times 10^{-6}.$$

Substitution of $\varepsilon_{1,\text{nominal}}$ and $\varepsilon_{20,\text{nominal}}$ in Equation (10.7) and solution for η gives $\eta = 457$. Successive applications of Equations (10.11) to (10.13) with $i = 1, 2, \ldots, 20$ gives coordinates of the points plotted in Figure 10.2b, representing (N/N_{r1}) versus (D/l).

Substitution of the values of N_{r1} and N_{rn} in Equation (10.16) and the result in Equation (10.15) gives the mean crack width at the start and at the end of the state of crack formation:

- At start

$$\varepsilon_{s2} = \frac{527 \times 10^3}{30 \times 10^9 (16 \times 10^{-3})} = 1098 \times 10^{-6},$$

$$w_m = 0.40(0.5)(1098 \times 10^{-6}) = 0.22 \text{ mm} (8.7 \times 10^{-3} \text{ inch}),$$

- At end

$$\varepsilon_{s2} = \frac{685 \times 10^3}{30 \times 10^9 (16 \times 10^{-3})} = 1426 \times 10^{-3},$$

$$w_m = 0.40(0.5)(1426 \times 10^{-6}) = 0.29 \text{ mm} (11 \times 10^{-3} \text{ inch}).$$

10.5 DISPLACEMENT-INDUCED CRACKING

Figure 10.3 represents the variation of N versus D for a member subjected to imposed elongation D producing an axial force N. The first crack occurs at the weakest section when stress in concrete reaches its tensile strength f_{ct1}. Successive cracks occur as D is increased. When the number of cracks, n, equals l/s_r, stabilized cracking is reached. Further increase of D increases the width of existing cracks but no new ones are developed. The values of N at which successive cracks occur, $N_{r1}, N_{r2}, \ldots, N_{rn}$, are given by Equation (10.8), which is the same as in force-induced cracking. But with displacement-induced cracking, a drop in the value of N occurs

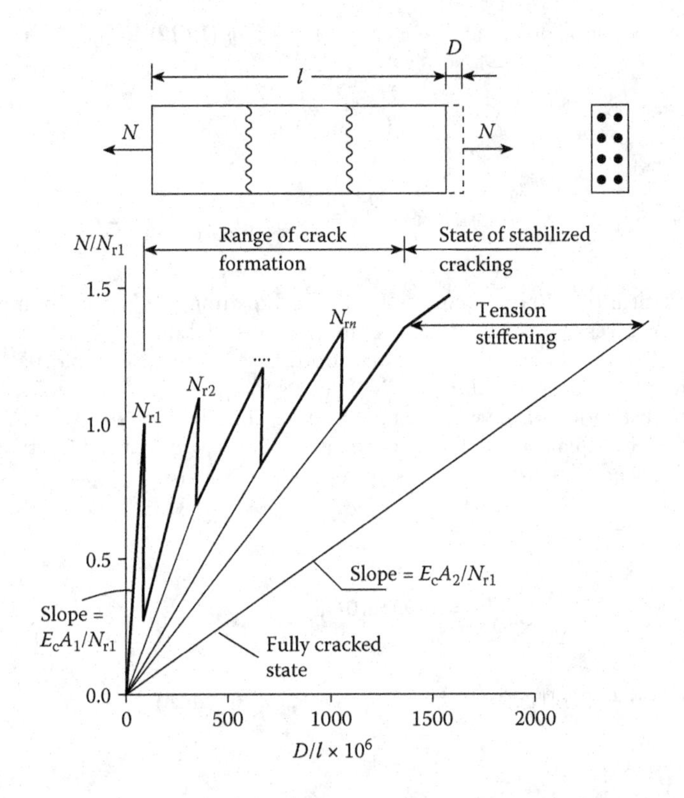

Figure 10.3 A reinforced concrete member subjected to imposed elongation D of increasing magnitude.

following each crack occurrence. Thus, displacement D_i at which the *i*th crack occurs corresponds to greater and lesser values of the normal forces N_i and $\bar{N}_i$, respectively.

Equations (10.9) to (10.11) give for the *i*th crack the upper value of N (N_{ri}) and D_i; the lower value $\bar{N}_{ri}$ is given by

$$\bar{N}_{ri} = E_c A_1 \left(\frac{D_i}{l} \right) \kappa_{i+1}^{-1}. \tag{10.21}$$

Figure 10.4b is a plot of (N/N_{r1}) versus (D/l) for the ring of Example 10.1, considering that cracking is caused by imposed displacement D.

With force-induced cracking (Figure 10.1), the stabilized crack state is reached when the applied force $N \geq N_{rn}$. The magnitude of applied axial forces that cause the first crack, N_{r1}, and the last crack, N_{r2}, are not substantially different. Thus, stabilized cracking can be expected in most cases when the cracking force N_{r1} is exceeded. However, with displacement-induced

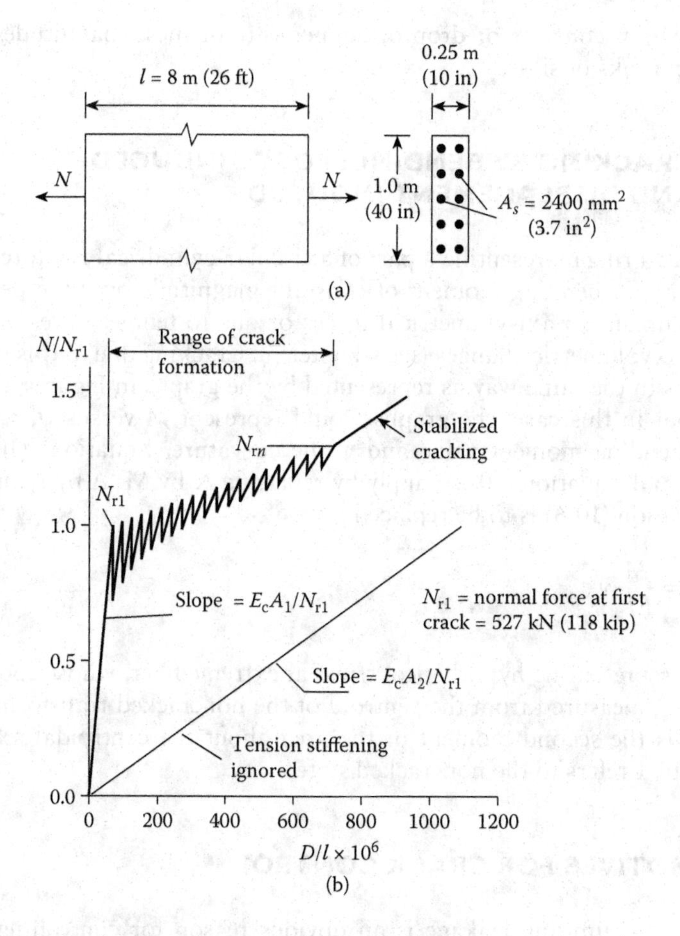

Figure 10.4 A circular ring subjected to imposed elongation D of increasing magnitude. Same data as for Example 10.1. (a) Cross-section and developed elevation. (b) (N/N_{r1}) versus (D/l).

cracking, the stabilized cracking state is not normally attained. This means that when cracking is caused only by temperature variation or by shrinkage, the stabilized crack condition rarely occurs and the internal forces do not exceed the values that can produce the last crack (N_{rn} for the member in Figure 10.3). The validity of this statement can be seen by application of Equation (10.13).

Consider the wall of Example 10.1 (Figure 10.2a); assume that the change in length is restrained and the member is subjected to shrinkage ε_{cs} or a temperature drop T degrees, with a coefficient of thermal expansion $\mu = 10 \times 10^{-6}$/°C (5.6×10^{-6}/°F). It can be verified that, using Equation (10.8) with $(D/l) = \varepsilon_{cs}$ or μT, the stabilized cracking state would be reached only when the values of ε_{cs} and T attain $\varepsilon_{cs} = -723 \times 10^{-6}$ and $T = -72$°C (−130°F).

Shrinkage of concrete or drop of temperature of these magnitudes rarely occur in tanks or silos.

10.6 CRACKING BY BENDING: FORCE INDUCED AND DISPLACEMENT INDUCED

Consider a ring representing a part of a circular-cylindrical reinforced concrete wall. A bending moment of constant magnitude over the perimeter can occur due to axisymmetrical forces or due to temperature variations or due to volumetric changes (see Chapter 7). Cracking due to this moment develops in the same way as represented by the graphs in Figures 10.1 and 10.3; but in this case, the graphs would represent M versus ψ, where M is the bending moment value and ψ the curvature. Equations (10.11) to (10.14) and Equation (10.21) apply by replacing N by M, A by I, and D by ψl. Equation (10.6) is to be replaced by

$$\sigma_c = \frac{M}{EI_1} y_{\text{extreme}} \tag{10.22}$$

with σ_c representing hypothetical stress at extreme fiber, whose coordinate is y_{extreme}, measured from the centroid of the noncracked transformed section; I is the second moment of the area about the centroidal axis; and subscript 1 refers to the noncracked state.

10.7 MOTIVES FOR CRACK CONTROL

Avoiding or limiting leakage is an obvious reason for controlling cracking in tanks. For a specified liquid, the rate of flow through a crack in a wall (mass per unit time per unit length) is proportional to the gradient of liquid pressure and inversely proportional to the wall thickness, but more important proportional to the width of the crack. It is well known that concrete in contact with water at a crack swells, particularly at the surface. This may close the cracks within a few days or weeks when their widths do not exceed 0.2 mm. Also, cracks of this width can become closed by fine material carried by water, or by cement or other material that can subsequently hydrate.[6]

The risk of corrosion is another reason for crack control. In fresh concrete, hydrating cement paste rapidly forms a thin layer of oxides strongly adhering to steel reinforcement and gives protection from corrosion. However, carbonation of concrete penetrating to steel reinforcement destroys the protective oxide film and corrosion starts. At a crack, the period before the initiation of corrosion is short (2 years). This initial corrosion does not occur at a crack that intersects reinforcing bars. But, a crack running along the line of a reinforcing bar makes it vulnerable to corrosion. Cracks of width not exceeding 0.30 mm (0.012 in) are considered to have no influence on the corrosion of steel reinforcement. The thickness and quality of concrete covering the reinforcement are the important factors that influence durability.

The risk of corrosion of prestressed steel is higher than the risk in non-prestressed bars. For this reason, a more restrictive crack width is allowed (0.1 to 0.2 mm, 0.004 to 0.008 in). Alternatively, a crack width that penetrates to the prestressed steel is not allowed.

There is no generally accepted value for allowable crack width. Crack width between 0.2 and 0.3 mm (0.008 and 0.012 in) is normally acceptable. A stricter requirement of 0.05 to 0.2 mm (0.002 to 0.008 in) is also common. The width of cracks in a member can vary from one crack to another; the width can also vary from point to point along the same crack. The allowable crack widths mentioned in this section represent the characteristic crack width w_K, of 5% probability of being exceeded; the mean crack width w_m should be smaller than w_K by a factor (e.g., 2/3).

10.8 MEANS OF CONTROLLING CRACKING

Quality of concrete and the care that should be taken in placement and curing of concrete to control cracking are not discussed here.[7] The following measures can be *taken in design* to control cracking:

1. Provision of sufficient prestressing level.
2. Provision of a minimum amount of non-prestressed reinforcement (Section 10.9). The bar spacings and the diameter of the bars should be limited in order to limit crack width.
3. Limiting the change in stress in the reinforcement when cracking occurs (as the section changes from the noncracked state 1 to the fully cracked state 2). Crack width is almost directly proportional to this stress change (Section 10.10).

The remaining sections of this chapter are intended to assist in structural design that controls cracking.

10.9 MINIMUM REINFORCEMENT TO AVOID YIELDING OF REINFORCEMENT

When a section cracks, the force carried by concrete in tension is transferred to the reinforcement. If the reinforcement is below a minimum ratio,

$\rho_{critical}$, it will yield at the formation of the first crack. Such a crack will widen excessively and formation of other cracks with limited width will not occur. This is true for force-induced cracking and for displacement-induced cracking. The minimum cross-sectional area of reinforcement $A_{s,critical}$ and the corresponding reinforcement ratio $\rho_{critical}$ that will avoid yielding are determined next.

Consider a reinforced concrete non-prestressed cross-section subjected to axial tension N_r just sufficient to produce cracking,

$$N_r = f_{ct} A_{gross} \left[1 + (\alpha - 1)\rho_{critical}\right], \tag{10.23}$$

where f_{ct} is the tensile strength of concrete; $\alpha = E_s/E_c$, the ratio of modulus of elasticity of reinforcement to that of concrete; A_{gross} is the gross area of a concrete cross-section; and $\rho_{critical} = A_{s,critical}/A_{gross}$, where $A_{s,critial}$ is the steel area. Immediately after cracking, the force N_r is resisted by reinforcement and produces yielding when $N_r = f_y A_{s,critical} = f_y \rho_{critical} A_{gross}$. Thus,

$$\rho_{critical} = \frac{A_{s,critical}}{A_{gross}} = \frac{f_{ct}/f_y}{1 - (\alpha - 1)(f_{ct}/f_y)}, \tag{10.24}$$

where f_y is the specified yield stress of reinforcement.

The preceding derivation implies that the magnitude of the axial force is the same just before and after cracking. This is not so in displacement-induced cracking, where crack occurrence is accompanied by a sudden drop in the value of N. The denominator in Equation (10.24) is approximately equal to 1.0; thus the equation is frequently written as

$$\rho_{critial} = A_{s,critical}/A_{gross} \cong f_{ct}/f_y. \tag{10.25}$$

In Equations (10.24) and (10.25), $\rho_{critical}$ and $A_{s,critical}$ refer to the total steel area in the section. If the section is prestressed, Equation (10.23) will give the excess of the axial force beyond the decompression force when cracking occurs. Equation (10.24) or (10.25) applies to give $\rho_{critical}$, which in this case will represent the ratio $[(A_{ns} + \alpha_y A_{ps})/A_{gross}]$, where A_{ns} and A_{ps} are the cross-sectional areas of the non-prestressed and the prestressed reinforcements. Here, it is implied that α is the same for the two reinforcement types. The coefficient α_y is the ratio of stress allowed in the prestressed reinforcement to f_y. Consider a reinforced concrete section subjected to a normal force N, at a reference point O, combined with a bending moment M (Figure 10.5). Assume that the reference point O is the centroid of the transformed noncracked section (state 1). When N and M are just sufficient to produce cracking,

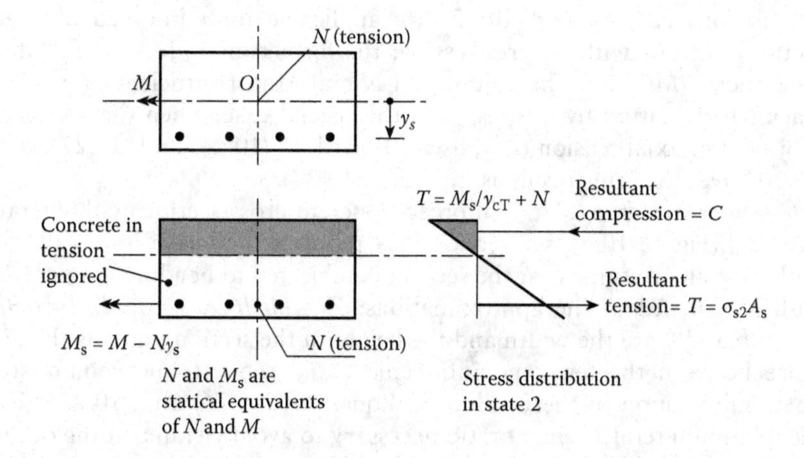

Figure 10.5 Analysis of tensile stress in the reinforcement in a fully cracked section (state 2).

$$f_{ct} = \frac{N}{A_1} + \frac{M}{Z_1}, \tag{10.26}$$

where A_1 and Z_1 are the cross-sectional area and section modulus of the transformed noncracked section (state 1). Again, assuming that the magnitudes of N and M are not changed by cracking, and equating the stress in the tension reinforcement to f_y, gives

$$A_{s,critical} = \frac{1}{f_y}\left(\frac{M_s}{y_{cT}} + N\right), \tag{10.27}$$

$$M_s = M - Ny_s. \tag{10.28}$$

Here M_s is the bending moment value when the pair M and N are replaced by statical equivalents composed of a force N at the centroid of the tension reinforcement combined with a moment; y_s is the coordinate of the tension reinforcement measured downward from the reference point O (Figure 10.5); and y_{cT} is the distance between the resultant compression and the tension reinforcement. Calculation of y_{cT} involves calculation of the depth c of the compression zone (see Chapter 9, Sections 9.4 and 9.5). When the cross-section is prestressed, Equation (10.27) gives $A_{s,critical}$, which is equal to the sum of the prestressed and the non-prestressed reinforcements. In this case, the values of N and M to be used in the equation are the values in excess of the decompression forces (defined in Chapter 9, Section 9.3).

Equations (10.26) and (10.27) are applicable for reinforced concrete sections with or without prestressing; the former gives, for any specified eccentricity (M/N), the magnitudes of N and M just sufficient to produce cracking; the latter gives $A_{s,critical}$. In the special case, when the section is subjected to axial tension only, use of Equations (10.26) and (10.27), with $M = 0$, gives the same result as Equation (10.24).

Consider a rectangular reinforced concrete cross-section without prestressing (Figure 10.5); the section may represent a section in a wall of a tank or a silo. Assume that the section is subjected to bending moment M, with $N = 0$. Adopt the approximations: $Z_1 \cong bh^2/6$, $y_{cT} \cong 0.9d$, $d \cong 0.9h$, where b and h are the width and the height of the section and d is the distance between the tension reinforcement and the extreme compression fiber. Substitution of these values in Equations (10.26) and (10.27) gives the minimum reinforcement ratio necessary to avoid yielding at the occurrence of the first crack in a rectangular section:

$$\rho_{critical} = \frac{A_{s,critical}}{bh} \cong 0.21 \frac{f_{ct}}{f_y}. \tag{10.29}$$

In this equation $\rho_{critical}$ and $A_{s,critical}$ refer to the cross-sectional area of the reinforcement at the tensile face of the section.

When cracking is caused by imposed deformation (e.g., due to temperature or shrinkage), the equations presented in this section can be employed to give $A_{s,critical}$. This is the minimum cross-sectional area of the reinforcement necessary to avoid yielding. With yielding, a single or a small number of isolated excessively wide cracks occurs. Provision of $A_{s,critical}$ ensures formation of a progressively increasing number of narrower cracks as the value of the imposed deformation increases. It is essential that the reinforcements do not yield after the first crack, but satisfying this condition is not sufficient to control crack width. The amount of reinforcement necessary to limit the crack width to a specified value is discussed in the following section.

10.10 AMOUNT OF REINFORCEMENT TO LIMIT CRACK WIDTH: DISPLACEMENT-INDUCED CRACKING

Provision of a sufficient amount of bonded reinforcement in the tension zone is a means of controlling crack width to any specified value. The same objective can be achieved by limiting the strain in the reinforcement at the cracked section. The mean crack width can be predicted by (Equation 9.29)

$$w_m = s_1 \zeta \varepsilon_{s2}, \tag{10.30}$$

where s_r is the average crack spacing; ζ is the interpolation coefficient defined by Equation (9.28); and ε_{s2} is the strain in the reinforcement in state 2. For a prestressed section, ε_{s2} is the strain in the tension reinforcement due to the forces N and M_s in excess of the forces that produce decompression (Section 9.3), where N is a normal force and M_s is a moment about an axis through the tension reinforcement. The interpolation coefficient ζ accounts for the contribution of concrete in resisting tensile stress in the vicinity of the cracked section. The value of ζ (Equation 9.28) depends upon the stress, $\sigma_{1,extreme}$, at the extreme tensile fiber immediately before cracking; thus in the present application $\sigma_{1,extreme} = f_{ct}$.

The strain change ε_{s2} at cracking is given by

$$\varepsilon_{s2} = \frac{M_s/y_{cT} + N}{E_s A_s}, \tag{10.31}$$

where A_s is the sum of prestressed steel and non-prestressed reinforcement in the tension zone; E_s is the modulus of elasticity of the reinforcement, a value assumed valid, for the two types of reinforcements; N is the normal force on the section; and M_s is given by Equation (10.28).

The change in stress in the reinforcement in state 2 is

$$\sigma_{s2} = E_c \varepsilon_{s2}. \tag{10.32}$$

The minimum area of steel required to limit the stress to a specified value σ_{s2} is

$$A_s = \frac{M_s/y_{cT} + N}{\sigma_{s2}}. \tag{10.33}$$

The crack spacing s_r can be predicted only by empirical equations. Thus, the use of Equation (10.30) may not accurately predict the crack width. However, the quantity (w_m/s_r), representing the sum of the width of cracks situated within a unit length, is a more reliable indicator of the expected crack width. The magnitude of w_m/s_r is almost directly proportional to ε_{s2} or σ_{s2} (see Equations 10.30 and 10.32). Thus, setting a limit to the change, σ_{s2}, in steel stress that occurs at cracking is a reliable means of controlling crack width.

EXAMPLE 10.2
Reinforcement Required to Limit the Width of Cracks Caused by Temperature Variation

Figure 10.6 exhibits a cross-section of unit width representing a wall of a circular-cylindrical tank or silo. Determine the reinforcement cross-sectional area required to limit the mean crack width to $w_m = 0.10$ mm.

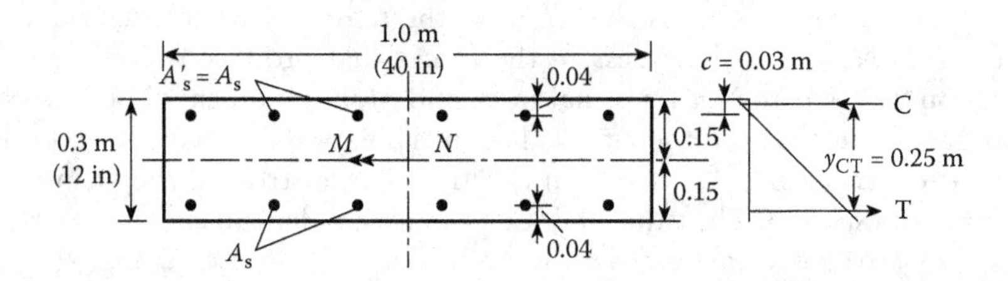

Figure 10.6 Wall cross-section cracked due to temperature variations (Example 10.2). Calculation of the reinforcement cross-sectional area required to control cracking.

The cracking is caused by temperature variation, producing normal force N combined with bending moment M, with eccentricity M/N = 0.25 m (10 in). Given data: f_{ct} = 3.00 MPa (435 psi); E_s = 200 GPa (29 000 ksi); E_c = 30.0 GPa (4350 ksi); mean crack spacing s_r = 200 mm (8 in); use the interpolation coefficient ζ = 0.5.

Limiting the crack width to a specified value can be achieved by limiting the stress σ_{s2} to (Equations 10.30 and 10.32):

$$\sigma_{s2} = \frac{E_s w_m}{\zeta s_r} = \frac{200\times10^9(0.10\times10^{-3})}{0.5(0.200)} = 200 \text{ MPa (29 ksi)}.$$

Assume a trial value, A_s – 1000 mm², for the reinforcement cross-sectional area at each face of the wall. With this assumption, the transformed section in the noncracked state has the following area properties:

$$A_1 = 0.3113 \text{ m}^2, \quad Z_1 = 15.91\times10^{-3} \text{ m}^3.$$

Application of Equation (10.26), with $N = M/0.256$, gives

$$3.0\times10^6 = \frac{(M/0.25)}{0.3113} + \frac{M}{15.91\times10^{-3}},$$

M = 39.63 kN m, N = 158.5 N.

The amount M and the normal force N at O (Figure 10.6) are statical equivalents to $M_s - M - N_{ys}$ and N at the centroid of the tension steel; where y_s (= 0.11 m) is the coordinate of this centroid with respect to the reference axis through O. Thus,

$$M_2 = 39.63 - 158.5(0.11) = 22.19 \text{ kN m},$$

$$c = 0.03 \text{ m}, \quad y_{cT} = 0.25 \text{ m}.$$

The minimum area of steel required to limit the crack width as specified is

$$A_s = \frac{22.19 \times 10^3/0.25 + 158.5 \times 10^3}{200 \times 10^6} = 1240 \text{ mm}^2 \text{ (1.92 inch}^2\text{)}.$$

The analysis may be repeated to reduce the difference between the assumed and the calculated values of A_s. However, this will have little effect on the answer to the present problem.

10.11 CHANGE IN STEEL STRESS AT CRACKING

The width of cracks, whether they are force induced or displacement induced, can be controlled by setting a limiting value on the change in stress σ_{s2} that occurs at cracking. This is considered here equal to the stress in a fully cracked section (state 2) subjected to known normal force N and bending moment M, with N and M being the forces on the section immediately before cracking. In the case of a prestressed section, N and M represent the values of the normal force and moment after deduction of the decompression forces (see Chapter 9, Section 9.3). The change in steel stress can be calculated by (Equation 10.33)

$$\sigma_{s2} = \frac{(M_s/y_{cT}) + N}{A_s}, \tag{10.34}$$

$$M_s = M - Ny_s. \tag{10.35}$$

N is assumed to act at a reference point O; y_s is the coordinate of the centroid of the tension reinforcement measured downward from O; and y_{cT} is the distance between the resultant compressive and tensile forces. As mentioned earlier, determination of y_{cT} involves calculation of the depth of the compression zone. The graphs in Figure 10.7 can be used in lieu of Equation (10.34) to give σ_{s2}. The graphs[8] can be used for prestressed or non-prestressed rectangular sections. In the former case, A_s represents the sum of the cross-sectional areas of the prestressed and the non-prestressed reinforcement; this implies the assumption that the resultant tension is at the centroid of A_s.

To use the graph, determine $\sigma_{1,extreme}$. This is the value of stress at the extreme fiber of the transformed noncracked section (state 1) subjected to N combined with M. Then enter the graphs with the dimensionless parameters $\rho = A_s/bd$ and M/Nd and read $\sigma_{s2}/\sigma_{1,extreme}$. Here N is a normal force

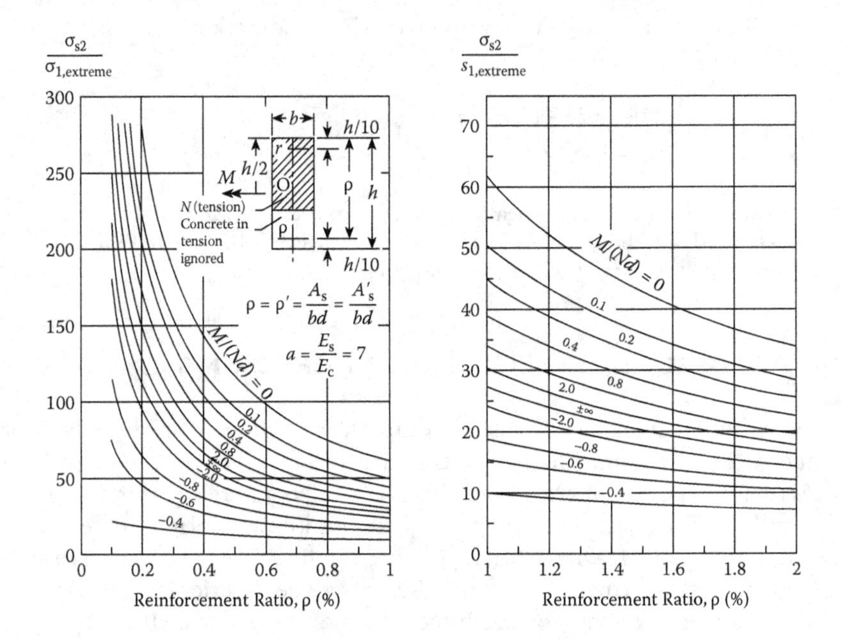

Figure 10.7 Change in stress in reinforcement σ_{s2} in a rectangular cracked section due to a normal force N and a bending moment M.

at O, the center of the rectangle; M is a bending moment about a horizontal axis through O; b is the width of the section; and d is the distance between the extreme compression fiber and the centroid of the tension reinforcement. The graphs are dimensionless, thus can be used when working with any system of units. In preparation of the graphs, it is assumed that A_s, the bottom reinforcement area, is the same as the top reinforcement area A_s'; the distance between the centroid of A_s or A_s' and the bottom or top face of the section is $h/10$, where h is the height of the section (Figure 10.7); the ratio $\alpha = E_s/E_c - 7$.

Obviously, the same graph may be used to determine ρ when σ_{s2} is given. Thus, the graphs can be used for checking the steel stress or for the design of the amount of reinforcement required to limit σ_{s2} to a specified value.

To demonstrate the use of the graphs in Figure 10.7 to check the value of σ_{s2}, consider the cross-section of Example 10.2 (Figure 10.6), with N = 158.5 kN and M = 39.63 kN m. A_s = 1240 mm² is the value determined in Example 10.2 to limit σ_{s2} to 200 MPa; thus, σ_{s2} determined by the graphs should be close to 200 MPa. The dimensionless parameters to be entered in the graph are

$$\rho = \frac{A_s}{bd} = \frac{1240}{1000(260)} = 0.0048 = 0.48\%,$$

$$\frac{M}{Nd} = \frac{39.63}{158.5(0.26)} = 0.96.$$

The graph gives

$$\frac{\sigma_{s2}}{\sigma_{1,\text{extreme}}} = 68.$$

The noncracked transformed section (state 1) has the following properties:

$$A_1 = 0.3149 \text{ m}^2, \quad Z_1 = 16.20 \times 10^{-3} \text{ m}^3.$$

The tensile stress at the extreme fiber in state 1 is (Equation 10.26)

$$\sigma_{1,\text{extreme}} = \frac{N}{A_1} + \frac{M}{Z_1}$$

$$= \frac{158.5 \times 10^3}{0.3149} + \frac{39.63 \times 10^3}{16.20 \times 10^{-3}} = 2.950 \text{ MPa}$$

$$\sigma_{s2} = 68(2.950) = 201 \text{ MPa (as expected)}.$$

NOTES

1. Neville, A. M. (2012), *Properties of Concrete*, 5th ed., Pearson Education, Harlow, England.
2. Fédération Internationale de Béton (fib), fib CEP-FIP (2012), *Model Code for Concrete Structures, MC2010, fib* bulletin No.65. fib, Case Postale 88, CH-1015, Lausanne, Switzerland.
3. ACI Committee 209 (2008), *Guide for Modeling and Calculating Shrinkage and Creep in Hardened Concrete*, ACI 209.2R-08, American Concrete Institute, Farmington, MI. ACI Committee 318 (2011), *Building Code Requirements for Structural Concrete*, ACI 318-11 and Commentary, American Concrete Institute, Farmington, MI.
4. Laurencet, P., Rotilio, J.-D., Jaccoud, J.-P., and Favre, R. (1999). *Influences des Actions Variables sur l'État Permanent des Ponts en Béton Précontraint*, Ecole Polytechnique Fédérale de Lausanne, Institut de Statique et Structures, Béton Armé et Précontraint, 1015 Lausanne, Switzerland, May.
5. Equation (10.7) (with η = 350) is derived from experiments (on specimens having l = 1.2 m) reported in Jaccoud, J. P., 1987, Armature minimale pour le contrôle de la fissuration des structures en beton, PhD thesis, Département de Génie Civil, École Polytechnique Fédérale de Lausanne, Switzerland.

6. For more references on this topic and means of control of cracking, see Laurencet et al., 1999. Also see Beeby, A. W., and Narayanan, R. S., 1995, *Designers' Handbook to Eurocode 2, Part 1.1: Design of Concrete Structures*, Thomas Telford, London.
7. See Neville, 2012.
8. Taken from Ghali, A., Elbadry, M., and Favre R., 2012, *Concrete Structures Stresses and Deformations*, 4th ed., Spon Press, London.

Chapter 11

Design and construction of concrete tanks

Recommended practice

This chapter is written in collaboration with Robert Bates, Bates Engineering, Lakewood, Colorado, for the sections on water tanks; and with Josef Roetzer, Strabag International, Dywidag LNG Technology, Munich, Germany, for the sections on liquefied natural gas (LNG) tanks.

11.1 INTRODUCTION

The recommendations presented in this chapter are based on the authors' experience and on guides[1] for design and construction of circular concrete tanks. The chapter concerns mainly water tanks; specialized techniques necessary for liquefied natural gas tanks are also considered. The techniques in design and construction of tank components, such as roof, floor slab, and wall, aim at constructability and satisfactory serviceability, sustainable for a lifespan exceeding 100 years. The objectives include: watertightness, absence of defects, and control of cracking. Tanks in which, at least, the walls are prestressed concrete may best achieve these objectives. This chapter reflects the experience of firms specialized in conventionally reinforced and prestressed concrete tank projects. The considered prestressed water tanks may have capacities ranging from 2000 to 120,000 m^3 (0.5 to 30 million U.S. gallons [MG]). Acceptance criteria for the water loss test are specified in the second reference in Note 1. Infiltration of nonpotable water into a tank can also be critical.

Concrete pumpers were introduced in the early 1980s as a substitute for the use of crane and bucket. Modern pumpers can deliver 90 m^3 (120 cubic yards) of concrete per hour, with a reach of 55 m (180 ft) from a 60 m (200 ft) boom. This has enhanced safety, increased efficiency, and reduced the risk of defects in concrete placement. It enables the monolithic placement of large slabs, for example, for a floor slab of 120 m (400 ft) diameter tank (Figure 11.1); 2900 m^3 (3800 cubic yards) of concrete can be placed in 13 hours.

Figure 11.1 Picture taken during the casting of a flat-plate roof for a 120,000 m^3 (30 MG) water tank.

11.2 WATER TANKS

The recommendations for details and construction of water tanks given here represent satisfactory practice. Alternative systems may also produce adequate serviceability.

11.2.1 Requirements

A water tank must maintain the good quality of the stored water (subject to periodic monitoring). Water loss through joints must be limited. Protection against corrosion of steel reinforcements must be adequate. Maintenance, including periodic cleaning, must be easy and safe; adequate access ladders (or stairs), ventilation, and light are necessary.

11.2.2 Concrete materials

Advances in concrete mixes and workmanship will no doubt continue to achieve high performance, improved constructability, and reduced risk of defects. Materials in use include super-plasticizers for high-slump high-strength concrete, hydration stabilizers to delay cement hydration for longer transit time, and fly ash to enhance impermeability.

Constructing slabs for large-capacity tanks, without construction joints, can be a complex proposition in locations of low humidity and windy conditions. Internally curing concrete (ICC) can be used to mitigate adverse concrete curing conditions. An ICC mix design includes the substitution of a small percentage of fine sand aggregate with expanded-shale fines. The ICC mix typically results in less shrinkage cracking, more complete cement hydration, and higher strength at early age, allowing faster formwork cycling and shorter construction schedules. For strength, the ICC is considered normal-weight concrete, because its specific weight is 22 kN/m^3 (140 lb/ft^3). Whether ICC or conventional concrete is used, curing remains a means of mitigating shrinkage cracks in the slab. Due to more complete cement hydration, fewer shrinkage cracks and remarkably consistent compressive strength test results can be achieved.

The so-called bug holes, resulting from air trapped along the form surface, increase opportunity for bacterial growth. When placement lift depths are limited to 0.3 m to 0.6 m, the air trapped at the form face can escape, leaving a smoother wall surface with fewer voids. Advances in concrete mix design and workmanship will no doubt continue to improve constructability and lower risk of defects.

11.2.3 Roof exposure

The majority of water tanks are buried with exposed roofs. Between 5% and 10% of the structure's cost can be saved by eliminating soil backfill on the roof. Exposed roofs are easier to maintain. Aesthetics can be a reason to bury a roof slab. Buried concrete roof slabs should be fenced to avoid unforeseen overloading.

11.2.4 Roof slab features

For ease of construction, the roof can be a flat plate. This is a two-way slab supported directly on columns, without beams or column capitals. Based on records of satisfactory serviceability, including deflection limitations, ACI 318 provides analysis, design, and reinforcement details of two-way slabs.[2]

Columns are laid out symmetrically with the tank's drain at the center (Figure 11.2). This choice of drain location facilitates washing of the floor (annually or biannually). Column spacing is commonly between 6 m and 12 m (20 ft and 40 ft). The span-to-thickness ratio of the roof slab depends on the applied loads; the long-term deflection often governs the thickness. Fewer columns can lead to economy and ease of maintenance. The cost of the roof of a tank is normally about 40% of the total of the structure.

After losses, the recommended compressive stress due to prestressing averaged over the slab thickness is ≥1.4 MPa (200 psi). The prestressing controls cracking, thus, reduces the risk of infiltration of pollutants to the stored water. Nonbonded 15 mm (0.6 in) strands, coated with corrosion-preventing grease in high-density 1 mm thick polyethylene extruded sheaths, are recommended as prestressing tendons. The tendons have rectangular

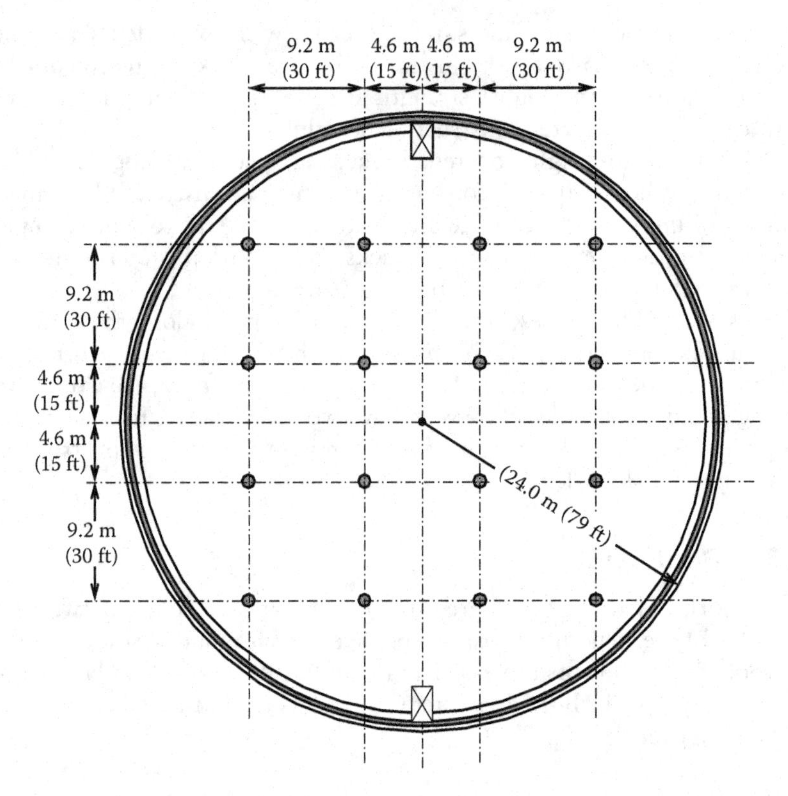

Figure 11.2 Layout of columns supporting a flat-plate roof of a water tank.

anchor plates that can fit vertically within the thickness of 190 mm (7.5 in) slab. Limiting the number of bundled strands to two reduces the risk of shrinkage cracking over the tendons. A minimum cover of 40 mm (1.5 in) is recommended for the tendons. In addition to the prestressing, bonded reinforcement above the columns is necessary; at the middle of the panels, bonded reinforcement may also be needed.[3] Reinforced concrete flat plates, with or without prestressing, often need shear reinforcement to resist punching shear at the columns.[4] Bars and tendons must be secured in their design locations and profiles until the concrete is cast. The reinforcement should be supported by sturdy chairs, such that the cover of bars or tendons at the top or the bottom is not too large or too small. A maximum chair spacing of 1.2 m (4 ft) is adequate to support the tendons and maintain appropriate profiles.

Inadequate reinforcement at the anchor zones of prestressing tendons can result in cracks that hamper durability. At slab edges, with the action of freeze-and-thaw, the cracks can damage the concrete. To reduce the

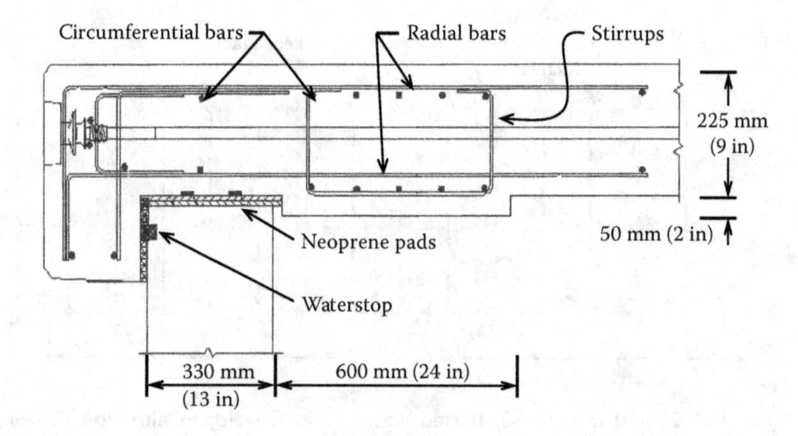

Figure 11.3 Radial section at the edge of the roof in Figure 11.2.

number of anchors at a slab edge, tendons that are shorter than 30 m (98 ft) can be stressed from one end, with the stressing ends alternating.

Breaches in the sheath of the tendons can diminish durability, thus, they should be repaired; damage in the sheath allows grease to travel on its exterior and may reach the surface. Leaking grease should not occur at properly sealed anchors; grease can be toxic. Thus, it must not reach stored water.

The roof slab is commonly supported on the wall, with a connection that does not restrain the inward movement of the slab edge caused by shrinkage and prestressing. Figure 11.3 is a radial sectional elevation at the joint between the wall and the roof slab in Figure 11.2. This particular design provides a small increase in slab thickness in a circumferential strip of width 0.6 m adjacent to the wall; a rubber waterstop is provided between the roof overhang and the wall, needed only when the roof is buried.

11.2.5 Crack repair

Cracks inevitably occur and their repair can be necessary for durability. Cost of repair, when included in the price of the structure, is a further incentive for the contractor to minimize cracking. Shrinkage cracks in slabs, occurring before prestressing, close at the time of prestressing. However, some cracks remain and should be repaired,[5] particularly when they traverse the full thickness.

11.2.6 Prestressed wall features

Walls can be post-tensioned by a combination of circumferential horizontal and vertical bonded tendons (Figures 11.4 and 11.5). Half the

Horizontal:
#4 @ 400 mm
each face
#4 each face
6.7 m
(22 ft)
24 m (79 ft)

Figure 11.4 20,000 m³ (5 MG) buried water tank. Developed elevation showing the reinforcement in one of four segments having no pilasters. Horizontal tendons are anchored at pilasters centered in adjacent segments. Wall sections are shown in Figure 11.5.

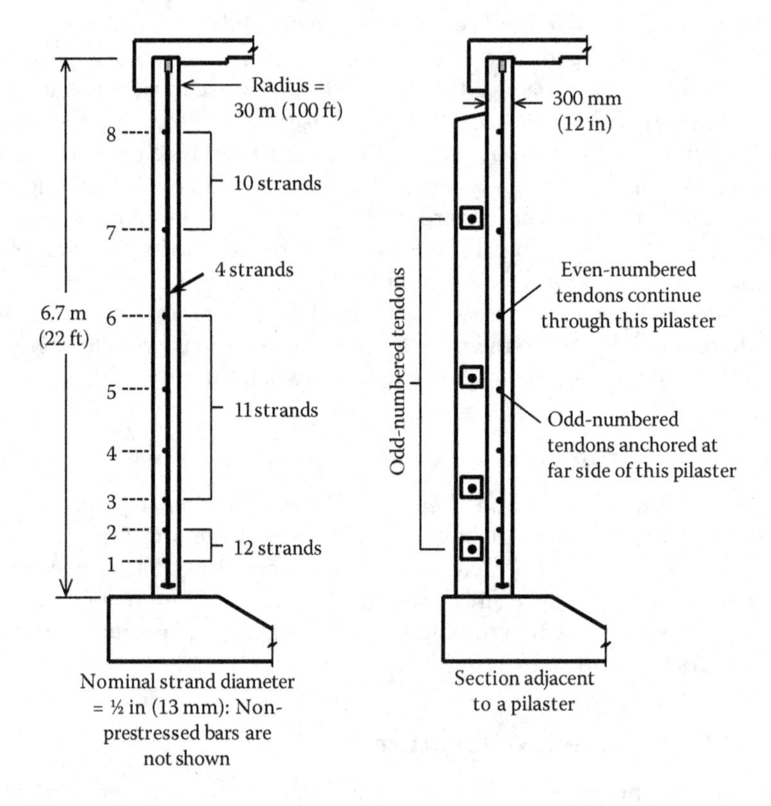

Figure 11.5 Vertical sections in the wall shown in Figure 11.4. Odd-numbered tendons overlap at the shown pilaster.

circumferential tendons are anchored at alternate pilasters; thus, an even number of circumferential tendons and pilasters are required. The distance between anchors, the friction between the strands and their ducts, the angular change in the profile of the tendons, the loss of prestressing at the seating of anchors, and the long-term losses due to relaxation of prestressed steel and creep and shrinkage of concrete are factors to be considered in computing the value of the average effective prestress. The recommended limits for the distance between adjacent pilasters and for the length of circumferential tendons are 40 m and 82 m, respectively. Longer limits result in greater prestress loss due to friction and increase the risk of vertical shrinkage cracking in the wall. Generally, the strands are pulled through their ducts after casting the concrete; for easier pull and improved quality of grout, larger duct diameters are preferred. Adequate reinforcement has to be provided to prevent bursting of the concrete at the tendon anchors. Generally, the circumferential tendons are stressed from their two ends; the vertical tendons are stressed from their top ends.

Figure 11.6 is a developed elevation of a post-tensioned wall, in which the profile of each tendon is in the shape of the letter W; the picture in Figure 11.7 is taken during the installation of the reinforcement. The tendons are stressed from their two ends at the top of the wall. No pilasters are needed for anchoring the tendons; thus, the wall thickness is constant at any section in the circumference. Because of this advantage, combined with the ease of stressing, the W-shaped tendons' profile is preferred over their substitute requiring pilasters and a combination of circumferential and vertical tendons. The W-shaped profile enhances productivity in construction; the improvement easily compensates for the increase in friction loss of prestress.

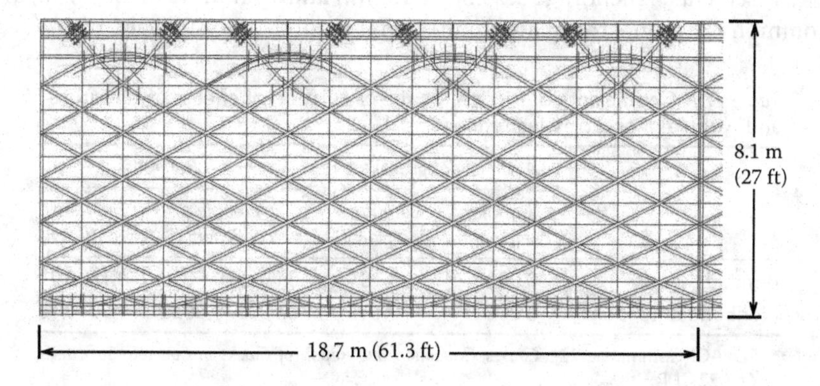

Figure 11.6 Developed elevation of post-tensioning tendons. The profile of each tendon is in the shape of the letter W. Typical segments of a wall cast in eight identical parts.

Figure 11.7 Picture taken during the installation of the reinforcements for the wall in Figure 11.6.

The manufacturers of prestressing tendons should provide the curvature and wobble coefficients to be used in calculating the variation of the force in the tendons due to friction. Unrealistic low estimates of the coefficients can induce cracking at sections far away from the jacking end. ACI 318 code requires determining curvature and wobble coefficients by measurements.[6] The measurements can also give the displacement at the seating of the anchorages. Testing the first tensioned tendon and adjusting accordingly the design forces are recommended. Table 11.1 compares a range of parameters given in ACI 318[7] with values adopted in design, based on extensive measurements. Anchorage seating of 10 mm and 13 mm (⅜ in and ½ in) are common for monostrand and multistrand tendons, respectively.

Table 11.1 Coefficients of friction for the calculation of effects of curvature and wobble of prestressing tendons

	Authors' design values		*Range*[a]	
	Wobble coefficient	*Curvature coefficient*[b]	*Wobble coefficient*	*Curvature coefficient*
Bonded tendons	0.0004	0.18	0.0005–0.0020	0.15–0.25
Nonbonded tendons	0.002	0.14	0.0003–0.0020	0.05–0.15

[a] See ACI Committee 224, *Causes, Evaluation, and Repair of Cracks in Concrete Structures*, ACI 224.1 R.

[b] For tanks with a diameter ≤ 45 m (150 ft), a higher coefficient for curvature is recommended.

11.2.7 Casting joints in wall

To reduce the risk of vertical shrinkage cracks, the wall is divided into an even number of segments of length not exceeding 24 m. Casting of the segments can take several weeks. Sixteen wall segments can be cast in about 5 weeks, at the rate of three segments per week. With W-shaped prestressing tendons (Figure 11.6), all segments can be identical (enhancing the efficiency and quality of workmanship). With vertical and horizontal prestressing, some of the segments have pilasters for anchorage of the horizontal tendons (Figures 11.4 and 11.5).

Casting of the segments is done in lifts, 0.3 m (1 ft) deep. Deeper lifts tend to trap air on the form surface, resulting in "bug holes" (potential spots for bacteria). The vertical wall joints between the segments have waterstops adhered to the concrete. These are similar to the one shown in Figure 11.3.

11.2.8 Wall bearing pads and waterstop

Figure 11.8a is a radial section at the connection of a tank wall and base; a waterstop between rubber pads at its sides is suitable for tanks of capacity 4000 m^3 to 40,000 m^3 (1 MG to 10 MG). For small tanks, it is possible to use rubber pads only at the inner side of the waterstop. The stiffness of the bearing pad depends upon the mechanical properties of the rubber, including its modulus of elasticity in shear and its hardness measurement (by a device known as durometer[8]). The thickness and the area of the bearing pads are designed to maintain their consistency through the vertical and horizontal displacements[9] expected in the structure's lifespan. The choice of width–thickness ratio of the pads can avoid their excessive deformability due to the design bearing stress. The force restraining radial horizontal displacement of the wall must not exceed the force considered in wall design; 30-durometer rubber is recommended because it provides little restraint.

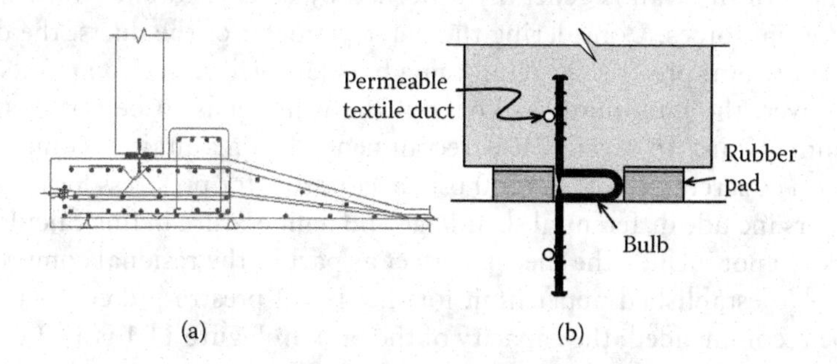

Figure 11.8 Radial section at a wall-base connection of a water tank. (a) Arrangement of the reinforcement in the base, including post-tensioned tendons. (b) Waterstop detail.

The tank–wall connection in Figure 11.8a contains a polyvinyl chloride (PVC) waterstop. Excessive wall translation can tear the PVC at its thin point; the bulb enables further displacement at the waterstop without loss of its function. The manufacturer of waterstops may conduct tests to determine their capacity to undergo displacement without tear. A waterstop having 75 mm (3 in) displacement capacity may be suitable for long-term displacement ≤40 mm (suitable for a 60 m diameter tank).

An intermittent curb of length 2.5 m to 3.0 m (9 ft to 10 ft), with gaps of the same length can control the drift of the wall relative to the base in an earthquake. The space between the curb and the wall is designed to allow for the long-term movement of the wall due to prestressing and service loads. The curb is commonly located inside the tank.

Segments of permeable textile ducts (up to 8 m [26 ft] long) are attached to the top and bottom of the ribbed PVC waterstop (Figure 11.8b) to pump chemical grout under pressure (≤ 1.4 MPa [200 psi]). The injected chemical is a low viscosity fluid, water reactive, resulting in flexible polyurethane grout for filling and repairing potential leakage through cracks or other imperfections.

11.2.9 Design of water tanks

The design requirements for prestressed concrete tanks are specified in references.[10] The analysis of internal forces in the walls of cylindrical tanks is discussed in other chapters; the tables in Chapter 12 give the displacements and internal forces due to chosen loadings; the solution can also be determined using the computer program CTW (Cylindrical Tank Walls). The computer program SOR (Shells of Revolution), based on the finite-element method, can analyze wall, roof, and base as a continuous shell. CTW and SOR can be downloaded from an annex Web site (see Appendix B). The thickness of the wall is generally governed by constructability rather than the internal forces. Considering the outer diameter of the ducts, the diameters of the non-prestressed reinforcing bars (at the intersection points) and their cover, the minimum thickness of the wall needs to be not less than 300 mm (12 in). In service, it is recommended to maintain a compressive stress on concrete ≥1.4 MPa (200 psi), accounting for prestress losses; some designers include differential shrinkage and temperature in this calculation; others, do not include the thermal effect as part of the residual compressive stress. No established upper limit for the size of prestressed concrete tank can be recommended; the capacity of the tank in Figure 11.1 is 120,000 m^3 (30 MG). The parameters that can limit a tank size include the translation at the waterstop and the bearing pads, and the variation of the earth pressure over the circumference.

11.2.10 Protection against corrosion

To sustain satisfactory service over a long lifespan, high level of corrosion protection is essential. Thicker concrete wall, facilitating placement of high-quality concrete, is the main line of defense. Maintaining compressive stress in the entire wall thickness eliminates cracking. Use of ducts made of noncorrosive and nonconductive HDPE (high-density polyethylene) with welded joints is recommended. High-quality grout can provide further corrosion protection for the strands.

A tendon's anchorage components can be vulnerable to corrosion. The duct containing the prestressing strands has sealed transition to the anchor block. Wedges, which seat-in an anchor head transfer stress from the strands to the anchor head then to the bearing plate and concrete. A bolted cap seals the anchor block and the cavity is filled with grout. Techniques of corrosion protection will no doubt continue to develop.

11.3 LIQUEFIED NATURAL GAS (LNG) TANKS

Liquefied natural gas is commonly contained in a steel tank suitable for extremely low temperatures. Leakage of the contained liquid and vapor must not harm the environment; measures of alleviating this risk are discussed in the following sections.

11.3.1 Function of LNG tanks

Until the second half of the 20th century, natural gas was considered a worthless by-product of oil production. Nowadays, natural gas represents about 21% of the global energy source. The reasons for this change are manifold and diverse, but the key factor is that natural gas is releasing significantly less greenhouse gas; thus, it is the most environmentally friendly source of fossil fuels.

The world consumption of natural gas is about 2.5 trillion cubic meters per year in 2013 and is rapidly increasing. Energy-lacking economies in Asia (such as Japan, Korea, China, and India) and some regions in Europe have little or no natural gas resources. Other regions in the world are newly becoming suppliers of natural gas. In Australia, huge sources have been developed. The increase in utilization of natural gas has changed the United States from an importer to an exporter of gas. All these countries and regions require infrastructure for transportation and storage of natural gas.

It is difficult to transport natural gas in its natural physical state. Therefore, since the mid-20th century the method of liquefying natural gas and then transporting by ship over long distances has been established. The abbreviation LNG has arisen from the English term "liquefied natural gas" and has become established in international usage. The application of LNG technology is based on the physical behavior of LNG. By cooling to −165°C, the gas is liquefied with a consecutive reduction in volume of 1/600. Only by this extreme volume reduction, transporting by ship is economically viable. All components that are necessary to liquefy and to transport gas from one location to another are referred to as the LNG chain. The LNG chain consists of an export terminal with a liquefaction plant in the producing countries, the specialized ships for transport in the ocean, and an import terminal with a jetty and vaporization unit in the recipient countries. For the temporary storage of liquefied gas in the liquefaction or the regasification plant, the usual size is 160,000 m^3 to 200,000 m^3; but volumes exceeding 200,000 m^3 are built in Korea and Japan.

11.3.2 Storage concept and tank types

The designation "containment tank system" in codes[11] is adopted here. Each containment tank system is a combination of several components. Codes distinguish between single-, double-, and full-containment tank systems. The choice of a containment tank system is based upon assessment of the risks for the inside and the outside of the boundary of the LNG plant.

The evolution of storage concepts and appropriate selection of materials is driven by hazard scenarios, like failure of the inner tank (primary container), fire, blast, and impact. The choice of the containment system depends upon the consequences of failure of the primary containment tank.

11.3.2.1 Single-containment tank system

A single-containment tank system consists of a primary container surrounded by a boundary wall (Figure 11.9). The primary container has to be liquid tight and vapor tight. It can have a liquid tight and vapor tight single wall subject to the cryogenic conditions, or a double wall, in which the inner wall has to be liquid tight to contain the liquid and the outer wall has to be vapor tight. The liquid can flow in the area surrounded by a boundary barrier. The distance between the primary container and the boundary wall can be up to 20 m. A large area is generated, where liquid and vapor are located at grade level. Evaporation and, most probably, pool fire affect this large area. The radiation to adjacent buildings, structures, and plant equipment cannot be prevented.

11.3.2.2 Double-containment tank system

The liquid can flow into a secondary containment, commonly a concrete wall. The distance between the primary and the secondary containments

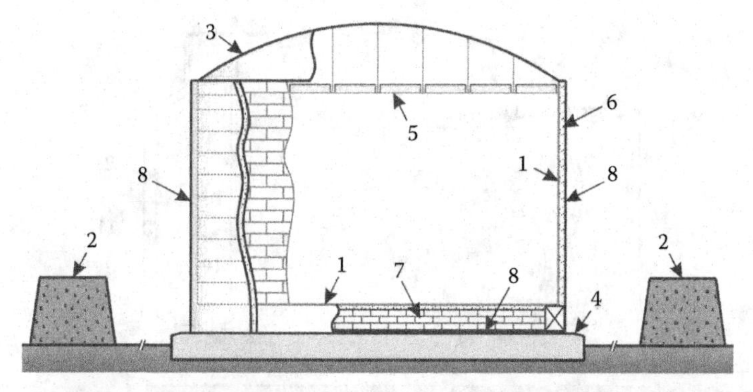

1 Primary liquid container (low temp steel)
2 Secondary liquid container dike
3 Warm vapor container (roof)
4 Concrete foundation
5 Suspended deck with insulation
6 Insulation (external)
7 Bottom insulation
8 Moisture vapor barrier

Figure 11.9 Single-containment LNG tank system.

is limited to 6 m, thus restricting the area where liquid and vapor can be located. The wall serves as protection to the adjacent plant area. The evaporation area is reduced and a subsequent pool fire, oriented upward only, cannot affect the tank area. The radiation to adjacent structures and plant equipment is reduced.

A double-containment tank system consists of a liquid-tight and vapor-tight primary container, which acts itself as a single-containment tank and is surrounded by a liquid-tight secondary container. The secondary container is open on top and cannot prevent the escape of vapor to the environment.

11.3.2.3 Full-containment tank system

The liquid can flow out of the primary tank into a secondary containment, which is able to contain the liquid and the vapor. This event cannot generate a pool fire. Vapor can exhaust from a relief valve. The effect of the event on an adjacent plant is significantly alleviated.

A full-containment tank system consists of a primary and a secondary container (Figure 11.10). The primary container is a self-standing steel tank holding liquid product; it can be open at the top (thus, does not contain the vapor) or it can have a dome roof.

The secondary container is a self-supporting steel or concrete tank having a dome roof able to contain, in normal operation, the vapor products and protect the thermal insulation from weather. In case of leakage of the inner tank, the secondary container contains all liquid products and remains structurally vapor tight.

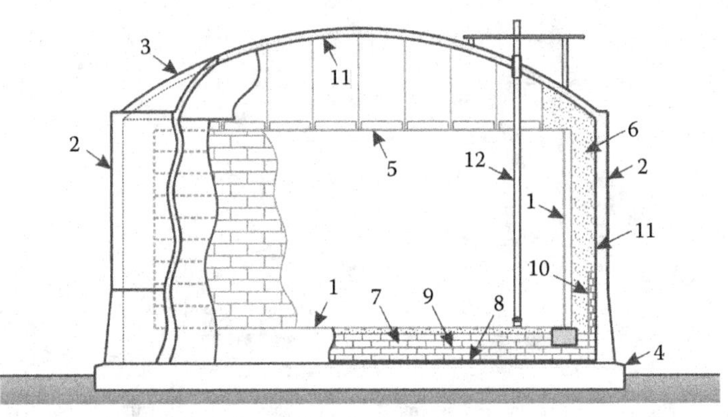

Figure 11.10 Full-containment LNG tank system.

The concrete secondary tank in Figure 11.10 has a monolithic connection between the wall and base slab. Failure of the inner tank would subject the wall to a temperature drop in the range of 100°C. A protective system composed of insulation and a collection basin made of 9% nickel steel (items 9 and 10 in Figure 11.10) allows smoother temperature variation, thus alleviates the stresses and controls cracking of the concrete. Figure 11.11 depicts dimensions of a typical LNG full-containment tank system. The remaining sections are concerned with the design and construction of full-containment tank systems.

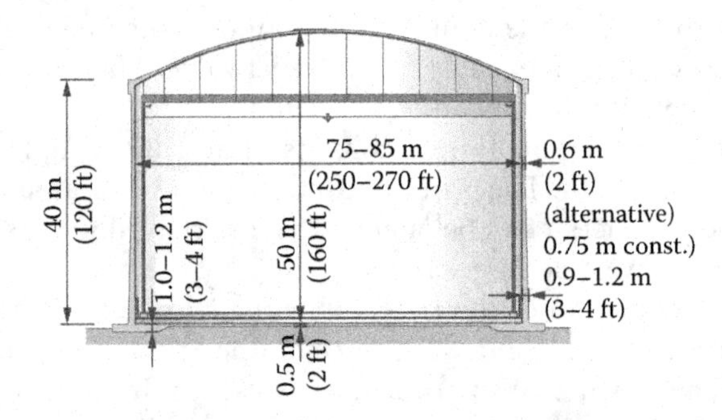

Figure 11.11 Dimensions of a typical full-containment LNG tank system. (Image also appears on book cover.)

11.3.3 Design of LNG tanks (full containment)

A state-of-the-art design requires a prestressed concrete outer tank that is able to operate as a catch basin in case of failure of the inner tank (liquid spill). This requires a closed container that can prevent the uncontrolled escape of gaseous clouds into the environment. The concrete outer tank is composed of a reinforced concrete bottom slab, a prestressed wall shell, often with a ring beam on top of the wall, and a reinforced concrete roof. The hydrostatic and internal pressure requires prestressing of the wall in the hoop (horizontal) direction. The need and the size of the vertical prestressing depend on the internal gas overpressure and the design requirements of the cracked concrete wall section.

The concrete outer tank protects the steel inner tank against external hazards, such as fire, impact, blast wave, and earthquake, and protects the environment from possible emergency situations in case of leakage of the inner tank. It is essential to consider the effect of extreme temperature in the design of the outer tank.

The liquid gas is stored in the steel inner tank. Adequate ductility of steel at –165°C requires a nickel content of 9%. The inner steel tank is open on top. The LNG is pumped into and out the tank by cryogenic pumps suspended on the concrete roof. At –165°C, LNG is close to its boiling point. Insulation limits the flow of heat and reduces evaporation.

The design loads occurring in operation are hydrostatic pressure acting on the inner tank and vapor pressure acting on the outer tank. The internal pressure is necessary to sustain the LNG in the liquid state. The design value of the vapor pressure is different for export and import terminals; its value is 10 kPa to 30 kPa above the atmospheric pressure. Normally, the lower value does not require vertical prestressing. The upper value requires vertical prestressing in the wall to control the thickness of its compressive zone (a criterion for tightness control).

Before commissioning the tank, it is tested for hydrostatic and pneumatic pressure. In the hydrostatic test, the tank is filled with water producing a maximum pressure equal to 125% of that of LNG. Based on an LNG density of 4.8 kN/m^3 and a water density of 10.0 kN/m^3, this results in a fill height of water equal to 60% of the filling height of LNG. The hydrostatic test certifies the tank foundation and the lower part of the steel tank. In the pneumatic test, the openings are temporarily closed with steel plates and the concrete tank is pressurized with 125% of the design pressure. The pneumatic test verifies the tightness of the carbon steel liner at the inner face of the wall and roof. It also tests all critical concrete sections. Codes permit carrying out the tests simultaneously or separately. Combining the hydrostatic and pneumatic pressures requires slightly more reinforcement but saves construction time; because of the presence of water, less volume needs to be pressurized.

The seismic design considers two characteristic cases. The Operating Basis Earthquake (OBE) defines the maximum earthquake event that causes no damage, and enables a restart and continuation of safe operation. The OBE is defined as the earthquake response spectrum with a damping of 5% with excess probability of 10% within a period of 50 years. This corresponds to a return period of 475 years.

The Safe Shutdown Earthquake (SSE) defines the largest earthquake in which the essential safety functions are still in operation. The SSE considers a response spectrum with a damping of 5% with excess probability of 1% or 2% within a period of 50 years; this corresponds to a return period of 4975 or 2475 years, respectively.

In operation, the crack width is usually limited to 0.30 mm for the concrete base slab and roof, and to 0.20 mm for the prestressed wall. Limitation of crack width does not apply for emergency load cases. In the event of liquid spill, the outer tank wall is designed for containing the liquid product and controlling the release of vapor to the environment. For vapor tightness, critical wall sections are verified to have a residual compression zone of thickness ≥100 mm and compressive stress ≥1 MPa (145 psi).

11.3.4 Construction features

The bottom slab may require more than 4000 m^3 concrete. Therefore, casting joints are arranged considering the amount of concrete that can be poured in a day shift. For construction of the wall, two principal methods are possible: climbing or sliding.

Casting the wall with a climbing formwork requires a smaller amount of preparation and less specialized team; it allows for breaks for bad weather, for placement of extensive built-in parts or for other reasons; it allows the casting in sections and the use of reinforcing mats or prefabricated reinforcement cages. Execution with climbing formwork is less sensitive to mistakes or tolerance deviations.

Wall casting with a sliding formwork requires a higher amount of preparation and an indispensable specialized team; it allows no breaks of the sliding process, excludes the use of reinforcement mats or prefabricated reinforcement cages; it also needs slightly more reinforcement to alleviate the effect of staggering of bars and inferior bond condition. Execution with sliding formwork is more sensitive to mistakes or tolerance deviations. The advantage of sliding is the significant reduction of construction time. In some cases, such as arctic climate, this advantage is decisive in decision making.

Design considering the hazard of blast wave, fire, or impact favors a concrete over steel tank roof. A concrete tank roof requires steel sheeting as a vapor and moisture barrier, and a means for fastening an aluminum suspended deck. The concrete roof in Figure 11.12 is cast on nonremovable

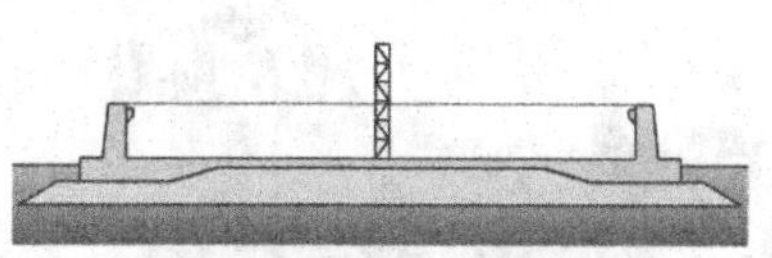

a) Erection of tank pad
b) Construction of bottom slab and outer concrete tank wall
c) Erection support structure for steel roof

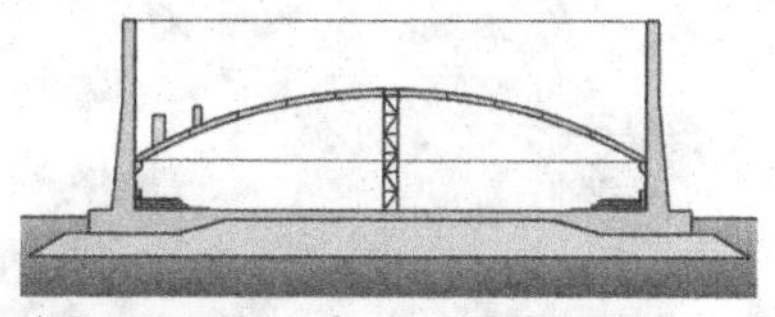

a) Continuation of outer concrete tank wall
b) Construction of steel roof including manholes and nozzles
c) Start insulation work

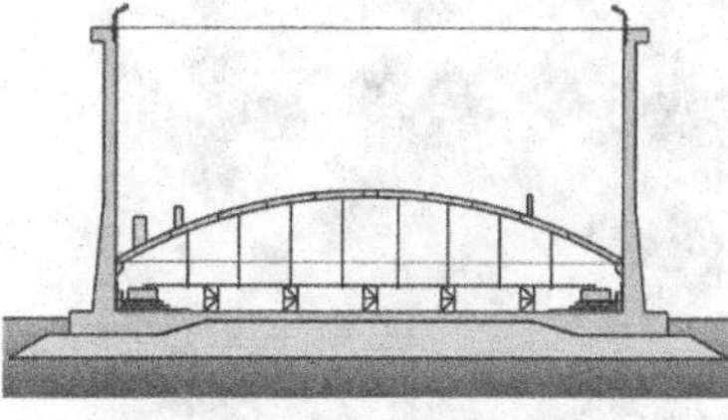

a) Construction of inner ring beam
b) Construction of concrete wall and concrete ring beam with compression ring
c) Suspended deck

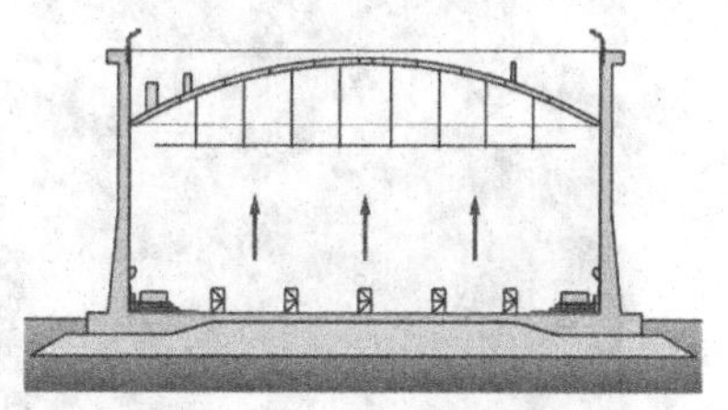

Air raising steel roof with suspended deck

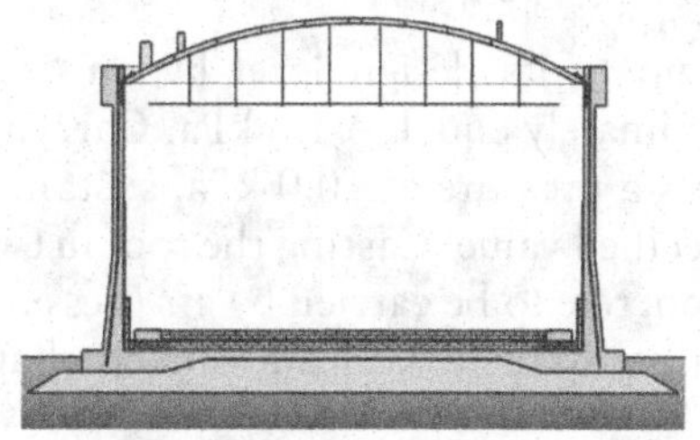

a) Fixing steel roof
b) Finalization of concrete ring beam
c) Bottom and wall insulation
d) Reinforcement works on concrete roof

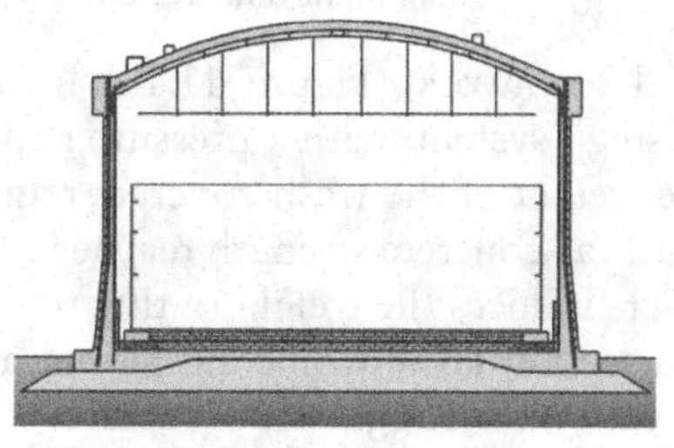

a) Concreting roof
b) Start erection of inner steel tank

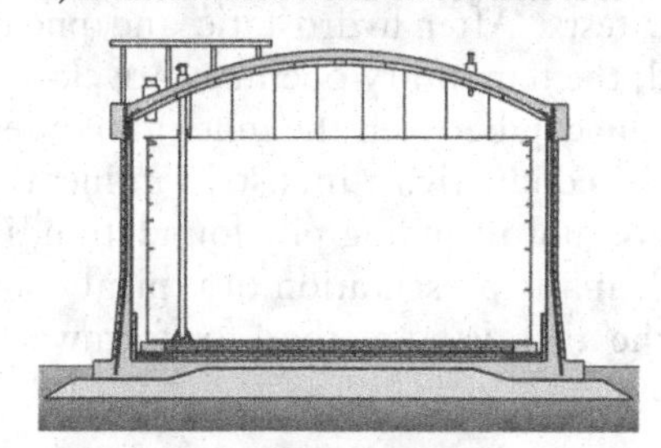

a) Finalization of inner steel tank
b) Installation of roof platform
c) Installation pump columns, instrumentation, etc.

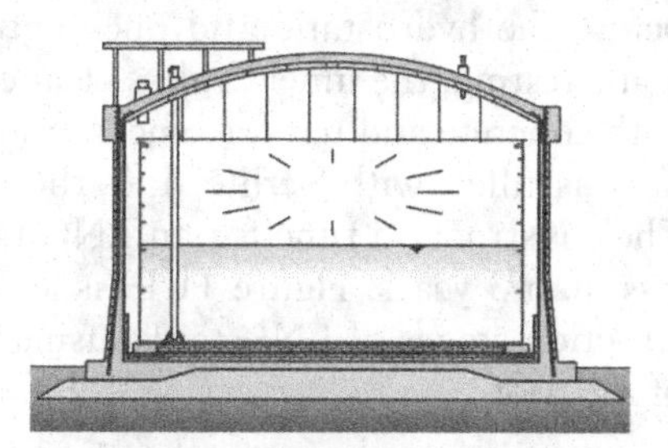

a) Hydraulic and pneumatic test
b) Drying and perlite filling
c) Purging
d) Cool down

Figure 11.12 Construction stages for an LNG full-containment tank system.

Figure 11.13 Steel formwork, permanently attached to a concrete dome in an LNG full-containment tank system.

steel formwork (Figure 11.13) lifted up to its design level by an "air-raising" system, with a pressure approximately equal to 1.0 kPa. Carrying the weight of the fresh concrete requires a pressure of 10.0 kPa, sustained until the concrete strength reaches a specified value. Casting the roof in two layers reduces the weight of the fresh concrete to be carried by air pressure; the second layer would be placed against hardened, clean—and perhaps intentionally roughened—surface.

After casting the outer tank and closing the structure, the inside construction work of the moisture sensitive insulation and steel tank can be started. Temporary openings in the wall allow access to the inner tank before the hydrostatic and pneumatic tests. After hydrostatic and pneumatic testing, the inner tank is cleaned, the temporary openings are closed with concrete and reinforcement, and the gap between the inner and outer tank is filled with perlite (low thermal-conductivity industrial mineral). The construction time for an LNG tank, not including pile foundation, is less than 3 years. Figure 11.13 is a schematic presentation of typical construction stages of LNG tanks using the air-raising method for formwork of the roof.

NOTES

1. American Concrete Institute Committee 350, *Code Requirements for Environmental Engineering Structures,* ACI 350-06; American Water Works Association, AWWA Standard, *Tendon-Prestressed Concrete Water Tanks,* ANSI/AWWA D115-06.
2. American Concrete Institute Committee 318, *Building Code Requirements for Structural Concrete and Commentary*, ACI 318 and ACI 318-R.
3. Ibid.
4. Joint ACI-ASCE Committee 421, Guide to Shear Reinforcements for slabs, ACI 421-1R; Joint ACI-ASCE Committee 421, Guide to Seismic Design of Punching Shear Reinforcement in Flat Plates, ACI 421-2R.
5. ACI Committee 224, *Causes, Evaluation, and Repair of Cracks in Concrete Structures,* ACI 224.1 R; ACI Committee 546, *Concrete Repair Guide,* ACI 546 R.
6. See note 2.
7. Ibid.
8. American Society for Testing and Materials, *Standard Test Method for Rubber Property—Durometer Hardness*, ASTM International, Designation: D 2240-05.
9. Lindley, P. B. (1974). *Engineering Design with Natural Rubber*, Malaysian Rubber Producer's Research Association, NR Technical Bulletin, London.
10. ACI Committee 373, *Design and Construction of Circular Prestressed Structures with Circumferential Tendons,* ACI 373R-97; ACI 350-06; ANSI/AWWA D115-06.
11. EN Standards-Euro Code, EN 14620, *Design and Manufacture of Site Built, Vertical, Cylindrical, Flat-Bottomed Steel Tanks for the Storage of Refrigerated, Liquefied Gases with Operating Temperatures between 0 Degrees and –165 DEGREES* C; American Petroleum Institute, API Standard 625, *Tank Systems for Refrigerated Liquefied Gas Storage.*
12. ACI Committee 376, *Code Requirements for Design and Construction of Concrete Structures for Refrigerated Liquefied Gases*, ACI 376-11.

Part II

Design tables

Key to Chapter 12 tables

Group	Edge conditions: Bottom	Edge conditions: Top	Load	Values given by	Table number
1	Free	Free	Triangular	N	1
			Intensity 0 at top and q at bottom	M, D_2, and D_4	2
			Uniform of	N	3
			Intensity q	M, D_2, and D_4	4
			Outward normal line load of	Influence coefficient N	5
			Unit intensity applied at 1 of 11 nodes	Influence coefficient M	6
			Spaced at one-tenth intervals	Influence coefficients D_2 and D_4	7
2	Free	Free	$F_1 = 1$	N	8
				M, D_2, and D_4	9
			$F_2 = 1$	N	10
				M, D_2, and D_4	11
			$F_3 = 1$	N	12
				M, D_2, and D_4	13
			$F_4 = 1$	N	14
				M, D_2, and D_4	15
3		Flexibility coefficients		f_{ij}	16
		Stiffness coefficients		S_{ij}	17
4	Encastré	Encastré	Triangular	N, $(V)_{x=0}$, and $(V)_{x=l}$	18
			Intensity 0 at top and q at bottom	M	19
			Uniform of	N, $(V)_{x=0}$, and $(V)_{x=0}$	20

continued

Key to Chapter 12 tables (continued)

Group	*Edge conditions*		*Load*	*Values given by*	*Table number*
	Bottom	*Top*			
			Intensity q	M	21
	Encastré	Free	Triangular	N and $(V)_{x=0}$	22
			Intensity 0 at top and q at bottom	M and D_4	23
			Uniform of	N and $(V)_{x=0}$	24
			Intensity q	M and D_4	25
	Hinged	Free	Triangular	N and $(V)_{x=0}$	26
			Intensity 0 at top and q at bottom	M, D_2, and D_4	27
			Uniform of	N and $(V)_{x=0}$	28
			Intensity q	M, D_2, and D_4	29

Chapter 12

Tables for analysis of circular walls of thickness varying linearly from top to bottom

A summary of the contents of this chapter is given in Figure 12.1 and in the table of the Part II page opener (which is a repeat of Table 4.1). Discussion of the tables and examples of their use are presented in Chapter 4. The symbols and the sign convention adopted are defined in Figure 12.1a,b.

The value of Poisson's ratio used in preparation of the tables is $v = 1/6$. For different values of v, correction may be made according to Chapter 4, Section 4.7.

The ratio of the thickness at the top and bottom, h_t/h_b, is taken as 1.0, 0.75, 0.5, 0.25, or 0.05, with the first value representing the case of constant thickness.

Tabulated values are for circular walls having the dimensionless parameter $\eta = l^2/(dh_b)$ between 0.4 and 24. For a higher value of η the tables may be used according to the procedure in Chapter 4, Section 4.6.

The tables in this section are usable for beams on elastic foundation as discussed in Chapter 4, Section 4.8.

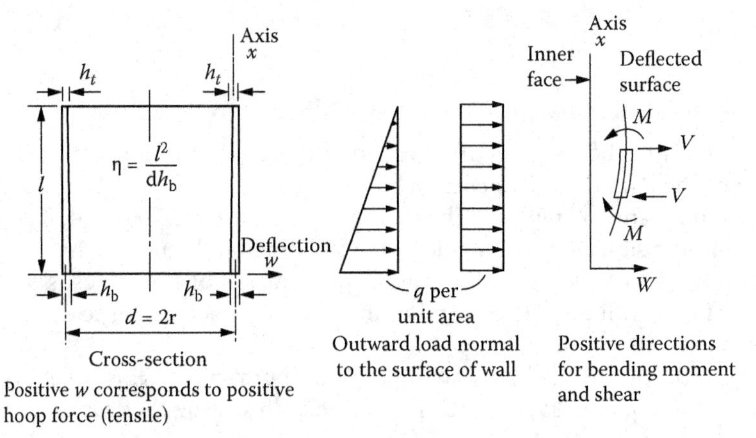

(a) Circular wall dimensions and positive directions for load and internal forces

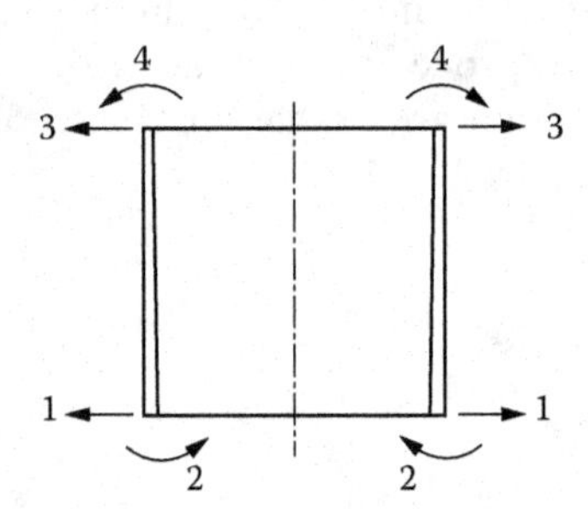

(b) Coordinate system indicating positive direction of forces $\{F\}$ or displacements $\{D\}$

Figure 12.1 Symbols and sign convention used in the tables in Chapter 12.

Table 12.1 Hoop force variation; triangular load of intensity q per unit area at bottom edge and zero at top; both edges free

η	h_t/h_b	x/l										
		0.0	*.1*	*.2*	*.3*	*.4*	*.5*	*.6*	*.7*	*.8*	*.9*	*1.0*
.4	1.00	.9997	.8998	.7998	.6999	.5999	.5000	.4001	.3001	.2002	.1002	.0003
	.75	1.0401	.9183	.8014	.6894	.5822	.4799	.3823	.2896	.2017	.1187	.0407
	.50	1.0746	.9341	.8028	.6804	.5671	.4626	.3671	.2806	.2030	.1346	.0755
	.25	1.0870	.9398	.8033	.6772	.5617	.4565	.3616	.2772	.2033	.1403	.0886
	.05	.0368	.9168	.8013	.6903	.5837	.4815	.3836	.2902	.2013	.1172	.0387
.8	1.00	.9997	.8998	.7998	.6999	.5999	.5000	.4001	.3001	.2002	.1002	.0003
	.75	1.0380	.9172	.8012	.6900	.5833	.4811	.3833	.2901	.2015	.1177	.0388
	.50	1.0694	.9315	.8024	.6819	.5697	.4656	.3696	.2818	.2024	.1320	.0710
	.25	1.0780	.9355	.8028	.6798	.5661	.4614	.3656	.2790	.2020	.1359	.0821
	.05	1.0314	.9143	.8012	.6920	.5864	.4844	.3859	.2910	.2001	.1145	.0367
1.2	1.00	.9997	.8998	.7998	.6999	.5999	.5000	.4001	.3001	.2002	.1002	.0003
	.75	1.0351	.9157	.8010	.6908	.5847	.4828	.3848	.2909	.2012	.1162	.0360
	.50	1.0623	.9279	.8019	.6839	.5732	.4697	.3730	.2834	.2016	.1283	.0649
	.25	1.0666	.9299	.8023	.6830	.5717	.4677	.3708	.2813	.2004	.1302	.0738
	.05	1.0249	.9113	.8010	.6939	.5895	.4878	.3886	.2920	.1988	.1113	.0341
1.6	1.00	.9997	.8998	.7998	.6999	.5999	.5000	.4001	.3001	.2002	.1002	.0003
	.75	1.0317	.9140	.8007	.6917	.5865	.4847	.3865	.2917	.2009	.1144	.0329
	.50	1.0545	.9240	.8014	.6860	.5771	.4741	.3767	.2853	.2007	.1244	.0581
	.25	1.0552	.9244	.8017	.6862	.5773	.4739	.3759	.2837	.1989	.1246	.0653
	.05	1.0191	.9086	.8008	.6955	.5924	.4910	.3911	.2930	.1978	.1082	.0315

continued

Table 12.1 (continued) Hoop force variation; triangular load of intensity q per unit area at bottom edge and zero at top; both edges free

		x/l										
η	h_t/h_b	*0.0*	*.1*	*.2*	*.3*	*.4*	*.5*	*.6*	*.7*	*.8*	*.9*	*1.0*
2.0	1.00	.9997	.8998	.7998	.6999	.6000	.5000	.4000	.3001	.2002	.1002	.0003
	.75	1.0283	.9122	.8005	.6926	.5882	.4867	.3882	.2926	.2005	.1126	.0297
	.50	1.0471	.9203	.8009	.6881	.5808	.4783	.3803	.2870	.1998	.1207	.0517
	.25	1.0453	.9196	.8011	.6890	.5821	.4794	.3805	.2859	.1976	.1196	.0578
	.05	1.0144	.9064	.8006	.6968	.5946	.4935	.3933	.2941	.1970	.1057	.0292
3.0	1.00	.9997	.8998	.7999	.6999	.6000	.5000	.4000	.3001	.2001	.1002	.0003
	.75	1.0209	.9084	.7999	.6947	.5919	.4911	.3918	.2945	.1998	.1088	.0227
	.50	1.0325	.9129	.7999	.6920	.5880	.4866	.3873	.2906	.1983	.1132	.0387
	.25	1.0282	.9112	.8000	.6936	.5903	.4889	.3887	.2902	.1957	.1108	.0437
	.05	1.0076	.9030	.8001	.6985	.5978	.4973	.3968	.2961	.1964	.1015	.0247
4.0	1.00	.9997	.8998	.7999	.6999	.6000	.5000	.4000	.3001	.2001	.1002	.0003
	.75	1.0157	.9057	.7995	.6961	.5945	.4941	.3944	.2959	.1993	.1061	.0177
	.50	1.0233	.9083	.7992	.6945	.5925	.4918	.3918	.2931	.1975	.1084	.0302
	.25	1.0190	.9066	.7994	.6959	.5946	.4940	.3934	.2930	.1952	.1057	.0350
	.05	1.0046	.9015	.7998	.6991	.5990	.4989	.3985	.2976	.1967	.0995	.0217
5.0	1.00	.9996	.8998	.7999	.6999	.6000	.5000	.4000	.3001	.2001	.1002	.0004
	.75	1.0121	.9039	.7992	.6970	.5963	.4961	.3961	.2968	.1990	.1043	.0143
	.50	1.0176	.9055	.7988	.6960	.5951	.4949	.3946	.2947	.1971	.1055	.0246
	.25	1.0139	.9041	.7990	.6971	.5968	.4967	.3961	.2949	.1952	.1028	.0293
	.05	1.0032	.9008	.7996	.6993	.5995	.4996	.3994	.2986	.1973	.0984	.0194
6.0	1.00	.9996	.8998	.7999	.7000	.6000	.5000	.4000	.3000	.2001	.1002	.0004
	.75	1.0097	.9027	.7991	.6977	.5974	.4973	.3973	.2975	.1989	.1031	.0120

	.50	1.0140	.9037	.7986	.6969	.5968	.4968	.3964	.2959	.1971	.1036	.0208
	.25	1.0109	.9026	.7988	.6977	.5980	.4982	.3977	.2963	.1956	.1011	.0253
	.05	1.0025	.9004	.7995	.6994	.5997	.4999	.3999	.2992	.1978	.0980	.0176
8.0	1.00	.9995	.8998	.7999	.7000	.6000	.5000	.4000	.3000	.2001	.1002	.0005
	.75	1.0068	.9013	.7990	.6985	.5987	.4988	.3986	.2983	.1988	.1017	.0091
	.50	1.0098	.9017	.7985	.6980	.5984	.4987	.3983	.2974	.1972	.1015	.0160
	.25	1.0076	.9011	.7986	.6984	.5991	.4995	.3993	.2980	.1964	.0993	.0201
	.05	1.0016	.9001	.7995	.6996	.5998	.5000	.4001	.2998	.1987	.0978	.0151
10.0	1.00	.9995	.8998	.7999	.7000	.6000	.5000	.4000	.3000	.2001	.1002	.0005
	.75	1.0052	.9006	.7990	.6989	.5993	.4995	.3992	.2988	.1988	.1009	.0074
	.50	1.0075	.9008	.7985	.6985	.5992	.4996	.3992	.2982	.1976	.1004	.0132
	.25	1.0058	.9005	.7987	.6989	.5995	.5000	.3999	.2989	.1972	.0986	.0168
	.05	1.0011	.8999	.7996	.6997	.5999	.5000	.4001	.3000	.1993	.0980	.0133
12.0	1.00	.9994	.8998	.8000	.7000	.6000	.5000	.4000	.3000	.2000	.1002	.0006
	.75	1.0041	.9002	.7991	.6992	.5996	.4998	.3996	.2991	.1989	.1005	.0064
	.50	1.0060	.9003	.7986	.6989	.5996	.4999	.3997	.2988	.1979	.0999	.0113
	.25	1.0046	.9001	.7989	.6992	.5997	.5001	.4001	.2994	.1979	.0983	.0146
	.05	1.0007	.8998	.7997	.6998	.5999	.5000	.4001	.3001	.1997	.0982	.0120
14.0	1.00	.9994	.8998	.8000	.7000	.6000	.5000	.4000	.3000	.2000	.1002	.0006
	.75	1.0034	.9000	.7992	.6994	.5998	.4999	.3998	.2993	.1990	.1003	.0056
	.50	1.0050	.9000	.7988	.6992	.5998	.5000	.3999	.2992	.1982	.0995	.0099
	.25	1.0038	.8999	.7990	.6994	.5998	.5001	.4001	.2997	.1984	.0982	.0129
	.05	1.0004	.8998	.7997	.6999	.6000	.5000	.4000	.3001	.1999	.0985	.0110

continued

Table 12.1 (continued) Hoop force variation; triangular load of intensity q per unit area at bottom edge and zero at top; both edges free

η	h_t/h_b	x/l										
		0.0	*.1*	*.2*	*.3*	*.4*	*.5*	*.6*	*.7*	*.8*	*.9*	*1.0*
16.0	1.00	.9993	.8998	.8000	.7000	.6000	.5000	.4000	.3000	.2000	.1002	.0007
	.75	1.0028	.8998	.7993	.6996	.5999	.5000	.3999	.2995	.1991	.1001	.0051
	.50	1.0042	.8998	.7989	.6994	.5999	.5001	.4000	.2994	.1985	.0994	.0089
	.25	1.0031	.8997	.7992	.6995	.5999	.5001	.4001	.2999	.1988	.0982	.0116
	.05	1.0002	.8998	.7998	.6999	.6000	.5000	.4000	.3001	.2000	.0988	.0102
18.0	1.00	.9993	.8998	.8000	.7001	.6000	.5000	.4000	.2999	.2000	.1002	.0007
	.75	1.0024	.8997	.7993	.6997	.5999	.5000	.3999	.2996	.1992	.1000	.0046
	.50	1.0036	.8997	.7991	.6995	.5999	.5001	.4000	.2996	.1988	.0993	.0081
	.25	1.0026	.8997	.7993	.6997	.6000	.5001	.4001	.3000	.1991	.0983	.0106
	.05	1.0001	.8997	.7998	.7000	.6000	.5000	.4000	.3000	.2000	.0990	.0096
20.0	1.00	.9992	.8998	.8000	.7001	.6000	.5000	.4000	.2999	.2000	.1002	.0008
	.75	1.0020	.8997	.7994	.6997	.6000	.5001	.4000	.2997	.1993	.0999	.0043
	.50	1.0031	.8996	.7992	.6997	.6000	.5001	.4001	.2997	.1990	.0992	.0075
	.25	1.0022	.8996	.7994	.6998	.6000	.5000	.4001	.3000	.1993	.0984	.0098
	.05	.9999	.8997	.7999	.7000	.6000	.5000	.4000	.3000	.2000	.0992	.0091
24.0	1.00	.9992	.8998	.8000	.7001	.6000	.5000	.4000	.2999	.2000	.1002	.0008
	.75	1.0015	.8996	.7996	.6999	.6000	.5000	.4000	.2998	.1994	.0998	.0038
	.50	1.0024	.8995	.7994	.6998	.6000	.5001	.4001	.2999	.1993	.0992	.0065
	.25	1.0016	.8996	.7996	.6999	.6000	.5000	.4000	.3001	.1996	.0986	.0086
	.05	.9997	.8997	.7999	.7000	.6000	.5000	.4000	.3000	.2001	.0994	.0083

Note: N = Coefficient × qr.

Table 12.2 Bending moment variation and rotations D_2 and D_4 at the bottom and top edges, respectively; triangular load of intensity q per unit area at bottom edge and zero at top; both edges free

η	h_t/h_b	x/l										
		.1	.2	.3	.4	.5	.6	.7	.8	.9	D_2	D_4
.4	1.00	.000003	.000004	.000004	.000002	–.000000	–.000002	–.000004	–.000004	–.000003	–.9994	–.9994
	.75	.000168	.000522	.000894	.001165	.001261	.001161	.000887	.000513	.000161	–.9823	–.9899
	.50	.000309	.000966	.001658	.002162	.002343	.002159	.001652	.000959	.000303	–.9133	–.9377
	.25	.000359	.001125	.001932	.002520	.002734	.002520	.001931	.001122	.000355	–.7098	–.7742
	.05	.000154	.000481	.000824	.001074	.001165	.001075	.000825	.000480	.000152	–.2373	–.3470
.8	1.00	.000003	.000004	.000004	.000002	.000000	.000002	–.000004	–.000004	–.000003	–.9994	–.9991
	.75	.000159	.000494	.000846	.001101	.001192	.001097	.000839	.000487	.000153	–.9728	–1.0016
	.50	.000287	.000896	.001536	.002002	.002170	.002002	.001536	.000894	.000283	–.8887	–.9795
	.25	.000322	.001007	.001728	.002254	.002449	.002264	.001744	.001020	.000326	–.6669	–.8994
	.05	.000132	.000411	.000704	.000920	.001004	.000933	.000725	.000429	.000139	–.2107	–.6038
1.2	1.00	.000003	.000004	.000004	.000002	–.000000	–.000002	–.000004	–.000004	–.000003	–.9994	–.9994
	.75	.000147	.000455	.000776	.001008	.001091	.001005	.000771	.000448	.000141	–.9591	–1.0187
	.50	.000257	.000799	.001368	.001781	.001932	.001785	.001374	.000804	.000256	–.8548	–1.0374
	.25	.000274	.000855	.001466	.001913	.002083	.001936	.001502	.000889	.000288	–.6122	–1.0630
	.05	.000106	.000327	.000561	.000737	.000809	.000762	.000603	.000367	.000123	–.1793	–.9387
1.6	1.00	.000003	.000004	.000004	.000002	.000000	–.000002	.000004	–.000004	–.000003	–.9993	–.9993
	.75	.000133	.000409	.000695	.000902	.000976	.000900	.000692	.000404	.000128	–.9433	–1.0383
	.50	.000224	.000695	.001186	.001541	.001673	.001550	.001199	.000706	.000227	–.8182	–1.1007
	.25	.000227	.000705	.001206	.001574	.001719	.001608	.001260	.000756	.000249	–.5581	–1.2308
	.05	.000082	.000251	.000431	.000568	.000630	.000603	.000489	.000307	.000107	–.1511	–1.2892

continued

Table 12.2 (continued) Bending moment variation and rotations D_2 and D_4 at the bottom and top edges, respectively; triangular load of intensity q per unit area at bottom edge and zero at top; both edges free

η	h_t/h_b	x/l										
		.1	.2	.3	.4	.5	.6	.7	.8	.9	D_2	D_4
2.0	1.00	.000003	.000004	.000004	.000002	.000000	–.000002	–.000004	–.000004	–.000003	–.9993	–.9993
	.75	.000118	.000362	.000614	.000794	.000859	.000793	.000611	.000358	.000114	–.9272	–1.0583
	.50	.000193	.000595	.001012	.001313	.001426	.001325	.001032	.000613	.000199	–.7832	–1.1628
	.25	.000186	.000574	.000978	.001277	.001399	.001318	.001046	.000638	.000215	–.5111	–1.3843
	.05	.000063	.000191	.000327	.000432	.000484	.000472	.000393	.000256	.000094	–.1292	–1.6256
3.0	1.00	.000003	.000004	.000003	.000002	–.000000	–.000002	–.000003	–.000004	–.000003	–.9991	–.9991
	.75	.000087	.000261	.000436	.000560	.000604	.000561	.000437	.000260	.000084	–.8922	–1.1022
	.50	.000132	.000399	.000669	.000862	.000937	.000880	.000698	.000426	.000143	–.7144	–1.2873
	.25	.000115	.000348	.000584	.000758	.000837	.000806	.000663	.000425	.000151	–.4308	–1.6822
	.05	.000034	.000100	.000169	.000223	.000255	.000260	.000232	.000167	.000069	–.0976	–2.3663
4.0	1.00	.000003	.000004	.000003	.000002	.000000	–.000002	–.000003	–.000004	–.000003	–.9988	–.9988
	.75	.000065	.000190	.000311	.000395	.000425	.000397	.000313	.000190	.000063	–.8672	–1.1340
	.50	.000094	.000276	.000453	.000578	.000628	.000597	.000484	.000305	.000106	–.6709	–1.3727
	.25	.000077	.000225	.000369	.000474	.000526	.000517	.000442	.000298	.000113	–.3880	–1.8853
	.05	.000022	.000060	.000098	.000126	.000145	.000153	.000146	.000114	.000053	–.0846	–2.9837
5.0	1.00	.000003	.000004	.000003	.000002	.000000	–.000002	–.000003	–.000004	–.000003	–.9985	–.9985
	.75	.000050	.000142	.000228	.000286	.000306	.000288	.000231	.000143	.000048	–.8502	–1.1560
	.50	.000070	.000200	.000321	.000403	.000437	.000420	.000349	.000228	.000082	–.6438	–1.4314
	.25	.000056	.000157	.000251	.000316	.000349	.000350	.000310	.000220	.000088	–.3644	–2.0324
	.05	.000016	.000042	.000064	.000080	.000090	.000096	.000097	.000082	.000042	–.0785	–3.5105
6.0	1.00	.000003	.000003	.000003	.000001	.000000	–.000001	–.000003	–.000003	–.000003	–.9982	–.9982

	.75	.000040	.000110	.000173	.000213	.000227	.000215	.000175	.000112	.000039	−.8384	−1.1716
	.50	.000055	.000152	.000237	.000293	.000316	.000307	.000262	.000176	.000066	−.6261	−1.4740
	.25	.000044	.000118	.000181	.000223	.000245	.000248	.000227	.000169	.000072	−.3500	−2.1460
	.05	.000013	.000031	.000046	.000055	.000060	.000065	.000067	.000061	.000034	−.0749	−3.9677
8.0	1.00	.000003	.000003	.000002	.000001	−.000000	−.000001	−.000002	−.000003	−.000003	−.9975	−.9975
	.75	.000028	.000072	.000107	.000128	.000135	.000129	.000110	.000073	.000027	−.8234	−1.1919
	.50	.000038	.000098	.000144	.000171	.000182	.000180	.000161	.000116	.000046	−.6046	−1.5325
	.25	.000030	.000075	.000109	.000127	.000135	.000139	.000134	.000109	.000051	−.3328	−2.3145
	.05	.000010	.000021	.000028	.000031	.000032	.000034	.000036	.000036	.000024	−.0700	−4.7257
10.0	1.00	.000002	.000002	.000001	.000001	0.000000	−.000001	−.000001	−.000002	−.000002	−.9969	−.9969
	.75	.000022	.000052	.000073	.000084	.000087	.000085	.000075	.000052	.000020	−.8139	−1.2048
	.50	.000029	.000070	.000097	.000111	.000116	.000116	.000108	.000082	.000035	−.5913	−1.5722
	.25	.000023	.000054	.000073	.000081	.000085	.000087	.000087	.000076	.000038	−.3219	−2.4368
	.05	.000008	.000015	.000019	.000020	.000020	.000021	.000022	.000024	.000017	−.0663	−5.3305
12.0	1.00	.000002	.000002	.000001	.000000	.000000	−.000000	−.000001	−.000002	−.000002	−.9962	−.9962
	.75	.000018	.000039	.000053	.000059	.000060	.000059	.000054	.000039	.000015	−.8072	−1.2139
	.50	.000023	.000053	.000070	.000077	.000079	.000080	.000077	.000062	.000028	−.5817	−1.6017
	.25	.000019	.000041	.000053	.000057	.000058	.000059	.000060	.000055	.000030	−.3137	−2.5308
	.05	.000006	.000012	.000014	.000014	.000014	.000014	.000015	.000016	.000013	−.0635	−5.8246
14.0	1.00	.000002	.000002	.000001	.000000	−.000000	−.000000	−.000001	−.000002	−.000002	−.9956	−.9956
	.75	.000015	.000031	.000040	.000043	.000044	.000044	.000041	.000031	.000013	−.8018	−1.2207
	.50	.000019	.000042	.000053	.000057	.000057	.000058	.000057	.000048	.000023	−.5742	−1.6248
	.25	.000016	.000032	.000040	.000042	.000042	.000042	.000044	.000042	.000024	−.3073	−2.6059
	.05	.000005	.000009	.000011	.000011	.000010	.000010	.000011	.000012	.000010	−.0612	−6.2352

continued

Table 12.2 (continued) Bending moment variation and rotations D_2 and D_4 at the bottom and top edges, respectively; triangular load of intensity q per unit area at bottom edge and zero at top; both edges free

η	h_t/h_b	x/l										
		.1	.2	.3	.4	.5	.6	.7	.8	.9	D_2	D_4
16.0	1.00	.000002	.000001	.000001	.000000	–.000000	–.000000	–.000001	–.000001	–.000002	–.9950	–.9950
	.75	.000013	.000025	.000031	.000033	.000033	.000033	.000032	.000025	.000010	–.7974	–1.2261
	.50	.000017	.000034	.000042	.000043	.000043	.000044	.000044	.000038	.000019	–.5680	–1.6434
	.25	.000013	.000026	.000031	.000032	.000032	.000032	.000033	.000033	.000020	–.3020	–2.6673
	.05	.000005	.000007	.000008	.000008	.000008	.000008	.000008	.000009	.000008	–.0593	–6.5812
18.0	1.00	.000002	.000001	.000000	.000000	.000000	–.000000	–.000000	–.000001	–.000002	–.9943	–.9943
	.75	.000011	.000021	.000025	.000026	.000026	.000026	.000025	.000021	.000009	–.7937	–1.2304
	.50	.000014	.000028	.000033	.000034	.000034	.000034	.000035	.000031	.000016	–.5628	–1.6587
	.25	.000011	.000022	.000025	.000025	.000025	.000025	.000026	.000026	.000017	–.2976	–2.7185
	.05	.000004	.000006	.000007	.000006	.000006	.000006	.000006	.000007	.000007	–.0576	–6.8758
20.0	1.00	.000002	.000001	.000000	.000000	.000000	–.000000	–.000000	–.000001	–.000002	–.9937	–.9937
	.75	.000010	.000018	.000021	.000021	.000021	.000021	.000021	.000017	.000008	–.7904	–1.2340
	.50	.000013	.000024	.000027	.000027	.000027	.000028	.000028	.000026	.000014	–.5583	–1.6715
	.25	.000010	.000018	.000021	.000020	.000020	.000020	.000021	.000021	.000014	–.2938	–2.7618
	.05	.000004	.000005	.000005	.000005	.000005	.000005	.000005	.000005	.000005	–.0560	–7.1286
24.0	1.00	.000002	.000001	.000000	–.000000	–.000000	.000000	–.000000	–.000001	–.000002	–.9925	–.9925
	.75	.000008	.000013	.000014	.000014	.000014	.000014	.000015	.000013	.000006	–.7849	–1.2394
	.50	.000010	.000017	.000019	.000019	.000019	.000019	.000020	.000019	.000011	–.5509	–1.6916
	.25	.000008	.000013	.000014	.000014	.000014	.000014	.000014	.000015	.000011	–.2875	–2.8310
	.05	.000003	.000004	.000004	.000004	.000004	.000004	.000004	.000004	.000004	–.0534	–7.5369

Notes: M = Coefficient × ql^2; D_2 or D_4 = Coefficient × $qr^2/(Eh_b l)$.

Table 12.3 Hoop force variation; uniform load of intensity q per unit area; both edges free

η	h_t/h_b	x/l										
		0.0	*.1*	*.2*	*.3*	*.4*	*.5*	*.6*	*.7*	*.8*	*.9*	*1.0*
.4	1.00	1.0000	1.0000	1.0000	1.0000	1.0000	1.0000	1.0000	1.0000	1.0000	1.0000	1.0000
	.75	.9865	.9938	.9995	1.0035	1.0059	1.0067	1.0059	1.0035	.9995	.9938	.9865
	.50	.9250	.9656	.9971	1.0195	1.0329	1.0374	1.0330	1.0196	.9972	.9656	.9247
	.25	.7377	.8796	.9897	1.0680	1.1150	1.1308	1.1155	1.0689	.9907	.8797	.7348
	.05	.2916	.6743	.9711	1.1827	1.3099	1.3533	1.3135	1.1896	.9791	.6755	.2652
.8	1.00	1.0000	1.0000	1.0000	1.0000	1.0000	1.0000	1.0000	1.0000	1.0000	1.0000	1.0000
	.75	.9872	.9942	.9995	1.0033	1.0056	1.0063	1.0056	1.0033	.9996	.9942	.9872
	.50	.9502	.9682	.9974	1.0180	1.0303	1.0344	1.0305	1.0184	.9978	.9682	.9292
	.25	.7647	.8927	.9910	1.0604	1.1017	1.1160	1.1034	1.0635	.9945	.8929	.7542
	.05	.3957	.7225	.9738	1.1516	1.2588	1.2981	1.2707	1.1746	1.0007	.7271	.3052
1.2	1.00	1.0000	1.0000	1.0000	1.0000	1.0000	1.0000	1.0000	1.0000	1.0000	1.0000	1.0000
	.75	.9882	.99447	.9996	1.0030	1.0051	1.0057	1.0051	1.0031	.9997	.9947	.9881
	.50	.9374	.9718	.9979	1.0160	1.0267	1.0303	1.0271	1.0167	.9986	.9719	.9354
	.25	.7992	.9094	.9928	1.0506	1.0849	1.0970	1.0879	1.0565	.9993	.9099	.7793
	.05	.5191	.7800	.9775	1.1153	1.1986	1.2323	1.2188	1.1548	1.0252	.7895	.3554
1.6	1.00	1.0000	1.0000	1.0000	1.0000	1.0000	1.0000	1.0000	1.0000	1.0000	1.0000	1.0000
	.75	.9893	.9953	.9997	1.0027	1.0045	1.0051	1.0045	1.0028	.9998	.9953	.9891
	.50	.9452	.9758	.9985	1.0139	1.0229	1.0259	1.0233	1.0148	.9995	.9758	.9421
	.25	.8333	.9260	.9946	1.0411	1.0682	1.0783	1.0724	1.0492	1.0039	.9268	.8047
	.05	.6305	.8324	.9817	1.0834	1.1446	1.1721	1.1698	1.1342	1.0455	.8481	.4052

continued

Table 12.3 (continued) Hoop force variation; uniform load of intensity q per unit area; both edges free

η	h_t/h_b	*x/l*										
		0.0	*.1*	*.2*	*.3*	*.4*	*.5*	*.6*	*.7*	*.8*	*.9*	*1.0*
2.0	1.00	1.0000	1.0000	1.0000	1.0000	1.0000	1.0000	1.0000	1.0000	1.0000	1.0000	1.0000
	.75	.9905	.9959	.9998	1.0024	1.0039	1.0044	1.0040	1.0025	.9999	.9959	.9902
	.50	.9526	.9795	.9990	1.0119	1.0192	1.0217	1.0198	1.0131	1.0003	.9796	.9486
	.25	.8631	.9405	.9962	1.0328	1.0537	1.0618	1.0586	1.0426	1.0077	.9418	.8276
	.05	.7187	.8745	.9859	1.0592	1.1024	1.1239	1.1289	1.1146	1.0592	.8967	.4499
3.0	1.00	1.0000	1.0000	1.0000	1.0000	1.0000	1.0000	1.0000	1.0000	1.0000	1.0000	1.0000
	.75	.9929	.9971	1.0000	1.0018	1.0027	1.0030	1.0027	1.0019	1.0001	.9971	.9926
	.50	.9672	.9869	1.0000	1.0079	1.0120	1.0134	1.0127	1.0094	1.0018	.9870	.9616
	.25	.9143	.9657	.9995	1.0191	1.0291	1.0333	1.0340	1.0297	1.0132	.9684	.8699
	.05	.8435	.9382	.9954	1.0270	1.0425	1.0515	1.0616	1.0739	1.0702	.9757	.5371
4.0	1.00	1.0000	1.0000	1.0000	1.0000	1.0000	1.0000	1.0000	1.0000	1.0000	1.0000	1.0000
	.75	.9947	.9980	1.0001	1.0013	1.0018	1.0020	1.0019	1.0014	1.0003	.9981	.9942
	.50	.9764	.9915	1.0007	1.0054	1.0075	1.0082	1.0082	1.0070	1.0026	.9918	.9702
	.25	.9418	.9794	1.0015	1.0122	1.0163	1.0180	1.0199	1.0211	1.0148	.9836	.8964
	.05	.9040	.9668	1.0018	1.0163	1.0194	1.0204	1.0275	1.0452	1.0636	1.0150	.5984
5.0	1.00	1.0000	1.0000	1.0000	1.0000	1.0000	1.0000	1.0000	1.0000	1.0000	1.0000	1.0000
	.75	.9958	.9986	1.0002	1.0010	1.0012	1.0013	1.0013	1.0011	1.0004	.9987	.9954
	.50	.9820	.9943	1.0011	1.0040	1.0049	1.0051	1.0054	1.0053	1.0030	.9948	.9758
	.25	.9569	.9870	1.0028	1.0087	1.0097	1.0098	1.0118	1.0153	1.0145	.9923	.9137
	.05	.9295	.9802	1.0055	1.0128	1.0107	1.0075	1.0107	1.0262	1.0521	1.0339	.6441

6.0	1.00	1.0000	1.0000	1.0000	1.0000	1.0000	1.0000	1.0000	1.0000	1.0000	1.0000	1.0000
	.75	.9966	.9990	1.0003	1.0008	1.0009	1.0009	1.0009	1.0009	1.0004	.9991	.9962
	.50	.9856	.9961	1.0013	1.0031	1.0033	1.0032	1.0036	1.0041	1.0030	.9967	.9797
	.25	.9658	.9913	1.0035	1.0068	1.0061	1.0054	1.0069	1.0111	1.0135	.9975	.9259
	.05	.9434	.9872	1.0072	1.0111	1.0072	1.0024	1.0026	1.0142	1.0406	1.0423	.6798
8.0	1.00	1.0000	1.0000	1.0000	1.0000	1.0000	1.0000	1.0000	1.0000	1.0000	1.0000	1.0000
	.75	.9976	.9995	1.0003	1.0005	1.0005	1.0004	1.0005	1.0006	1.0004	.9995	.9971
	.50	.9897	.9980	1.0015	1.0021	1.0016	1.0013	1.0016	1.0026	1.0028	.9988	.9845
	.25	.9755	.9958	1.0039	1.0048	1.0029	1.0014	1.0022	1.0060	1.0107	1.0027	.9419
	.05	.9586	.9938	1.0076	1.0086	1.0044	.9998	.9978	1.0025	1.0226	1.0448	.7326
10.0	1.00	1.0000	1.0000	1.0000	1.0000	1.0000	1.0000	1.0000	1.0000	1.0000	1.0000	1.0000
	.75	.9981	.9997	1.0003	1.0004	1.0002	1.0002	1.0003	1.0004	1.0004	.9998	.9977
	.50	.9919	.9990	1.0014	1.0015	1.0008	1.0004	1.0007	1.0017	1.0025	.9998	.9875
	.25	.9808	.9979	1.0037	1.0036	1.0016	1.0002	1.0004	1.0032	1.0082	1.0049	.9521
	.05	.9675	.9969	1.0067	1.0063	1.0028	.9997	.9978	.9988	1.0115	1.0401	.7701
12.0	1.00	1.0000	1.0000	1.0000	1.0000	1.0000	1.0000	1.0000	1.0000	1.0000	1.0000	1.0000
	.75	.9984	.9998	1.0003	1.0003	1.0001	1.0001	1.0001	1.0003	1.0004	.9999	.9981
	.50	.9933	.9995	1.0013	1.0011	1.0005	1.0001	1.0003	1.0011	1.0021	1.0004	.9895
	.25	.9842	.9990	1.0033	1.0027	1.0010	.9998	.9998	1.0017	1.0061	1.0056	.9591
	.05	.9734	.9986	1.0058	1.0046	1.0018	.9999	.9986	.9980	1.0052	1.0337	.7985
14.0	1.00	1.0000	1.0000	1.0000	1.0000	1.0000	1.0000	1.0000	1.0000	1.0000	1.0000	1.0000
	.75	.9986	.9999	1.0003	1.0002	1.0001	1.0000	1.0001	1.0002	1.0003	1.0000	.9984
	.50	.9943	.9998	1.0012	1.0009	1.0003	1.0000	1.0001	1.0008	1.0017	1.0007	.9909
	.25	.9866	.9997	1.0029	1.0020	1.0006	.9997	.9996	1.0008	1.0046	1.0058	.9644
	.05	.9776	.9997	1.0049	1.0033	1.0010	.9999	.9993	.9983	1.0017	1.0275	.8207

continued

Table 12.3 (continued) Hoop force variation; uniform load of intensity q per unit area; both edges free

η	h_t/h_b	x/l										
		0.0	*.1*	*.2*	*.3*	*.4*	*.5*	*.6*	*.7*	*.8*	*.9*	*1.0*
16.0	1.00	1.0000	1.0000	1.0000	1.0000	1.0000	1.0000	1.0000	1.0000	1.0000	1.0000	1.0000
	.75	.9988	1.0000	1.0002	1.0002	1.0000	1.0000	1.0000	1.0002	1.0003	1.0000	.9986
	.50	.9951	1.0000	1.0011	1.0007	1.0002	.9999	1.0000	1.0005	1.0014	1.0008	.9920
	.25	.9884	1.0001	1.0025	1.0015	1.0004	.9998	.9996	1.0003	1.0034	1.0057	.9684
	.05	.9807	1.0003	1.0042	1.0024	1.0006	.9999	.9997	.9988	.9999	1.0221	.8388
18.0	1.00	1.0000	1.0000	1.0000	1.0000	1.0000	1.0000	1.0000	1.0000	1.0000	1.0000	1.0000
	.75	.9990	1.0000	1.0002	1.0001	1.0000	1.0000	1.0000	1.0001	1.0003	1.0001	.9987
	.50	.9956	1.0001	1.0009	1.0005	1.0001	.9999	.9999	1.0003	1.0012	1.0009	.9928
	.25	.9897	1.0004	1.0022	1.0011	1.0002	.9998	.9997	1.0000	1.0025	1.0054	.9717
	.05	.9830	1.0007	1.0036	1.0018	1.0003	.9999	.9999	.9992	.9991	1.0175	.8537
20.0	1.00	1.0000	1.0000	1.0000	1.0000	1.0000	1.0000	1.0000	1.0000	1.0000	1.0000	1.0000
	.75	.9991	1.0000	1.0002	1.0001	1.0000	1.0000	1.0000	1.0001	1.0002	1.0001	.9988
	.50	.9961	1.0002	1.0008	1.0004	1.0000	.9999	.9999	1.0002	1.0010	1.0009	.9935
	.25	.9908	1.0005	1.0019	1.0009	1.0001	.9999	.9998	.9999	1.0018	1.0051	.9743
	.05	.9848	1.0010	1.0031	1.0013	1.0001	.9999	1.0000	.9995	.9988	1.0138	.8663
24.0	1.00	1.0000	1.0000	1.0000	1.0000	1.0000	1.0000	1.0000	1.0000	1.0000	1.0000	1.0000
	.75	.9992	1.0001	1.0002	1.0001	1.0000	1.0000	1.0000	1.0000	1.0002	1.0001	.9990
	.50	.9968	1.0003	1.0006	1.0002	1.0000	.9999	.9999	1.0001	1.0007	1.0009	.9946
	.25	.9925	1.0007	1.0015	1.0005	1.0000	.9999	.9999	.9998	1.0009	1.0043	.9783
	.05	.9876	1.0013	1.0024	1.0007	1.0000	.9999	1.0000	.9999	.9989	1.0083	.8864

Note: N = Coefficient × qr.

Table 12.4 Bending moment variation and rotations D_2 and D_4 at bottom and top edges, respectively; uniform load of intensity q per unit area; both edges free

η	h_t/h_b	x/l .1	.2	.3	.4	.5	.6	.7	.8	.9	D_2	D_4
.4	1.00	–.000000	–.000000	–.000000	–.000000	–.000000	–.000000	–.000000	–.000000	–.000000	–.0000	–.0000
	.75	–.000055	–.000173	–.000297	–.000388	–.000421	–.000388	–.000297	–.000173	–.000055	.3276	.3302
	.50	–.000305	–.000962	–.001656	–.002161	–.002345	–.002163	–.001658	–.000964	–.000306	.9140	.9385
	.25	–.001068	–.003366	–.005793	–.007565	–.008212	–.007577	–.005813	–.003385	–.001077	2.1323	2.3262
	.05	–.002887	–.009101	–.015673	–.020485	–.022267	–.020583	–.015832	–.009255	–.002962	4.5350	6.6687
.8	1.00	.000000	–.000000	–.000000	.000000	.000000	–.000000	–.000000	–.000000	–.000000	.0000	.0000
	.75	–.000.052	–.000163	–.000281	–.000366	–.000398	–.000367	–.000281	–.000164	–.000052	.3245	.3341
	.50	–.000284	–.000892	–.001533	–.002001	–.002172	–.002005	–.001541	–.000899	–.000287	.8894	.9804
	.25	–.000957	–.003011	–.005178	–.006764	–.007355	–.006808	–.005249	–.003078	–.000988	2.0035	2.7039
	.05	–.002461	–.007758	–.013377	–.017538	–.019165	–.017865	–.013913	–.008277	–.002715	4.0257	11.6855
1.2	1.00	–.000000	–.000000	–.000000	–.000000	–.000000	–.000000	–.000000	–.000000	–.000000	–.0000	–.0000
	.75	–.000048	–.000150	–.000258	–.000336	–.000364	–.000336	–.000258	–.000151	–.000048	.3199	.3398
	.50	–.000254	–.000796	–.001365	–.001780	–.001933	–.001789	–.001379	–.000809	–.000260	.8556	1.0386
	.25	–.000814	–.002556	–.004391	–.005739	–.006256	–.005820	–.004523	–.002682	–.000873	1.8392	3.1974
	.05	–.001956	–.006163	–.010645	–.014017	–.015443	–.014585	–.011579	–.007078	–.002408	3.4264	18.2523
1.6	1.00	.000000	.000000	.000000	.000000	.000000	.000000	.000000	.000000	.000000	.0000	.0000
	.75	–.000043	–.000135	–.000231	–.000300	–.000325	–.000301	–.000232	–.000136	–.000044	.3147	.3464
	.50	–.000221	–.000691	–.001183	–.001540	–.001674	–.001553	–.001204	–.000711	–.000230	.8190	1.1022
	.25	–.000673	–.002105	–.003610	–.004720	–.005161	–.004833	–.003795	–.002283	–.000757	1.6770	3.7041
	.05	–.001498	–.004715	–.008154	–.010791	–.012010	–.011531	–.009380	–.005930	–.002109	2.8905	25.1679

continued

Table 12.4 (continued) Bending moment variation and rotations D_2 and D_4 at bottom and top edges, respectively; uniform load of intensity q per unit area; both edges free

η	h_t/h_b	x/l									D_2	D_4
		.1	.2	.3	.4	.5	.6	.7	.8	.9		
2.0	1.00	.000000	.000000	.000000	.000000	.000000	.000000	.000000	.000000	.000000	.0000	.0000
	.75	–.000038	–.000119	–.000203	–.000264	–.000286	–.000265	–.000205	–.000121	–.000039	.3093	.3531
	.50	–.000190	–.000592	–.001009	–.001312	–.001427	–.001328	–.001036	–.000617	–.000202	.7841	1.1639
	.25	–.000550	–.001712	–.002927	–.003827	–.004200	–.003963	–.003151	–.001928	–.000654	1.5362	4.1685
	.05	–.001135	–.003562	–.006161	–.008192	–.009216	–.009014	–.007537	–.004947	–.001846	2.4731	31.8559
3.0	1.00	–.000000	–.000000	–.000000	–.000000	–.000000	–.000000	.000000	.000000	.000000	–.0000	.0000
	.75	–.000028	–.000086	–.000144	–.000186	–.000202	–.000188	–.000147	–.000088	–.000029	.2977	.3679
	.50	–.000129	–.000396	–.000666	–.000861	–.000938	–.000882	–.000701	–.000430	–.000146	.7155	1.2901
	.25	–.000337	–.001032	–.001743	–.002271	–.002512	–.002423	–.001998	–.001285	–.000463	1.2959	5.0728
	.05	–.000598	–.001841	–.003150	–.004203	–.004836	–.004954	–.004453	–.003224	–.001360	1.8788	46.7968
4.0	1.00	.000000	.000000	.000000	.000000	.000000	–.000000	.000000	.000000	.000000	.0000	–.0000
	.75	–.000021	–.000062	–.000103	–.000131	–.000142	–.000133	–.000105	–.000065	–.000022	.2895	.3786
	.50	–.000091	–.000272	–.000451	–.000576	–.000628	–.000598	–.000487	–.000308	–.000109	.6723	1.3766
	.25	–.000223	–.000665	–.001100	–.001419	–.001576	–.001553	–.001331	–.000903	–.000347	1.1686	5.6933
	.05	–.000366	–.001037	–.001806	–.002371	–.002743	–.002906	–.002786	–.002211	–.001046	1.6395	59.5265
5.0	1.00	.000000	.000000	.000000	.000000	–.000000	–.000000	–.000000	–.000000	0.000000	.0000	.0000
	.75	–.000016	–.000046	–.000075	–.000095	–.000102	–.000096	–.000078	–.000049	–.000017	.2839	.3862
	.50	–.000068	–.000197	–.000318	–.000402	–.000437	–.000421	–.000351	–.000231	–.000085	.6456	1.4365
	.25	–.000160	–.000463	–.000745	–.000945	–.001047	–.001050	–.000933	–.000667	–.000272	1.0989	6.1464
	.05	–.000259	–.000736	–.001174	–.001490	–.001700	–.001830	–.001843	–.001583	–.000830	1.5321	70.6221

6.0	1.00	.000000	–.000000	.000000	.000000	.000000	–.000000	–.000000	–.000000	–.000000	.0000	.0000
	.75	–.000012	–.000036	–.000057	–.000071	–.000076	–.000072	–.000059	–.000038	–.000014	.2801	.3915
	.50	–.000053	–.000149	–.000235	–.000292	–.000316	–.000308	–.000263	–.000179	–.000069	.6283	1.4802
	.25	–.000123	–.000345	–.000538	–.000666	–.000733	–.000744	–.000684	–.000513	–.000222	1.0569	6.4992
	.05	–.000201	–.000551	–.000841	–.001027	–.001141	–.001227	–.001278	–.001174	–.000675	1.4704	80.4619
8.0	1.00	0.000000	.000000	.000000	0.000000	–.000000	.000000	.000000	.000000	.000000	.0000	–.0000
	.75	–.000009	–.000023	–.000035	–.000042	–.000045	–.000043	–.000037	–.000025	–.000010	.2753	.3987
	.50	–.000036	–.000095	–.000143	–.000170	–.000182	–.000180	–.000162	–.000118	–.000049	.6074	1.5411
	.25	–.000084	–.000219	–.000322	–.000378	–.000406	–.000417	–.000404	–.000330	–.000158	1.0075	7.0293
	.05	–.000140	–.000360	–.000514	–.000536	–.000615	–.000644	–.000689	–.000697	–.000470	1.3902	97.3066
10.0	1.00	.000000	.000000	.000000	–.000000	.000000	.000000	–.000000	–.000000	–.000000	.0000	.0000
	.75	–.000006	–.000017	–.000024	–.000028	–.000029	–.000028	–.000025	–.000018	–.000007	.2724	.4033
	.50	–.000027	–.000068	–.000096	–.000110	–.000116	–.000116	–.000108	–.000084	–.000037	.5948	1.5831
	.25	–.000063	–.000156	–.000217	–.000244	–.000254	–.000261	–.000262	–.000229	–.000120	.9768	7.4207
	.05	–.000105	–.000258	–.000352	–.000384	–.000388	–.000394	–.000420	–.000449	–.000344	1.3343	111.3550
12.0	1.00	–.000000	–.000000	.000000	.000000	.000000	.000000	–.000000	–.000000	–.000000	.0000	–.0000
	.75	–.000005	–.000013	–.000017	–.000019	–.000020	–.000020	–.000018	–.000014	–.000006	.2704	.4067
	.50	–.000021	–.000051	–.000070	–.000077	–.000079	–.000080	–.000077	–.000063	–.000030	.5859	1.6149
	.25	–.000050	–.000118	–.000157	–.000170	–.000173	–.000177	–.000181	–.000167	–.000094	.9544	7.7279
	.05	–.000083	–.000195	–.000255	–.000271	–.000269	–.000268	–.000280	–.000307	–.000260	1.2933	123.3640
14.0	1.00	.000000	.000000	.000000	.000000	–.000000	.000000	–.000000	–.000000	–.000000	.0000	.0000
	.75	–.000004	–.000010	–.000013	–.000014	–.000015	–.000015	–.000014	–.000011	–.000005	.2688	.4094
	.50	–.000017	–.000040	–.000053	–.000056	–.000057	–.000058	–.000057	–.000049	–.000024	.5790	1.6403
	.25	–.000041	–.000093	–.000119	–.000125	–.000125	–.000127	–.000132	–.000127	–.000077	.9370	7.9783
	.05	–.000067	–.000152	–.000193	–.000201	–.000198	–.000195	–.000201	–.000220	–.000202	1.2624	133.8203

continued

Table 12.4 (continued) Bending moment variation and rotations D_2 and D_4 at bottom and top edges, respectively; uniform load of intensity q per unit area; both edges free

η	h_t/h_b	x/l										
		.1	.2	.3	.4	.5	.6	.7	.8	.9	D_2	D_4
16.0	1.00	.000000	–.000000	.000000	–.000000	–.000000	–.000000	.000000	–.000000	–.000000	–.0000	.0000
	.75	–.000004	–.000008	–.000010	–.000011	–.000011	–.000011	–.000011	–.000009	–.000004	.2676	.4115
	.50	–.000015	–.000033	–.000041	–.000043	–.000043	–.000044	–.000044	–.000039	–.000020	.5735	1.6612
	.25	–.000034	–.000075	–.000093	–.000096	–.000095	–.000096	–.000100	–.000099	–.000064	.9232	8.1877
	.05	–.000056	–.000122	–.000150	–.000155	–.000152	–.000150	–.000151	–.000164	–.000161	1.2383	143.0548
18.0	1.00	–.000000	.000000	.000000	.000000	.000000	.000000	–.000000	.000000	–.000000	.0000	–.0000
	.75	–.000003	–.000007	–.000008	–.000009	–.000009	–.000009	–.000009	–.000007	–.000003	.2665	.4133
	.50	–.000013	–.000027	–.000033	–.000034	–.000034	–.000034	–.000035	–.000032	–.000017	.5689	1.6788
	.25	–.000029	–.000062	–.000075	–.000076	–.000075	–.000075	–.000078	–.000079	–.000054	.9119	8.3666
	.05	–.000048	–.000101	–.000120	–.000122	–.000120	–.000119	–.000119	–.000127	–.000130	1.2189	151.3023
20.0	1.00	.000000	–.000000	–.000000	–.000000	.000000	.000000	.000000	.000000	–.000000	–.0000	–.0000
	.75	–.000003	–.00000b	–.000007	–.000007	–.000007	–.000007	–.000007	–.000006	–.000003	.2656	.4149
	.50	–.000011	–.000023	–.000027	–.000027	–.000027	–.000027	–.000023	.000026	–.000015	.5651	1.6939
	.25	–.000025	–.000052	–.000061	–.000062	–.000060	–.000060	–.000063	–.000064	–.000046	.9024	8.5219
	.05	–.000041	–.000084	–.000098	–.000099	–.000097	–.000096	–.000096	–.000101	–.000107	1.2027	158.7353
24.0	1.00	.000000	.000000	.000000	–.000000	–.000000	.000000	–.000000	.000000	–.000000	.0000	.0000
	.75	–.000002	–.000004	–.000005	–.000005	–.000005	–.000005	–.000005	–.000004	–.000002	.2642	.4174
	.50	–.000009	–.000017	–.000019	–.000019	–.000019	–.000019	–.000019	–.000019	–.000012	.5589	1.7186
	.25	–.000020	–.000038	–.000043	–.000043	–.000042	–.000042	–.000043	–.000045	–.000035	.8875	8.7800
	.05	–.000032	–.000061	–.000069	–.000068	–.000067	–.000067	–.000067	.000068	–.000075	1.1773	171.6496

Notes: M = Coefficient × ql^2; D_2 or D_4 = Coefficient × $qr^2/(Eh_bl)$.

Table 12.5 Influence coefficients for hoop force; outward normal line load of unit intensity applied at 1 of 11 nodes spaced at intervals one-tenth the height; both edges free

η	h_t/h_b	x/ℓ	Load position x/l										
			0.0	*.1*	*.2*	*.3*	*.4*	*.5*	*.6*	*.7*	*.8*	*.9*	*1.0*
.4	1.00	0.0	4.0685	3.4313	2.7999	2.1778	1.5664	.9655	.3738	–.2110	–.7911	–1.3689	–1.9459
		0.2	2.7999	2.4417	2.0825	1.7204	1.3576	.9957	.6356	.2773	–.0795	–.4354	–.7911
		0.4	1.5664	1.4623	1.3576	1.2505	1.1384	1.0190	.8940	.7656	.6356	.5048	.3738
		0.6	.3738	.5048	.6356	.7656	.8940	1.0190	1.1384	1.2505	1.3576	1.4623	1.5664
		0.8	–.7911	–.4354	–.0795	.2773	.6356	.9957	1.3576	1.7204	2.0825	2.4417	2.7999
		1.0	–1.9459	–1.3689	–.7911	–.2110	.3738	.9655	1.5664	2.1778	2.7999	3.4313	4.0685
	.50	0.0	4.7178	3.9235	3.1359	2.3603	1.6006	.8587	.1348	–.5727	–1.2674	–1.9542	–2.6379
		0.2	2.8223	2.4602	2.0967	1.7295	1.3602	.9915	.6250	.2615	–.0992	–.4577	–.8154
		0.4	1.2805	1.2451	1.2091	1.1704	1.1250	1.0679	1.0002	.9247	.8439	.7605	.6761
		0.6	.0944	.2903	.4861	.6814	.8752	1.0650	1.2463	1.4136	1.5693	1.7185	1.8654
		0.8	–.7605	–.4135	–.0661	.2823	.6329	.9870	1.3451	1.7067	2.0679	2.4228	2.7743
		1.0	–1.3190	–.8863	–.4530	–.0176	.4225	.8711	1.3324	1.8113	2.3119	2.8354	3.3766
2.0	1.00	0.0	5.3600	4.0063	2.7873	1.7707	.9715	.3709	–.0686	–.3916	–.6423	–.8575	–1.0620
		0.2	2.7873	2.4833	2.1504	1.7480	1.3217	.9147	.5456	.2158	–.0836	–.3658	–.6423
		0.4	.9715	1.1522	1.3217	1.4483	1.4764	1.3535	1.1262	.8465	.5456	.2389	–.0686
		0.6	–.0686	.2389	.5456	.8465	1.1262	1.3535	1.4764	1.4483	1.3217	1.1522	.9715
		0.8	–.6423	–.3658	–.0836	.2158	.5456	.9147	1.3217	1.7480	2.1504	2.4833	2.7873
		1.0	–1.0620	–.8575	–.6423	–.3916	–.0686	.3709	.9715	1.7707	2.7873	4.0063	5.3600
	.50	0.0	6.3671	4.6358	3.0554	1.7232	.6853	–.0550	–.5271	–.7833	–.8884	–.9092	–.9013
		0.2	2.7499	2.4833	2.1826	1.7872	1.3413	.9024	.5067	.1699	–.1112	–.3536	–.5807
		0.4	.5483	.8747	1.1922	1.4686	1.6270	1.5676	1.3352	1.0037	.6274	.2388	–.1502

continued

Table 12.5 (continued) Influence coefficients for hoop force; outward normal line load of unit intensity applied at 1 of 11 nodes spaced at intervals one-tenth the height; both edges free

η	h_t/h_b	x/ℓ	Load position x/l										
			0.0	*.1*	*.2*	*.3*	*.4*	*.5*	*.6*	*.7*	*.8*	*.9*	*1.0*
		0.6	–.3690	.0112	.3941	.7823	1.1683	1.5169	1.7366	1.7019	1.4723	1.1453	.7885
		0.8	–.5330	–.3064	–.0741	.1779	.4706	.8271	1.2620	1.7598	2.2331	2.5393	2.7610
		1.0	–.4506	–.3894	–.3226	–.2343	–.0939	.1481	.5632	1.2458	2.3008	3.8039	5.6928
4.0	1.00	0.0	7.3916	4.7606	2.6042	1.0802	.1446	–.3345	–.5055	–.4953	–.3953	–.2612	–.1193
		0.2	2.6042	2.5211	2.3266	1.8588	1.2905	.7779	.3799	.0964	–.1030	–.2565	–.3953
		0.4	.1446	.7260	1.2905	1.7682	1.9939	1.7941	1.3568	.8597	.3799	–.0698	–.5055
		0.6	–.5055	–.0698	.3799	.8597	1.3568	1.7941	1.9939	1.7682	1.2905	.7260	.1446
		0.8	–.3953	–.2565	–.1030	.0964	.3799	.7779	1.2905	1.8588	2.3266	2.5211	2.6042
		1.0	–.1193	–.2612	–.3953	–.4953	–.5055	–.3345	.1446	1.0802	2.6042	4.7606	7.3916
	.50	0.0	8.4655	5.3177	2.6958	.8306	–.2796	–.7733	–.8430	–.6745	–.4085	–.1244	.1549
		0.2	2.4262	2.4773	2.4013	1.9692	1.3545	.7643	.3068	.0112	–.1469	–.2197	–.2616
		0.4	–.2237	.4921	1.2040	1.8562	2.2401	2.0466	1.4987	.8733	.3150	–.1476	–.5609
		0.6	–.5901	–.1867	.2386	.7280	1.3114	1.9355	2.3329	2.0738	1.3919	.5786	–.2472
		0.8	–.2451	–.1771	–.0979	.0228	.2363	.6085	1.1930	1.9567	2.6050	2.6154	2.3162
		1.0	.0774	–.0331	–.1453	–.2574	–.3506	–.3671	–.1765	.4641	1.9302	4.6431	8.6431
6.0	1.00	0.0	9.0465	5.1713	2.2550	.4845	–.3634	–.6186	–.5658	–.3943	–.2016	–.0193	.1549
		0.2	2.2550	2.5240	2.5617	2.0216	1.2793	.6483	.2240	–.0126	–.1235	–.1718	–.2016
		0.4	–.3634	.4569	1.2793	2.0239	2.3981	2.0513	1.3864	.7422	.2240	–.1909	–.5658
		0.6	–.5658	–.1909	.2240	.7422	1.3864	2.0513	2.3981	2.0239	1.2793	.4569	–.3634

		0.8	–.2016	–.1718	–.1235	–.0126	.2240	.6483	1.2793	2.0216	2.5617	2.5240	2.2550
		1.0	.1549	–.0193	–.2016	–.3943	–.5658	–.6186	–.3634	.4845	2.2550	5.1713	9.0465
	.50	0.0	10.1007	5.6157	2.1907	.1265	–.7775	–.9205	–.6947	–.3804	–.1197	.0606	.2004
		0.2	1.9717	2.4544	2.6830	2.1855	1.3571	.6146	.1258	–.1071	–.1608	–.1251	–.0639
		0.4	–.6220	.2778	1.2063	2.1271	2.6890	2.2861	1.4196	.6183	.0796	–.2272	–.4344
		0.6	–.4863	–.2161	.0978	.5507	1.2422	2.1414	2.7855	2.3304	1.3148	.2791	–.6770
		0.8	–.0718	–.0942	–.1072	–.0787	.0597	.4199	1.1269	2.1781	3.0402	2.6302	1.6341
		1.0	.1002	.0366	–.0355	–.1336	–.2715	–.4281	–.4836	–.1094	1.3618	4.9018	10.9446
10.0	1.00	0.0	11.6614	5.4207	1.4193	–.3637	–.7798	–.6141	–.3312	–.1180	–.0013	.0542	.0894
		0.2	1.4193	2.5122	3.0949	2.3319	1.2067	.3966	–.0081	–.1333	–.1220	–.0655	–.0013
		0.4	–.7798	.1608	1.2067	2.3507	3.0148	2.3148	1.2247	.4223	–.0081	–.2077	–.3312
		0.6	–.3312	–.2077	–.0081	.4223	1.2247	2.3148	3.0148	2.3507	1.2067	.1608	–.7798
		0.8	–.0013	–.0655	–.1220	–.1333	–.0081	.3966	1.2067	2.3319	3.0949	2.5122	1.4193
		1.0	.0894	.0542	–.0013	–.1180	–.3312	–.6141	–.7798	–.3637	1.4193	5.4207	11.6614
	.50	0.0	12.7005	5.7030	1.1685	–.7703	–1.0587	–.6856	–.2639	–.0184	.0583	.0481	.0149
		0.2	1.0516	2.4366	3.2912	2.5265	1.2309	.2999	–.1114	–.1766	–.1066	–.0236	.0458
		0.4	–.8469	.0465	1.0941	2.4122	3.3672	2.5207	1.1254	.1940	–.1590	–.1750	–.0905
		0.6	–.1847	–.1644	–.0866	.2109	.9847	2.3597	3.5261	2.5925	1.0443	–.0326	–.7483
		0.8	.0350	–.0158	–.0711	–.1247	–.1192	.1160	.8951	2.4548	3.8852	2.6451	.3585
		1.0	.0074	.0184	.0255	.0128	–.0565	–.2364	–.5345	–.6929	.2988	4.7112	14.6046
18.0	1.00	0.0	15.6153	5.0474	.0157	–1.0395	–.6753	–.2203	–.0015	.0457	.0304	.0077	–.0117
		0.2	.0157	2.5318	4.1058	2.6367	.8507	.0014	–.1768	–.1161	–.0387	.0046	.0304
		0.4	–.6753	–.1074	.8507	2.5738	3.9594	2.5331	.8128	–.0050	–.1768	–.1098	–.0015
		0.6	–.0015	–.1098	–.1768	–.0050	.8128	2.5331	3.9594	2.5738	.8507	–.1074	–.6753

continued

Table 12.5 (continued) Influence coefficients for hoop force; outward normal line load of unit intensity applied at 1 of 11 nodes spaced at intervals one-tenth the height; both edges free

η	h_t/h_b	x/ℓ	Load position x/l										
			0.0	*.1*	*.2*	*.3*	*.4*	*.5*	*.6*	*.7*	*.8*	*.9*	*1.0*
		0.8	.0304	.0046	–.0387	–.1161	–.1768	.0014	.8507	2.6367	4.1058	2.5318	.0157
		1.0	–.0117	.0077	.0304	.0457	–.0015	–.2203	–.6753	–1.0395	.0157	5.0474	15.6153
	.50	0.0	16.6349	5.1168	–.3616	–1.3072	–.6959	–.1345	.0578	.0517	.0139	–.0027	–.0064
		0.2	–.3254	2.4717	4.3794	2.7722	.7426	–.1234	–.2014	–.0768	–.0014	.0116	.0047
		0.4	–.5568	–.1555	.6601	2.5243	4.4214	2.6505	.5736	–.1833	–.1793	–.0413	.0550
		0.6	.0404	–.0585	–.1567	–.1291	.5019	2.4759	4.7143	2.6549	.4243	–.2540	–.2546
		0.8	.0084	.0081	–.0009	–.0424	–.1345	–.1719	.3637	2.4878	5.2260	2.6063	–1.1049
		1.0	–.0032	–.0017	.0026	.0151	.0344	.0170	–.1819	–.7288	–.9207	3.4794	20.1349
24.0	1.00	0.0	18.0053	4.4972	–.6877	–1.0764	–.4257	–.0372	.0508	.0317	.0077	–.0020	–.0054
		0.2	–.6877	2.5414	4.7409	2.6624	.5574	–.1554	–.1686	–.0590	–.0019	.0096	.0077
		0.4	–.4257	–.1981	.5574	2.5818	4.5584	2.5647	.5374	–.1529	–.1686	–.0523	.0508
		0.6	.0508	–.0523	–.1686	–.1529	.5374	2.5647	4.5584	2.5818	.5574	–.1981	–.4257
		0.8	.0077	.0096	–.0019	–.0590	–.1686	–.1554	.5574	2.6624	4.7409	2.5414	–.6877
		1.0	–.0054	–.0020	.0077	.0317	.0508	–.0372	–.4257	–1.0764	–.6877	4.4972	18.0053
	.50	0.0	19.0148	4.4578	–1.0662	–1.2219	–.3637	.0335	.0627	.0166	–.0027	–.0026	.0007
		0.2	–.9596	2.4791	5.0386	2.7349	.4020	–.2406	–.1444	–.0169	.0126	.0051	–.0031
		0.4	–.2909	–.2136	.3573	2.4697	5.0989	2.5965	.2514	–.2527	–.1060	.0018	.0270
		0.6	.0439	–.0159	–.1123	–.2024	.2200	2.4021	5.4534	2.5401	.0920	–.2644	–.0263
		0.8	–.0016	.0041	.0084	–.0048	–.0795	–.2093	.0788	2.3273	6.0076	2.4626	–1.5018
		1.0	.0003	–.0009	–.0017	.0011	.0169	.0438	–.0188	–.4937	–1.2515	2.4453	23.4702

Note: N = Coefficient × r/l.

Table 12.6 Influence coefficients for bending moment; outward normal line load of unit intensity applied at 1 of 11 nodes spaced at intervals one-tenth the height; both edges free

η	h_t/h_b	x/ℓ	Load position x/l										
			0.0	*.1*	*.2*	*.3*	*.4*	*.5*	*.6*	*.7*	*.8*	*.9*	*1.0*
.4	1.00	0.2	–.1271	–.0379	.0513	.0405	.0299	.0195	.0092	–.0010	–.0111	–.0212	–.0313
		0.4	–.1420	–.0781	–.0142	.0498	.1142	.0788	.0438	.0091	–.0254	–.0598	–.0942
		0.6	–.0942	–.0598	–.0254	.0091	.0438	.0788	.1142	.0498	–.0142	–.0781	–.1420
		0.8	–.0313	–.0212	–.0111	–.0010	.0092	.0195	.0299	.0405	.0513	–.0379	–.1271
	.50	0.2	–.1188	–.0316	.0556	.0429	.0304	.0181	.0061	–.0056	–.0172	–.0287	–.0401
		0.4	–.1234	–.0640	–.0045	.0552	.1152	.0757	.0369	–.0014	–.0392	–.0766	–.1140
		0.6	–.0757	–.0457	–.0156	.0146	.0450	.0758	.1072	.0393	–.0281	–.0950	–.1618
		0.8	–.0231	–.0150	–.0068	.0015	.0098	.0182	.0269	.0358	.0450	–.0455	–.1359
.8	1.00	0.2	–.1245	–.0366	.0514	.0397	.0287	.0181	.0081	–.0015	–.0108	–.0200	–.0292
		0.4	–.1364	–.0752	–.0138	.0482	.1112	.0755	.0411	.0078	–.0248	–.0569	–.0889
		0.6	–.0889	–.0569	–.0248	.0078	.0411	.0755	.1112	.0482	–.0138	–.0752	–.1364
		0.8	–.0292	–.0200	–.0108	–.0015	.0081	.0181	.0287	.0397	.0514	–.0366	–.1245
	.50	0.2	–.1151	–.0298	.0556	.0415	.0282	.0158	.0042	–.0064	–.0164	–.0260	–.0354
		0.4	–.1159	–.0601	–.0041	.0525	.1104	.0702	.0323	–.0035	–.0374	–.0702	–.1025
		0.6	–.0689	–.0420	–.0150	.0124	.0407	.0705	.1024	.0367	–.0267	–.0886	–.1498
		0.8	–.0205	–.0135	–.0065	.0007	.0082	.0161	.0248	.0345	.0454	–.0428	–.1305
1.2	1.00	0.2	–.1206	–.0347	.0515	.0385	.0267	.0161	.0065	–.0022	–.0104	–.0182	–.0260
		0.4	–.1280	–.0708	–.0131	.0458	.1069	.0706	.0370	.0058	–.0239	–.0527	–.0812
		0.6	–.0812	–.0527	–.0239	.0058	.0370	.0706	.1069	.0458	–.0131	–.0708	–.1280
		0.8	–.0260	–.0182	–.0104	–.0022	.0065	.0161	.0267	.0385	.0515	–.0347	–.1206

continued

Table 12.6 (continued) Influence coefficients for bending moment; outward normal line load of unit intensity applied at 1 of 11 nodes spaced at intervals one-tenth the height; both edges free

η	h_t/h_b	x/ℓ	*Load position x/l*										
			0.0	*.1*	*.2*	*.3*	*.4*	*.5*	*.6*	*.7*	*.8*	*.9*	*1.0*
	.50	0.2	–.1101	–.0274	.0556	.0396	.0252	.0125	.0017	–.0074	–.0152	–.0222	–.0289
		0.4	–.1053	–.0546	–.0036	.0488	.1038	.0626	.0260	–.0063	–.0350	–.0614	–.0869
		0.6	–.0595	–.0370	–.0142	.0094	.0348	.0631	.0957	.0331	–.0248	–.0797	–.1332
		0.8	–.0169	–.0115	–.0060	–.0003	.0060	.0132	.0220	.0327	.0458	–.0390	–.1229
2.0	1.00	0.2	–.1107	–.0299	.0517	.0354	.0218	.0110	.0027	–.0038	–.0092	–.0139	–.0185
		0.4	–.1072	–.0598	–.0114	.0398	.0959	.0584	.0271	.0011	–.0215	–.0422	–.0623
		0.6	–.0623	–.0422	–.0215	.0011	.0271	.0584	.0959	.0398	–.0114	–.0598	–.1072
		0.8	–.0185	–.0139	–.0092	–.0038	.0027	.0110	.0218	.0354	.0517	–.0299	–.1107
	.50	0.2	–.0986	–.0221	.0553	.0351	.0184	.0056	–.0034	–.0091	–.0124	–.0144	–.0159
		0.4	–.0824	–.0429	–.0027	.0405	.0893	.0464	.0129	–.0117	–.0294	–.0428	–.0547
		0.6	–.0399	–.0263	–.0123	.0033	.0224	.0475	.0813	.0254	–.0210	–.0607	–.0978
		0.8	–.0097	–.0074	–.0050	–.0022	.0015	.0072	.0157	.0284	.0463	–.0310	–.1059
4.0	1.00	0.2	–.0868	–.0192	.0509	.0271	.0106	.0006	–.0043	–.0061	–.0061	–.0053	–.0043
		0.4	–.0613	–.0360	–.0085	.0257	.0716	.0322	.0069	–.0077	–.0156	–.0202	–.0239
		0.6	–.0239	–.0202	–.0156	–.0077	.0069	.0322	.0716	.0257	–.0085	–.0360	–.0613
		0.8	–.0043	–.0053	–.0061	–.0061	–.0043	.0006	.0106	.0271	.0509	–.0192	–.0868
	.50	0.2	–.0757	–.0128	.0528	.0249	.0058	–.0049	–.0090	–.0089	–.0065	–.0032	.0002
		0.4	–.0426	–.0234	–.0025	.0246	.0638	.0202	–.0056	–.0167	–.0182	–.0147	–.0097
		0.6	–.0108	–.0099	–.0085	–.0049	.0038	.0226	.0572	.0117	–.0152	–.0300	–.0403
		0.8	–.0004	–.0016	–.0027	–.0036	–.0037	–.0016	.0048	.0190	.0446	–.0176	–.0734

6.0	1.00	0.2	–.0701	–.0130	.0485	.0204	.0036	–.0042	–.0063	–.0056	–.0036	–.0014	.0008
		0.4	–.0350	–.0228	–.0077	.0166	.0572	.0183	–.0024	–.0104	–.0114	–.0094	–.0066
		0.6	–.0066	–.0094	–.0114	–.0104	–.0024	.0183	.0572	.0166	–.0077	–.0228	–.0350
		0.8	.0008	–.0014	–.0036	–.0056	–.0063	.0042	.0036	.0204	.0485	–.0130	–.0701
	.50	0.2	–.0608	–.0082	.0490	.0173	–.0009	–.0081	–.0085	–.0059	–.0028	–.0001	.0023
		0.4	–.0231	–.0145	–.0036	.0153	.0507	.0091	–.0104	–.0147	–.0112	–.0049	.0016
		0.6	–.0016	–.0039	–.0059	–.0065	–.0024	.0122	.0458	.0049	–.0130	–.0171	–.0164
		0.8	.0011	–.0000	–.0012	–.0026	–.0039	–.0040	–.0003	.0123	.0408	–.0116	–.0537
10.0	1.00	0.2	–.0487	–.0074	.0426	.0109	–.0031	–.0063	–.0049	–.0027	–.0009	.0003	.0013
		0.4	–.0119	–.0114	–.0077	.0069	.0430	.0068	–.0072	–.0089	–.0060	–.0021	.0018
		0.6	.0018	–.0021	–.0060	–.0089	–.0072	.0068	.0430	.0069	–.0077	–.0114	–.0119
		0.8	.0013	.0003	–.0009	–.0027	–.0049	–.0063	–.0031	.0109	.0426	–.0074	–.0487
	.50	0.2	–.0418	–.0043	.0417	.0076	–.0063	–.0078	–.0047	–.0016	.0001	.0005	.0006
		0.4	–.0069	–.0072	–.0052	.0059	.0382	.0010	–.0104	–.0088	–.0039	–.0002	.0025
		0.6	.0014	–.0006	–.0028	–.0050	–.0049	.0043	.0353	–.0009	–.0105	–.0073	–.0012
		0.8	.0004	.0002	–.0000	–.0006	–.0019	–.0037	–.0033	.0044	.0335	–.0073	–.0311
14.0	1.00	0.2	–.0355	–.0052	.0373	.0052	–.0053	–.0054	–.0029	–.0009	.0001	.0003	.0004
		0.4	–.0033	–.0067	–.0075	.0022	.0360	.0023	–.0073	–.0063	–.0030	–.0003	.0019
		0.6	.0019	–.0003	–.0030	–.0063	–.0073	.0023	.0360	.0022	–.0075	–.0067	–.0033
		0.8	.0004	.0003	.0001	–.0009	–.0029	–.0054	–.0053	.0052	.0373	–.0052	–.0355
	.50	0.2	–.0301	–.0031	.0359	.0022	–.0074	–.0056	–.0020	–.0000	.0004	.0002	–.0001
		0.4	–.0014	–.0042	–.0054	.0018	.0320	–.0019	–.0086	–.0050	–.0012	.0004	.0009
		0.6	.0010	.0001	–.0012	–.0033	–.0048	.0009	.0299	–.0032	–.0082	–.0036	.0017
		0.8	.0000	.0001	.0002	.0000	–.0007	–.0024	–.0041	.0006	.0284	–.0060	–.0188

continued

Table 12.6 (continued) Influence coefficients for bending moment; outward normal line load of unit intensity applied at 1 of 11 nodes spaced at intervals one-tenth the height; both edges free

η	h_t/h_b	x/ℓ	Load position x/l										
			0.0	*.1*	*.2*	*.3*	*.4*	*.5*	*.6*	*.7*	*.8*	*.9*	*1.0*
18.0	1.00	0.2	–.0264	–.0043	.0331	.0017	–.0058	–.0041	–.0014	–.0001	.0003	.0002	.0000
		0.4	–.0001	–.0043	–.0069	–.0002	.0317	–.0000	–.0067	–.0043	–.0014	.0002	.0011
		0.6	.0011	.0002	–.0014	–.0043	–.0067	–.0000	.0317	–.0002	–.0069	–.0043	–.0001
		0.8	.0000	.0002	.0003	–.0001	–.0014	–.0041	–.0058	.0017	.0331	–.0043	–.0264
	.50	0.2	–.0221	–.0027	.0316	–.0008	–.0069	–.0037	–.0007	.0003	.0003	.0000	–.0001
		0.4	.0005	–.0026	–.0049	–.0003	.0282	–.0033	–.0069	–.0028	–.0001	.0004	.0002
		0.6	.0004	.0002	–.0004	–.0020	–.0042	–.0008	.0264	–.0041	–.0063	–.0018	.0018
		0.8	–.0000	.0000	.0001	.0001	–.0002	–.0014	–.0035	–.0012	.0248	–.0055	–.0116
24.0	1.00	0.2	–.0174	–.0039	.0284	–.0011	–.0052	–.0024	–.0004	.0002	.0002	.0000	–.0001
		0.4	.0013	–.0023	–.0057	–.0021	.0274	–.0019	–.0054	–.0024	–.0003	.0003	.0004
		0.6	.0004	.0003	–.0003	–.0024	–.0054	–.0019	.0274	–.0021	–.0057	–.0023	.0013
		0.8	–.0001	.0000	.0002	.0002	–.0004	–.0024	–.0052	–.0011	.0284	–.0039	–.0174
	.50	0.2	–.0143	–.0028	.0270	–.0030	–.0055	–.0018	.0001	.0003	.0001	–.0000	–.0000
		0.4	.0011	–.0013	–.0039	–.0019	.0244	–.0042	–.0049	–.0011	.0003	.0002	–.0001
		0.6	.0001	.0001	.0000	–.0009	–.0032	–.0021	.0228	–.0046	–.0042	–.0005	.0011
		0.8	–.0000	–.0000	.0000	.0001	.0001	–.0006	–.0026	–.0024	.0212	–.0052	–.0055

Table 12.7 Influence coefficients of rotations D_2 and D_4 at bottom and top edges; outward normal line load of unit intensity applied at 1 of 11 nodes spaced at intervals one-tenth the height; both edges free

		Coefficients for D_2 Load position x/l										
η	h_t/h_b	*0.0*	*.1*	*.2*	*.3*	*.4*	*.5*	*.6*	*.7*	*.8*	*.9*	*1.0*
.4	1.00	−6.3833	−4.9458	−3.5805	−2.2815	−1.0404	.1532	1.3101	2.4412	3.5566	4.6642	5.7695
	.50	−7.9544	−5.9465	−4.0222	−2.1869	−.4414	1.2178	2.7989	4.3150	5.7832	7.2230	8.6527
.8	1.00	−7.4954	−5.3662	−3.5242	−1.9437	−.5882	.5846	1.6189	2.5573	3.4373	4.2892	5.1328
	.50	−9.5540	−6.5636	−3.9054	−1.5951	.3702	2.0117	3.3671	4.4896	5.4443	6.3018	7.1250
1.2	1.00	−9.2077	−5.9928	−3.4194	−1.4209	.0926	1.2214	2.0653	2.7166	3.2534	3.7362	4.2033
	.50	−11.8863	−7.4200	−3.6931	−.7241	1.5153	3.0926	4.1074	4.6820	4.9520	5.0524	5.0949
1.6	1.00	−11.3603	−6.7425	−3.2544	−.7594	.9198	1.9713	2.5735	2.8820	3.0221	3.0853	3.1270
	.50	−14.6410	−8.3573	−3.3718	.3168	2.8010	4.2401	4.8354	4.8079	4.3759	3.7331	3.0231
2.0	1.00	−13.8008	−7.5369	−3.0197	−.0042	1.8153	2.7486	3.0740	3.0204	2.7604	2.4109	2.0377
	.50	−17.5893	−9.2621	−2.9367	1.4430	4.0866	5.2991	5.4278	4.8170	3.7702	2.5220	1.2180
3.0	1.00	−20.4657	−9.3745	−2.1038	2.0759	4.0032	4.4450	4.0101	3.1264	2.0568	.9339	−.1958
	.50	−25.1109	−11.0715	−1.3912	4.3360	6.8861	7.1740	6.0743	4.2978	2.3222	.3753	−1.5197
4.0	1.00	−27.3104	−10.7294	−.7519	4.2000	5.8210	5.5323	4.3440	2.8634	1.3773	−.0414	−1.4233
	.50	−32.5251	−12.1778	.6563	7.1111	8.9440	7.9779	5.7390	3.2755	1.1200	−.6425	−2.2079
5.0	1.00	−34.1244	−11.5731	.9350	6.2379	7.2007	6.0427	4.1829	2.3487	.7786	−.5678	−1.8260
	.50	−39.8422	−12.7151	3.0130	9.6513	10.3459	8.0307	4.8746	2.1266	.2209	−.9821	−1.8859
6.0	1.00	−40.9080	−11.9848	2.8470	8.1301	8.1888	6.1232	3.7152	1.7269	.2847	−.7901	−1.7340
	.50	−47.1050	−12.8080	5.5322	11.8986	11.2051	7.6083	3.8144	1.0856	−.3829	−.9838	−1.2576
8.0	1.00	−54.4528	−11.8065	7.0038	11.3512	9.2159	5.4997	2.4272	.5481	−.3707	−.7729	−1.0160
	.50	−61.5190	−11.9737	10.6945	15.4355	11.6713	6.0307	1.7737	−.3544	−.8886	−.6516	−.2129

continued

Table 12.7 (continued) Influence coefficients of rotations D_2 and D_4 at bottom and top edges; outward normal line load of unit intensity applied at 1 of 11 nodes spaced at intervals one-tenth the height; both edges free

		Coefficients for D_2 Load position x/l										
η	h_t/h_b	*0.0*	*.1*	*.2*	*.3*	*.4*	*.5*	*.6*	*.7*	*.8*	*.9*	*1.0*
10.0	1.00	−67.9905	−10.6187	11.2352	13.7426	9.2896	4.3706	1.1762	−.2746	−.6441	−.5481	−.3364
	.50	−75.8100	−10.1115	15.6858	17.7544	10.9914	4.1436	.2378	−.9936	−.8375	−.3057	.2232
12.0	1.00	−81.5103	−8.6758	15.2915	15.3365	8.7197	3.1090	.1958	−.7138	−.6633	−.3207	.0605
	.50	−90.0010	−7.4922	20.2998	19.0357	9.6573	2.3716	−.7340	−1.1168	−.5850	−.0789	.2768
14.0	1.00	−95.0054	−6.1524	19.0472	16.2352	7.7534	1.9101	−.4807	−.8664	−.5510	−.1485	.2188
	.50	−104.1094	−4.3040	24.4430	19.4829	8.0127	.8958	−1.2441	−.9787	−.3220	.0372	.1925
16.0	1.00	−108.4746	−3.1783	22.4413	16.5587	6.5772	.8698	−.8919	−.8412	−.3940	−.0361	.2359
	.50	−118.1468	−.6850	28.0803	19.2822	6.2829	−.2343	−1.4240	−.7425	−.1216	.0793	.0952
18.0	1.00	−121.9180	.1464	25.4485	16.4216	5.3240	.0253	−1.0960	−.7229	−.2424	.0280	.1896
	.50	−132.1213	3.2606	31.2088	18.5932	4.6079	−1.0355	−1.3911	−.4982	.0049	.0808	.0267
20.0	1.00	−135.3359	3.7427	28.0644	15.9257	4.0835	−.6207	−1.1504	−.5682	−.1188	.0580	.1268
	.50	−146.0389	7.4512	33.8443	17.5485	3.0697	−1.5533	−1.2362	−.2878	.0707	.0644	−.0096
24.0	1.00	−162.0952	11.5087	32.1648	14.1922	1.8479	−1.3898	−.9956	−.2707	.0285	.0618	.0282
	.50	−173.7199	16.3221	37.7456	14.8035	.5490	−1.9562	−.7972	−.0122	.0936	.0250	−.0235
.4	1.00	−5.7695	−4.6642	−3.5566	−2.4412	−1.3101	−.1532	1.0404	2.2815	3.5805	4.9458	6.3833
	.50	−6.8349	−5.4059	−5.9737	−2.5292	−1.0556	.4734	2.0960	3.8654	5.8537	8.1568	10.9000
.8	1.00	−5.1328	−4.2892	−3.4373	−2.5573	−1.6189	−.5846	.5882	1.9437	3.5242	5.3662	7.4954
	.50	−5.3871	−4.5662	−3.7348	−2.8627	−1.8924	−.7299	.7665	2.8022	5.6657	9.7513	15.5808

1.2	1.00	−4.2033	−3.7362	−3.2534	−2.7166	−2.0653	−1.2214	−.0926	1.4209	3.4194	5.9928	9.2077
	.50	−3.4759	−3.4371	−3.3821	−3.2616	−2.9744	−2.3446	−1.0879	1.2328	5.2646	11.9278	22.4607
1.6	1.00	−3.1270	−3.0853	−3.0221	−2.8820	−2.5735	−1.9713	−.9198	.7594	3.2544	6.7425	11.3603
	.50	−1.5472	−2.2616	−2.9606	−3.5903	−4.0245	−4.0175	−3.1334	−.6479	4.5798	14.2457	30.6792
2.0	1.00	−2.0377	−2.4109	−2.7604	−3.0204	−3.0740	−2.7486	−1.8153	.0042	3.0197	7.5369	13.8008
	.50	.1021	−1.2055	−2.5074	−3.7658	−4.8600	−5.5050	−5.1276	−2.6851	3.5770	16.3909	39.5964
5.0	1.00	.1958	−.9339	−2.0568	−3.1264	−4.0101	−4.4450	−4.0032	−2.0759	2.1038	9.3745	20.4657
	.50	2.4451	.5532	−1.3870	−3.4585	−5.6861	−7.8557	−9.1998	−7.8823	−.2330	20.1695	62.9472
4.0	1.00	1.4233	.0414	−1.3773	−2.8634	−4.3440	−5.5323	−5.8210	−4.2000	.7519	10.7294	27.3104
	.50	2.7818	1.2283	−.4359	−2.4543	−5.1036	−8.4406	−11.7808	−12.7064	−5.3973	21.6076	86.7442
5.0	1.00	1.8260	.5678	−.7786	−2.3487	−4.1829	−6.0427	−7.2007	−6.2379	−.9350	11.5731	34.1244
	.50	2.1818	1.2958	.2628	−1.2806	−3.8602	−7.9135	−13.1154	−16.8465	−11.2525	21.1816	110.8207
6.0	1.00	1.7340	.7901	−.2847	−1.7269	−3.7152	−6.1232	−8.1888	−8.1301	−2.8470	11.9848	40.9080
	.50	1.3560	1.1008	.6997	−.2559	−2.4742	−6.7994	−13.4881	−20.1644	−17.3095	19.3248	135.1479
8.0	1.00	1.0160	.7729	.3707	−.5481	−2.4272	−5.4997	−9.2159	−11.3512	−7.0038	11.8065	54.4528
	.50	.1185	.5668	.9514	.9714	−.1795	−4.0089	−12.2739	−24.2997	−28.9086	12.4562	184.3237
10.0	1.00	.3364	.5481	.6441	.2746	−1.1762	−4.3706	−9.2896	−13.7426	−11.2352	10.6187	67.9905
	.50	−.3455	.1800	.7465	1.2749	1.1427	−1.4341	−9.6410	−25.5827	−38.9436	2.6269	233.9076
12.0	1.00	−.0605	.3207	.6633	.7138	−.1958	−3.1090	−8.7197	−15.3365	−15.2915	8.6758	81.5103
	.50	.3648	−.0204	.4278	1.0933	1.6664	.4887	−6.5825	−24.7976	−47.0434	−9.1394	283.7155
14.0	1.00	−.2188	.1485	.5510	.8664	.4807	−1.9101	−7.7534	−16.2352	−19.0472	6.1524	95.0054
	.50	−.2380	−.0956	.1612	.7533	1.6867	1.7124	−3.6701	−22.6580	−53.1898	−22.1383	333.6487

continued

Table 12.7 (continued) Influence coefficients of rotations D_2 and D_4 at bottom and top edges; outward normal line load of unit intensity applied at 1 of 11 nodes spaced at intervals one-tenth the height; both edges free

		Coefficients for D_2 Load position x/l										
η	h_t/h_b	*0.0*	*.1*	*.2*	*.3*	*.4*	*.5*	*.6*	*.7*	*.8*	*.9*	*1.0*
16.0	1.00	−.2359	.0361	.3940	.8412	.8919	−.8698	−6.5772	−16.5587	−22.4413	3.1783	108.4746
	.50	−.1097	−.1039	−.0083	.4237	1.4438	2.3455	−1.1820	−19.7216	−57.5167	−35.8676	383.6428
18.0	1.00	−.1896	−.0280	.2424	.7229	1.0960	−.0253	−5.3240	−16.4216	−25.4485	−.1464	121.9180
	.50	−.0241	−.0837	−.0914	.1690	1.0996	2.5405	.7836	−16.3982	−60.2216	−49.9613	433.6525
20.0	1.00	−.1268	−.0580	.1188	.5682	1.1504	.6207	−4.0835	−15.9257	−28.0644	−3.7427	135.3359
	.50	.0191	−.0564	−.1158	−.0002	.7499	2.4431	2.2290	−12.9769	−61.5216	−64.1481	485.6446
24.0	1.00	−.0282	−.0618	−.0285	.2707	.9956	1.3898	−1.8479	−14.1922	−32.1648	−11.5087	162.0952
	.50	.0319	−.0136	−.0840	−.1396	.2022	1.8152	3.7907	−6.5544	−60.7526	−92.0400	583.4818

Note: D_2 or D_4 = Coefficient × r^2 $(Eh_b/l2)$.

Table 12.8 Hoop force variation; outward normal line load of intensity $F_1 = 1$ (force/unit length) at bottom edge; both edges free

η	h_t/h_b	*x/l*										
		0.0	*.1*	*.2*	*.3*	*.4*	*.5*	*.6*	*.7*	*.8*	*.9*	*1.0*
.4	1.00	4.0685	3.4313	2.7999	2.1778	1.5664	.9655	.3738	–.2110	–.7911	–1.3689	–1.9459
	.75	4.3572	3.5631	2.8101	2.1017	1.4393	.8224	.2491	–.2832	–.7778	–1.2378	–1.6656
	.50	4.7178	3.7273	2.8223	2.0063	1.2805	.6440	.0944	–.3723	–.7605	–1.0748	–1.3190
	.25	5.2160	3.9535	2.8383	1.8739	1.0611	.3984	–.1179	–.4933	–.7350	–.8505	–.8468
	.05	5.8662	4.2480	2.8585	1.7008	.7751	.0787	–.3937	–.6499	–.7006	–.5583	–.2370
.8	1.00	4.2690	3.5238	2.8016	2.1159	1.4718	.8681	.2991	–.2433	–.7685	–1.2847	–1.7982
	.75	4.5966	3.6706	2.8095	2.0270	1.3277	.7095	.1647	–.3173	–.7487	–1.1410	–1.5028
	.50	5.0059	3.8523	2.8175	1.9147	1.1480	.5133	.0003	–.4056	–.7214	–.9639	–1.1460
	.25	5.5565	4.0941	2.8249	1.7617	.9066	.2527	–.2148	–.5172	–.6798	–.7295	–.6897
	.05	6.2088	4.3775	2.8303	1.5783	.6211	–.0526	–.4637	–.6420	–.6248	–.4547	–.1772
1.2	1.00	4.5732	3.6621	2.8019	2.0211	1.3297	.7235	.1897	–.2897	–.7338	–1.1600	–1.5808
	.75	4.9508	3.8267	2.8053	1.9152	1.1647	.5471	.0454	–.3637	–.7052	–1.0021	–1.2714
	.50	5.4166	4.0263	2.8056	1.7822	.9620	.3339	–.1254	–.4470	–.6653	–.8124	–.9132
	.25	6.0173	4.2780	2.7993	1.6066	.7016	.0662	–.3330	–.5401	–.6047	–.5762	–.4956
	.05	6.6594	4.5402	2.7843	1.4137	.4241	–.2118	–.5402	–.6219	–.5276	–.3326	–.1101
1.6	1.00	4.9475	3.8285	2.7979	1.9030	1.1574	.5516	.0620	–.3417	–.6906	–1.0122	–1.3255
	.75	5.3730	4.0075	2.7944	1.7797	.9739	.3620	–.0870	–.4119	–.6526	–.8444	–1.0123
	.50	5.8858	4.2176	2.7837	1.6276	.7544	.1409	–.2548	–.4842	–.6003	–.6508	–.6701
	.25	6.5165	4.4669	2.7596	1.4339	.4864	–.1189	–.4409	–.5509	–.5233	–.4269	–.3142
	.05	7.1373	4.7013	2.7224	1.2345	.2240	–.3606	–.5989	–.5873	–.4289	–.2232	–.0549

continued

Table 12.8 (continued) Hoop force variation; outward normal line load of intensity F_l = 1 (force/unit length) at bottom edge; both edges free

η	h_t/h_b	x/l										
		0.0	*.1*	*.2*	*.3*	*.4*	*.5*	*.6*	*.7*	*.8*	*.9*	*1.0*
2.0	1.00	5.3600	4.0063	2.7873	1.7707	.9715	.3709	–.0686	–.3916	–.6423	–.8575	–1.0620
	.75	5.8236	4.1931	2.7743	1.6321	.7754	.1761	–.2146	–.4536	–.5957	–.6872	–.7591
	.50	6.3671	4.4041	2.7499	1.4647	.5483	–.0412	–.3690	–.5092	–.5330	–.5000	–.4506
	.25	7.0076	4.6400	2.7060	1.2587	.2839	–.2800	–.5227	–.5451	–.4439	–.3001	–.1695
	.05	7.6035	4.8449	2.6473	1.0554	.0401	–.4828	–.6318	–.5405	–.3388	–.1381	–.0175
3.0	1.00	6.4165	4.4277	2.7228	1.4199	.5195	–.0397	–.3436	–.4779	–.5152	–.5087	–.4899
	.75	6.9266	4.6055	2.6791	1.2560	.3197	–.2138	–.4533	–.5044	–.4535	–.3615	–.2628
	.50	7.4916	4.7890	2.6152	1.0667	.1039	–.3870	–.5464	–.5073	–.3752	–.2203	–.0786
	.25	8.1154	4.9724	2.5225	.8455	–.1275	–.5506	–.6076	–.4708	–.2731	–.0966	.0249
	.05	8.6626	5.1153	2.4195	.6388	–.3252	–.6679	–.6199	–.3976	–.1663	–.0208	.0168
4.0	1.00	7.3916	4.7606	2.6042	1.0802	.1446	–.3345	–.5055	–.4953	–.3953	–.2612	–.1193
	.75	7.9084	4.9087	2.5274	.9042	–.0366	–.4645	–.5622	–.4816	–.3269	–.1558	.0088
	.50	8.4655	5.0519	2.4262	.7060	–.2237	–.5800	–.5901	–.4384	–.2451	–.0684	.0774
	.25	9.0677	5.1862	2.2940	.4811	–.4158	–.6740	–.5787	–.3567	–.1478	–.0072	.0696
	.05	9.5886	5.2857	2.1612	.2785	–.5726	–.7290	–.5348	–.2600	–.0616	.0169	.0142
5.0	1.00	8.2616	5.0020	2.4434	.7676	–.1448	–.5179	–.5685	–.4595	–.2901	–.1074	.0754
	.75	8.7754	5.1179	2.3372	.5865	–.3014	–.6027	–.5774	–.4143	–.2215	–.0417	.1212
	.50	9.3240	5.2247	2.2060	.3866	–.4580	–.6679	–.5561	–.3429	–.1448	.0038	.1141
	.25	9.9128	5.3203	2.0456	.1659	–.6126	–.7083	–.4999	–.2443	–.0634	.0252	.0584
	.05	10.4178	5.3877	1.8938	–.0260	–.7315	–.7180	–.4280	–.1512	–.0054	.0227	.0062

Note: N = Coefficient × r/l.

Table 12.9 Bending moment variation and rotations D_2 and D_4 at bottom and top edges, respectively; outward normal line load of intensity F_1 = 1 (force/unit length) at bottom edge; both edges free

η	h_t/h_b	x/l									D_2	D_4
		.1	.2	.3	.4	.5	.6	.7	.8	.9		
.4	1.00	–.080691	–.127064	–.145431	–.142012	–.122920	–.094166	–.061668	–.031277	–.008794	–6.3833	–5.7695
	.75	–.079516	–.123369	–.139084	–.133745	–.113977	–.085950	–.055398	–.027648	–.007648	–7.0383	–6.2348
	.50	–.078049	–.118756	–.131168	–.123444	–.102842	–.075731	–.047608	–.023145	–.006227	–7.9544	–6.8349
	.25	–.076024	–.112394	–.120259	–.109261	–.087530	–.061697	–.036928	–.016983	–.004287	–9.4315	–7.6918
	.05	–.073383	–.104100	–.106044	–.090792	–.067604	–.043447	–.023051	–.008989	–.001775	–11.7347	–8.8169
.8	1.00	–.079868	–.124478	–.141043	–.136414	–.117035	–.088947	–.057845	–.029163	–.008159	–7.4954	–5.1328
	.75	–.078539	–.120320	–.133942	–.127226	–.107167	–.079954	–.051042	–.025263	–.006939	–8.3638	–5.3107
	.50	–.076884	–.115147	–.125129	–.115852	–.094989	–.068893	–.042708	–.020505	–.005458	–9.5540	–5.3871
	.25	–.074661	–.108225	–.113372	–.100734	–.078866	–.054320	–.031787	–.014313	–.003545	–11.3517	–5.1085
	.05	–.072034	–.100063	–.099549	–.083013	–.060036	–.037371	–.019155	–.007201	–.001370	–13.7652	–3.8065
1.2	1.00	–.078621	–.120579	–.134453	–.128043	–.108266	–.081196	–.052184	–.026042	–.007223	–9.2077	–4.2033
	.75	–.077100	–.115849	–.126437	–.117760	–.097327	–.071330	–.044803	–.021858	–.005929	–10.3596	–4.0135
	.50	–.075229	–.110057	–.116668	–.105292	–.084139	–.059510	–.036024	–.016923	–.004418	–11.8863	–3.4759
	.25	–.072828	–.102665	–.104276	–.089588	–.067664	–.044887	–.025285	–.010968	–.002624	–14.0294	–2.0401
	.05	–.070275	–.094852	–.091274	–.073249	–.050696	–.030011	–.014528	–.005121	–.000908	–16.5319	1.3110
1.6	1.00	–.077095	–.115832	–.126478	–.117972	–.097777	–.071973	–.045479	–.022358	–.006121	–11.3603	–3.1270
	.75	–.075393	–.110586	–.117673	–.106792	–.086015	–.061486	–.037726	–.018017	–.004796	–12.8023	–2.5904
	.50	–.073353	–.104338	–.107260	–.093675	–.072336	–.049411	–.028899	–.013138	–.003327	–14.6410	–1.5472
	.25	–.070862	–.096775	–.094781	–.078141	–.056359	–.035536	–.018950	–.007762	–.001753	–17.0577	.6330
	.05	–.068431	–.089477	–.082899	–.063588	–.041690	–.023117	–.010330	–.003298	–.000517	–19.6114	4.7725

continued

Table 12.9 (continued) Bending moment variation and rotations D_2 and D_4 at bottom and top edges, respectively; outward normal line load of intensity F_1 = 1 (force/unit length) at bottom edge; both edges free

η	h_t/h_b	x/l										
		.1	.2	.3	.4	.5	.6	.7	.8	.9	D_2	D_4
2.0	1.00	–.075424	–.110673	–.117883	–.107208	–.086657	–.062266	–.038467	–.018526	–.004980	–13.8008	–2.0377
	.75	–.073586	–.105067	–.108578	–.095535	–.074529	–.051592	–.030678	–.014222	–.003683	–15.5003	–1.2417
	.50	–.071447	–.098600	–.097953	–.082358	–.061014	–.039871	–.022265	–.009657	–.002335	–17.5893	.1021
	.25	–.068951	–.091148	–.085888	–.067656	–.046251	–.027386	–.013570	–.005105	–.001046	–20.1968	2.5124
	.05	–.066658	–.084408	–.075192	–.054952	–.033911	–.017395	–.007002	–.001926	–.000239	–22.7868	6.2432
3.0	1.00	–.071207	–.097899	–.097032	–.081638	–.060771	–.040094	–.022716	–.010040	–.002481	–20.4657	.1958
	.75	–.069240	–.092097	–.087750	–.070446	–.049617	–.030689	–.016141	–.006556	–.001473	–22.6210	1.1700
	.50	–.067089	–.085852	–.077950	–.058903	–.038435	–.021567	–.010005	–.003441	–.000613	–25.1109	2.4451
	.25	–.064752	–.079209	–.067808	–.047379	–.027779	–.013388	–.004917	–.001108	–.000048	–28.0155	3.9626
	.05	–.062737	–.073621	–.059562	–.038450	–.020085	–.008083	–.002141	–.000171	.000063	–30.7308	4.5094
4.0	1.00	–.067420	–.086835	–.079687	–.061257	–.041011	–.023860	–.011619	–.004260	–.000827	–27.3104	1.4233
	.75	–.065493	–.081383	–.071376	–.051770	–.032115	–.016839	–.007041	–.002001	–.000219	–29.7618	2.1296
	.50	–.063450	–.075726	–.062997	–.042564	–.023904	.010764	–.003397	–.000384	.000163	–32.5251	2.7818
	.25	–.061281	–.069882	–.054657	–.033875	–.016734	.006049	–.001052	.000365	.000244	–35.6807	2.9356
	.05	–.059439	–.065048	–.048027	–.027380	–.011895	–.003420	–.000232	.000290	.000094	–38.5688	1.1599
5.0	1.00	–.064148	–.077684	–.066059	–.046135	–.027223	–.013231	–.004797	–.000912	.000081	–34.1244	1.8260
	.75	–.062298	–.072682	–.058846	–.038442	–.020578	–.008480	–.002047	.000264	.000347	–36.8264	2.1437
	.50	–.060357	–.067566	–.051729	–.031231	–.014801	–.004793	–.000268	.000812	.000403	–39.8422	2.1818
	.25	–.058313	–.062327	–.044740	–.024597	–.010028	–.002302	.000458	.000710	.000251	–43.2459	1.4547
	.05	–.056590	–.058028	–.039238	–.019718	–.006934	–.001112	.000412	.000312	.000060	–46.3085	–.7449
6.0	1.00	–.061292	–.070058	–.055329	–.035004	–.017836	–.006608	–.000947	.000786	.000493	–40.9080	1.7340
	.75	–.059513	–.065459	–.049074	–.028823	–.013015	–.003622	.000442	.001188	.000524	–43.8478	1.7190

	.50	–.057653	–.060782	–.042961	–.023141	–.009010	–.001569	.001032	.001105	.000401	–47.1050	1.3560
	.25	–.055705	–.056018	–.037003	–.017992	–.005839	–.000409	.000924	.000648	.000178	–50.7404	.3263
	.05	–.054075	–.052140	–.032359	–.014267	–.003864	.000006	.000551	.000227	.000027	–53.9665	–1.2316
8.0	1.00	–.056459	–.057982	–.039712	–.020427	–.007083	–.000248	.001944	.001656	.000593	–54.4528	1.0160
	.75	–.054790	–.054018	–.034906	–.016392	–.004656	.000627	.001863	.001310	.000420	–57.8302	.6723
	.50	–.053059	–.050020	–.030266	–.012782	–.002800	.000969	.001427	.000806	.000215	–61.5190	.1185
	.25	–.051262	–.045993	–.025817	–.009609	–.001472	.000913	.000830	.000317	.000049	–65.5624	–.5643
	.05	–.049776	–.042760	–.022418	–.007402	–.000753	.000693	.000401	.000076	–.000001	–69.0878	–.5840
10.0	1.00	–.052435	–.048741	–.029028	–.011851	–.002015	.001764	.002168	.001311	.000392	–67.9905	.3364
	.75	–.050856	–.045272	–.025260	–.009187	–.000891	.001718	.001627	.000831	.000212	–71.7491	.0002
	.50	–.049230	–.041803	–.021673	–.006874	–.000141	.001432	.001024	.000393	.000068	–75.8100	–.3455
	.25	–.047555	–.038344	–.018291	–.004923	.000280	.001015	.000498	.000101	–.000003	–80.2110	–.4813
	.05	–.046179	–.035591	–.015749	–.003626	.000413	.000660	.000205	.000010	–.000005	–84.0080	–.0029
12.0	1.00	–.048980	–.041408	–.021411	–.006606	.000356	.002126	.001678	.000800	.000195	–81.5103	–.0605
	.75	–.047481	–.038349	–.018431	–.004861	.000762	.001735	.001106	.000413	.000071	–85.6072	–.2625
	.50	–.045945	–.035312	–.015636	–.003405	.000931	.001270	.000606	.000138	–.000000	–90.0010	–.3648
	.25	–.044371	–.032307	–.013040	–.002238	.000902	.000804	.000248	.000009	–.000013	–94.7262	–.2137
	.05	–.043084	–.029930	–.011116	–.001507	.000771	.000476	.000082	–.000009	–.000002	–98.7724	.1322
14.0	1.00	–.045954	–.035451	–.015840	–.003343	.001386	.001905	.001125	.000406	.000067	–95.0054	–.2188
	.75	–.044526	–.032739	–.013477	–.002230	.001399	.001428	.000662	.000156	.000001	–99.4107	–.2869
	.50	–.043069	–.030063	–.011289	–.001350	.001267	.000963	.000314	.000018	–.000020	–104.1094	–.2380
	.25	–.041582	–.027430	–.009286	–.000696	.001029	.000557	.000100	–.000019	–.000010	–109.1329	–.0324
	.05	–.040370	–.025360	–.007824	–.000323	.000795	.000302	.000020	–.000010	–.000001	–113.4096	.0870

continued

Table 12.9 (continued) Bending moment variation and rotations D_2 and D_4 at bottom and top edges, respectively; outward normal line load of intensity F_1 = 1 (force/unit length) at bottom edge; both edges free

η	h_t/h_b	*x/l*									D_2	D_4
		.1	.2	.3	.4	.5	.6	.7	.8	.9		
16.0	1.00	–.043265	–.030528	–.011697	–.001308	.001737	.001522	.000685	.000159	–.000000	–108.4746	–.2359
	.75	–.041902	–.028114	–.009820	–.000635	.001541	.001070	.000355	.000023	–.000025	–113.1654	–.2154
	.50	–.040515	–.025743	–.008106	–.000145	.001268	.000672	.000136	–.000028	–.000020	–118.1468	–.1097
	.25	–.039104	–.023423	–.006560	.000174	.000951	.000354	.000024	–.000022	–.000005	–123.4473	.0423
	.05	–.037957	–.021608	–.005448	.000320	.000693	.000174	–.000007	–.000007	–.000000	–127.9393	.0282
18.0	1.00	–.040850	–.026408	–.008577	–.000054	.001747	.001140	.000378	.000025	–.000028	–121.9180	–.1896
	.75	–.039546	–.024250	–.007089	.000313	.001453	.000755	.000162	–.000033	–.000028	–126.8757	–.1275
	.50	–.038221	–.022142	–.005748	.000539	.001127	.000439	.000038	–.000037	–.000014	–132.1214	–.0241
	.25	–.036878	–.020088	–.004558	.000640	.000799	.000208	–.000011	–.000017	–.000002	–137.6812	.0535
	.05	–.035788	–.018488	–.003717	.000643	.000556	.000088	–.000015	–.000004	.000000	–142.3755	–.0022
20.0	1.00	–.038661	–.022923	–.006209	.000696	.001597	.000813	.000180	–.000038	–.000034	–135.3359	–.1268
	.75	–.037410	–.020990	–.005032	.000852	.001269	.000505	.000050	–.000049	–.000023	–140.5447	–.0565
	.50	–.036144	–.019108	–.003989	.000902	.000941	.000268	–.000011	–.000032	–.000008	–146.0389	.0191
	.25	–.034861	–.017284	–.003079	.000863	.000635	.000109	–.000023	–.000011	–.000000	–151.8432	.0396
	.05	–.033823	–.015869	–.002447	.000780	.000423	.000036	–.000015	–.000002	.000000	–156.7285	–.0102
24.0	1.00	–.034829	–.017398	–.003015	.001329	.001155	.000360	–.000008	–.000060	–.000022	–162.0952	–.0282
	.75	–.033674	–.015836	–.002296	.001245	.000853	.000186	–.000037	–.000038	–.000009	–167.7669	.0147
	.50	–.032508	–.014328	–.001681	.001107	.000583	.000071	–.000036	–.000015	–.000001	–173.7199	.0319
	.25	–.031333	–.012879	–.001169	.000951	.000358	.000009	–.000021	–.000003	.000001	–179.9759	.0082
	.05	–.030385	–.011763	–.000831	.000774	.000216	–.000010	–.000008	.000000	.000000	–185.2152	–.0044

Notes: M = Coefficient × l; D_2 or D_4 = Coefficient × r^2 $(Eh_b/2)$.

Table 12.10 Hoop force variation; uniformly distributed moment of intensity $F_2 = 1$ (force × length/length) at bottom edge; both edges free

η	h_t/h_b	*x/l*										
		0.0	*.1*	*.2*	*.3*	*.4*	*.5*	*.6*	*.7*	*.8*	*.9*	*1.0*
.4	1.00	−6.3833	−4.9458	−3.5805	−2.2815	−1.0404	.1532	1.3101	2.4412	3.5566	4.6642	5.7695
	.75	−7.0383	−5.2394	−3.5975	−2.1057	−.7539	.4705	1.5817	2.5942	3.5215	4.3753	5.1636
	.50	−7.9544	−5.6492	−3.6200	−1.8588	−.3531	.9133	1.9592	2.8048	3.4699	3.9727	4.3263
	.25	−9.4315	−6.3093	−3.6553	−1.4596	.2940	1.6271	2.5660	3.1405	3.3827	3.3242	2.9914
	.05	−11.7347	−7.3413	−3.7135	−.8387	1.3044	2.7446	3.5183	3.6691	3.2467	2.3066	.9064
.8	1.00	−7.4954	−5.3662	−3.5242	−1.9437	−.5882	.5846	1.6189	2.5573	3.4373	4.2892	5.1328
	.75	−8.3638	−5.7345	−3.5224	−1.6974	−.2153	.9764	1.9344	2.7148	3.3690	3.9392	4.4543
	.50	−9.5540	−6.2355	−3.5149	−1.3559	.2962	1.5088	2.3570	2.9182	3.2666	3.4660	3.5625
	.25	−11.3517	−6.9873	−3.4965	−.8339	1.0708	2.3077	2.9816	3.2056	3.0953	2.7588	2.2829
	.05	−13.7652	−7.9959	−3.4698	−.1298	2.1144	3.3822	3.8167	3.5803	2.8490	1.8058	.6286
1.2	1.00	−9.2077	−5.9928	−3.4194	−1.4209	.0926	1.2214	2.0653	2.7166	3.2534	3.7362	4.2033
	.75	−10.3596	−6.4511	−3.3836	−1.0783	.5743	1.6990	2.4238	2.8690	3.1394	3.3174	3.4568
	.50	−11.8863	−7.0490	−3.3238	−.6155	1.2122	2.3195	2.8752	3.0433	2.9712	2.7788	2.5475
	.25	−14.0294	−7.8726	−3.2190	.0498	2.1087	3.1707	3.4686	3.2379	2.7008	2.0491	1.4239
	.05	−16.5319	−8.8170	−3.0730	.8456	3.1594	4.1451	4.1163	3.4036	2.3361	1.2218	.3253
1.6	1.00	−11.3603	−6.7425	−3.2544	−.7594	.9198	1.9713	2.5735	2.8820	3.0221	3.0853	3.1270
	.75	−12.8023	−7.2768	−3.1681	−.3139	1.5020	2.5141	2.9497	3.0099	2.8590	2.6175	2.3589
	.50	−14.6410	−7.9395	−3.0346	.2693	2.2408	3.1801	3.3848	3.1252	2.6255	2.0532	1.5115
	.25	−17.0577	−8.7784	−2.8184	1.0634	3.2085	4.0117	3.8784	3.1884	2.2678	1.3671	.6463
	.05	−19.6114	−9.6257	−2.5385	1.9397	4.2312	4.8398	4.3008	3.1358	1.8122	.7090	.0882

continued

Table 12.10 (continued) Hoop force variation; uniformly distributed moment of intensity $F_2 = 1$ (force × length/length) at bottom edge; both edges free

η	h_t/h_b	*x/l*										
		0.0	*.1*	*.2*	*.3*	*.4*	*.5*	*.6*	*.7*	*.8*	*.9*	*1.0*
2.0	1.00	−13.8008	−7.5369	−3.0197	−.0042	1.8153	2.7486	3.0740	3.0204	2.7604	2.4109	2.0377
	.75	−15.5003	−8.1168	−2.8674	.5380	2.4715	3.3184	3.4309	3.1016	2.5513	1.9282	1.3134
	.50	−17.5893	−8.7990	−2.6430	1.2265	3.2693	3.9743	3.7995	3.1310	2.2621	1.3871	.6090
	.25	−20.1968	−9.6002	−2.3005	2.1253	4.2542	4.7213	4.1403	3.0431	1.8397	.8004	.0594
	.05	−22.7868	−10.3380	−1.8863	3.0680	5.2257	5.3832	4.3356	2.7962	1.3323	.3232	−.0582
3.0	1.00	−20.4657	−9.3745	−2.1038	2.0759	4.0032	4.4450	4.0101	3.1264	2.0568	.9339	−.1958
	.75	−22.6210	−9.9332	−1.7410	2.8067	4.7147	4.9214	4.1814	3.0240	1.7653	.5503	−.5853
	.50	−25.1109	−10.5180	−1.2521	3.6856	5.5089	5.3805	4.2520	2.7936	1.3933	.2064	−.7598
	.25	−28.0155	−11.1175	−.5849	4.7615	6.4000	5.7900	4.1546	2.3645	.9101	−.0522	−.5877
	.05	−30.7308	−11.6046	.1256	5.8133	7.2015	6.0575	3.8988	1.8203	.4241	−.1555	−.1407
4.0	1.00	−27.3104	−10.7294	−.7519	4.2000	5.8210	5.5323	4.3440	2.8634	1.3773	−.0414	−1.4233
	.75	−29.7618	−11.1593	−.1587	5.0526	6.4702	5.7982	4.2560	2.5677	1.0528	−.2408	−1.3903
	.50	−32.5251	−11.5689	.5907	6.0444	7.1552	5.9835	4.0173	2.1291	.6720	−.3534	−1.1039
	.25	−35.6807	−11.9502	1.5390	7.2092	7.8771	6.0569	3.5858	1.5222	.2419	−.3506	−.5573
	.05	−38.5688	−12.2340	2.4728	8.2891	8.4759	6.0072	3.0794	.9288	−.0899	−.2448	−.0744
5.0	1.00	−34.1244	−11.5731	.9350	6.2379	7.2007	6.0427	4.1829	2.3487	.7786	−.5678	−1.8260
	.75	−36.8264	−11.8476	1.7405	7.1609	7.7394	6.0871	3.8711	1.9197	.4644	−.5983	−1.4745
	.50	−39.8422	−12.0794	2.7117	8.2036	8.2767	6.0230	3.4122	1.3823	.1325	−.5402	−.9429
	.25	−43.2459	−12.2615	3.8792	9.3799	8.7967	5.8213	2.7951	.7613	−.1742	−.3879	−.3350
	.05	−46.3085	−12.3660	4.9761	10.4190	9.1726	5.5262	2.1868	.2629	−.3175	−.1971	−.0138

6.0	1.00	−40.9080	−11.9848	2.8470	8.1301	8.1888	6.1232	3.7152	1.7269	.2847	−.7901	−1.7340
	.75	−43.8478	−12.1049	3.8317	9.0769	8.5950	5.9742	3.2506	1.2417	.0152	−.7000	−1.2149
	.50	−47.1050	−12.1676	4.9790	10.1138	8.9641	5.7062	2.6701	.7056	−.2297	−.5411	−.6288
	.25	−50.7404	−12.1633	6.3108	11.2375	9.2644	5.2899	1.9834	.1760	−.3876	−.3240	−.1315
	.05	−53.9665	−12.1008	7.5261	12.1893	9.4174	4.8151	1.3692	−.1720	−.3706	−.1219	.0139
8.0	1.00	−54.4528	−11.8065	7.0038	11.3512	9.2159	5.4997	2.4272	.5481	−.3707	−.7729	−1.0160
	.75	−57.8302	−11.6319	8.2433	12.2247	9.3260	5.0692	1.8528	.1308	−.4937	−.5748	−.5039
	.50	−61.5190	−11.3750	9.6251	13.1202	9.3371	4.5230	1.2416	−.2304	−.5332	−.3584	−.1065
	.25	−65.5624	−11.0199	11.1630	14.0182	9.2085	3.8403	.6221	−.4715	−.4422	−.1479	.0620
	.05	−69.0878	−10.6515	12.5183	14.7231	8.9727	3.1831	.1497	−.5309	−.2606	−.0149	.0146
10.0	1.00	−67.9905	−10.6187	11.2352	13.7426	9.2896	4.3706	1.1762	−.2746	−.6441	−.5481	−.3364
	.75	−71.7491	−10.1639	12.6171	14.4342	9.1072	3.7839	.6579	−.5136	−.6177	−.3492	−.0285
	.50	−75.8100	−9.6060	14.1172	15.0912	8.7931	3.1077	.1665	−.6458	−.5025	−.1681	.1116
	.25	−80.2110	−8.9280	15.7447	15.6911	8.3178	2.3430	−.2626	−.6376	−.2990	−.0295	.0739
	.05	−84.0080	−8.2862	17.1451	16.1122	7.8024	1.6811	−.5217	−.5161	−.1120	.0203	.0031
12.0	1.00	−81.5103	−8.6758	15.2915	15.3365	8.7197	3.1090	.1958	−.7138	−.6633	−.3207	.0605
	.75	−85.6072	−7.9558	16.7339	15.7959	8.2901	2.4678	−.1932	−.7735	−.5356	−.1641	.1654
	.50	−90.0010	−7.1176	18.2698	16.1803	7.7258	1.7787	−.5138	−.7259	−.3510	−.0434	.1384
	.25	−94.7262	−6.1454	19.9027	16.4667	7.0115	1.0634	−.7298	−.5676	−.1446	.0225	.0386
	.05	−98.7724	−5.2590	21.2805	16.6061	6.3241	.5004	−.7921	−.3717	−.0167	.0202	−.0014
14.0	1.00	−95.0054	−6.1524	19.0472	16.2352	7.7534	1.9101	−.4807	−.8664	−.5510	−.1485	.2188
	.75	−99.4107	−5.1847	20.4890	16.4490	7.1412	1.2922	−.7209	−.7922	−.3796	−.0436	.1910
	.50	−104.1094	−4.0888	21.9987	16.5604	6.4102	.6719	−.8709	−.6361	−.1932	.0204	.0962
	.25	−109.1329	−2.8499	23.5749	16.5466	5.5574	.0823	−.9010	−.4149	−.0372	.0341	.0099
	.05	−113.4096	−1.7443	24.8815	16.4269	4.7886	.3347	−.8175	−.2161	.0260	.0114	−.0016

continued

Table 12.10 (continued) Hoop force variation; uniformly distributed moment of intensity $F_2 = 1$ (force × length/length) at bottom edge; both edges free

η	h_t/h_b	*x/l*										
		0.0	*.1*	*.2*	*.3*	*.4*	*.5*	*.6*	*.7*	*.8*	*.9*	*1.0*
16.0	1.00	−108.4746	−3.1783	22.4413	16.5587	6.5772	.8698	−.3919	−.8412	−.3940	−.0361	.2359
	.75	−113.1654	−1.9818	23.8356	16.5341	5.8482	.3283	−.9923	−.6862	−.2239	.0224	.1468
	.50	−118.1468	−.6507	25.2723	16.3898	5.0263	−.1757	−.9968	−.4826	−.0730	.0436	.0476
	.25	−123.4473	.8292	26.7460	16.1045	4.1182	−.6060	−.8882	−.2579	.0208	.0281	−.0036
	.05	−127.9393	2.1311	27.9468	15.7584	3.3384	−.8667	−.7152	−.0938	.0361	.0040	−.0007
18.0	1.00	−121.9180	.1464	25.4485	16.4216	5.3240	.0253	−1.0960	−.7229	−.2424	.0280	.1896
	.75	−126.8757	1.5526	26.7592	16.1774	4.5365	−.4095	−1.0792	−.5318	−.0998	.0508	.0888
	.50	−132.1214	3.0975	28.0879	15.8042	3.6863	−.7766	−.9737	−.3238	.0029	.0444	.0134
	.25	−137.6812	4.7945	29.4264	15.2838	2.7873	−1.0431	−.7765	−.1294	.0437	.0176	−.0070
	.05	−142.3755	6.2716	30.4972	14.7484	2.0478	−1.1588	−.5618	−.0132	.0310	−.0000	−.0001
20.0	1.00	−135.3359	3.7427	28.0644	15.5257	4.0835	−.6207	−1.1504	−.5682	−.1188	.0580	.1268
	.75	−140.5447	5.3402	29.2645	15.4866	3.2857	−.9362	−1.0443	−.3734	−.0147	.0563	.0407
	.50	−146.0389	7.0787	30.4599	14.9162	2.4557	−1.1650	−.8654	−.1871	.0424	.0354	−.0048
	.25	−151.8432	8.9705	31.6400	14.2003	1.6130	−1.2806	−.6229	−.0377	.0459	.0085	−.0058
	.05	−156.7285	10.6035	32.5648	13.5131	.9484	−1.2738	−.4021	.0318	.0213	−.0015	.0001
24.0	1.00	−162.0952	11.5087	32.1648	14.1922	1.8479	−1.3898	−.9956	−.2707	.0285	.0618	.0282
	.75	−167.7669	13.4368	33.0900	13.4476	1.1324	−1.4765	−.7924	−.1207	.0607	.0386	−.0085
	.50	−173.7199	15.5060	33.9711	12.5830	.4392	−1.4671	−.5580	−.0079	.0562	.0138	−.0118
	.25	−179.9759	17.7266	34.7949	11.5925	−.2068	−1.3511	−.3165	.0515	.0278	−.0010	−.0015
	.05	−185.2152	19.6199	35.4025	10.7069	−.6683	−1.1790	−.1426	.0547	.0051	−.0012	.0001

Note: N = Coefficient × r/l^2.

Table 12.11 Bending moment variation and rotations D_2 and D_4 at bottom and top edges, respectively; uniformly distributed moment of intensity $F_2 = 1$ (force × length/length) at bottom edge; both edges free

		x/l											
η	h_t/h_b	*0.0*	*.1*	.2	.3	.4	.5	.6	.7	.8	.9	D_2	D_4
.4	1.00	1.0000	.9704	.8914	.7764	.6386	.4904	.3437	.2100	.1008	.0271	14.7464	11.0513
	.75	1.0000	.9678	.8831	.7622	.6202	.4706	.3255	.1963	.0928	.0246	17.0265	12.3892
	.50	1.0000	.9641	.8714	.7424	.5945	.4429	.3003	.1771	.0818	.0211	20.4696	14.2876
	.25	1.0000	.9581	.8527	.7104	.5531	.3985	.2598	.1464	.0642	.0156	26.5071	17.3372
	.05	1.0000	.9488	.8235	.6605	.4885	.3289	.1963	.0984	.0366	.0070	36.6365	21.8590
.8	1.00	1.0000	.9661	.8783	.7550	.6122	.4633	.3201	.1930	.0915	.0243	22.7773	8.4301
	.75	1.0000	.9626	.8675	.7369	.5891	.4388	.2982	.1767	.0823	.0215	26.3451	8.5502
	.50	1.0000	.9579	.8529	.7123	.5577	.4058	.2686	.1548	.0700	.0177	31.4634	8.2026
	.25	1.0000	.9507	.8308	.6752	.5107	.3564	.2247	.1224	.0519	.0122	39.5780	6.3150
	.05	1.0000	.9410	.8011	.6255	.4477	.2903	.1660	.0794	.0281	.0051	50.9310	.1970
1.2	1.00	1.0000	.9594	.8584	.7228	.5725	.4228	.2852	.1680	.0779	.0203	35.4863	4.6600
	.75	1.0000	.9549	.8446	.6998	.5437	.3930	.2589	.1488	.0673	.0171	40.8503	3.2641
	.50	1.0000	.9489	.8263	.6698	.5063	.3544	.2252	.1245	.0540	.0131	48.1676	.3777
	.25	1.0000	.9405	.8010	.6282	.4549	.3018	.1799	.0923	.0367	.0081	58.7770	−6.2603
	.05	1.0000	.9307	.7715	.5802	.3961	.2424	.1294	.0570	.0183	.0029	71.5563	−20.4170
1.6	1.00	1.0000	.9512	.8340	.6834	.5246	.3745	.2438	.1385	.0620	.0156	52.1024	.4011
	.75	1.0000	.9455	.8172	.6561	.4910	.3403	.2143	.1175	.0506	.0122	59.4579	−2.3476
	.50	1.0000	.9384	.7960	.6219	.4494	.2985	.1788	.0925	.0374	.0084	69.0554	−7.1526
	.25	1.0000	.9291	.7685	.5781	.3969	.2466	.1358	.0633	.0224	.0043	82.0476	−16.3583
	.05	1.0000	.9193	.7401	.5333	.3441	.1959	.0952	.0369	.0099	.0012	96.2516	−31.9784

continued

Table 12.11 (continued) Bending moment variation and rotations D_2 and D_4 at bottom and top edges, respectively; uniformly distributed moment of intensity F_2 = 1 (force × length/length) at bottom edge; both edges free

η	h_t/h_b	x/l											
		0.0	.1	.2	.3	.4	.5	.6	.7	.8	.9	D_2	D_4
2.0	1.00	1.0000	.9419	.8070	.6407	.4733	.3235	.2006	.1081	.0457	.0108	71.8799	−3.7483
	.75	1.0000	.9354	.7878	.6101	.4366	.2869	.1698	.0867	.0344	.0076	81.2181	−7.3917
	.50	1.0000	.9274	.7647	.5737	.3935	.2449	.1353	.0633	.0225	.0043	92.9677	−13.0776
	.25	1.0000	.9176	.7365	.5300	.3430	.1971	.0976	.0391	.0110	.0014	108.1161	−22.3023
	.05	1.0000	.9079	.7092	.4888	.2968	.1554	.0669	.0214	.0039	.0001	123.7656	−33.6270
3.0	1.00	1.0000	.9173	.7378	.5347	.3505	.2052	.1036	.0418	.0110	.0009	131.6096	−11.2972
	.75	1.0000	.9094	.7158	.5019	.3138	.1716	.0778	.0255	.0033	−.0011	145.5219	−14.8893
	.50	1.0000	.9004	.6913	.4661	.2753	.1380	.0537	.0117	−.0024	−.0023	162.0951	−18.8805
	.25	1.0000	.8901	.6637	.4274	.2356	.1060	.0336	.0025	−.0046	−.0022	182.1963	−21.5519
	.05	1.0000	.8805	.6390	.3938	.2032	.0826	.0218	.0001	−.0029	−.0009	201.7447	−14.5571
4.0	1.00	1.0000	.8932	.6738	.4428	.2511	.1160	.0356	−.0016	−.0101	−.0049	202.4396	−13.7840
	.75	1.0000	.8847	.6516	.4122	.2203	.0914	.0199	−.0093	−.0125	−.0051	220.6560	−15.3501
	.50	1.0000	.8752	.6276	.3804	.1902	.0696	.0083	−.0127	−.0121	−.0044	241.8587	−15.4529
	.25	1.0000	.8646	.6013	.3470	.1608	.0514	.0021	−.0110	−.0083	−.0025	266.9273	−10.8584
	.05	1.0000	.8549	.5781	.3187	.1377	.0393	.0006	.0066	−.0035	−.0007	290.5795	5.0870
5.0	1.00	1.0000	.8705	.6170	.3671	.1760	.0553	.0056	−.0245	−.0198	−.0071	282.4920	−12.5062
	.75	1.0000	.8615	.5952	.3397	.1519	.0398	−.0119	−.0247	−.0178	−.0060	305.0573	−11.8000
	.50	1.0000	.8516	.5719	.3117	.1292	.0276	−.0140	−.0209	−.0134	−.0041	330.9566	−8.7564
	.25	1.0000	.8407	.5466	.2827	.1076	.0187	−.0120	−.0139	−.0073	−.0018	361.0137	−.8121
	.05	1.0000	.8309	.5246	.2584	.0910	.0135	−.0085	−.0074	−.0025	−.0003	388.7317	12.3110

6.0	1.00	1.0000	.8488	.5663	.3045	.1199	.0154	–.0280	–.0339	–.0221	–.0072	370.9266	–9.3322
	.75	1.0000	.8395	.5450	.2802	.1015	.0072	–.0274	–.0289	–.0174	–0053	397.9084	–6.9950
	.50	1.0000	.8293	.5223	.2554	.0847	.0019	–.0235	–.0214	–.0113	–.0030	428.5091	–2.4546
	.25	1.0000	.8180	.4980	.2301	.0691	–.0008	–.0172	–.0125	–.0051	–.0010	463.4735	5.1450
	.05	1.0000	.8082	.4772	.2093	.0574	–.0017	–.0116	–.0062	–.0015	–.0001	495.1897	10.7490
8.0	1.00	1.0000	.8082	.4789	.2080	.0455	–.0260	–.0420	–.0326	–.0170	–.0047	570.4752	–2.3133
	.75	1.0000	.7983	.4587	.1888	.0357	–.0251	–.0343	–.0237	–.0109	–.0026	606.2838	.8379
	.50	1.0000	.7875	.4375	.1696	.0272	–.0225	–.0256	–.0148	–.0054	–.0009	646.1262	4.5087
	.25	1.0000	.7760	.4153	.1506	.0201	–.0185	–.0169	–.0073	–.0015	.0000	690.6734	6.8664
	.05	1.0000	.7660	.3966	.1356	.0155	–.0147	–.0106	–.0031	–.0002	.0001	730.2629	1.9220
10.0	1.00	1.0000	.7706	.4059	.1382	.0025	–.0405	–.0384	–.0230	–.0095	–.0020	796.8925	2.1874
	.75	1.0000	.7602	.3871	.1233	–.0016	–.0352	–.0291	–.0148	–.0048	–.0007	841.4142	4.1615
	.50	1.0000	.7492	.3676	.1089	–.0044	–.0290	–.0202	–.0080	–.0015	.0001	890.3264	5.2036
	.25	1.0000	.7374	.3474	.0952	–.0059	–.0222	–.0123	–.0032	.0000	.0002	944.2909	3.3963
	.05	1.0000	.7274	.3308	.0845	–.0062	–.0168	–.0072	–.0011	.0002	.0001	991.6398	–1.9135
12.0	1.00	1.0000	.7354	.3442	.0868	–.0221	–.0427	–.0300	–.0137	–.0038	–.0003	1047.1472	3.8142
	.75	1.0000	.7248	.3268	.0758	–.0221	–.0354	–.0215	–.0077	–.0011	.0003	1100.3140	4.2559
	.50	1.0000	.7136	.3090	.0654	–.0211	–.0279	–.0139	–.0034	.0003	.0004	1158.2140	3.3544
	.25	1.0000	.7018	.2909	.0557	–.0190	–.0205	–.0078	–.0009	.0005	.0002	1221.4965	.4359
	.05	1.0000	.6918	.2762	.0486	–.0168	–.0149	–.0042	–.0001	.0002	.0000	1276.5101	–1.6800
14.0	1.00	1.0000	.7024	.2915	.0489	–.0352	–.0394	–.0216	–.0070	–.0005	.0005	1319.0667	3.6157
	.75	1.0000	.6917	.2756	.0411	–.0325	–.0316	–.0145	–.0031	.0007	.0006	1380.8492	2.9830
	.50	1.0000	.6804	.2596	.0340	–.0290	–.0240	–.0087	–.0008	.0009	.0004	1447.6839	1.3384
	.25	1.0000	.6686	.2434	.0277	–.0248	–.0168	–.0044	.0002	.0005	.0001	1520.2025	–.8651
	.05	1.0000	.6586	.2303	.0233	–.0211	–.0118	–.0020	.0003	.0001	.0000	1582.8022	–.16135

continued

Table 12.11 (continued) Bending moment variation and rotations D_2 and D_4 at bottom and top edges, respectively; uniformly distributed moment of intensity F_2 = 1 (force × length/length) at bottom edge; both edges free

η	h_t/h_b	*x/l*											
		0.0	*.1*	.2	.3	.4	.5	.6	.7	.8	.9	D_2	D_4
16.0	1.00	1.0000	.6714	.2462	.0210	–.0412	–.0338	–.0145	–.0025	.0011	.0008	1610.9972	2.6285
	.75	1.0000	.6606	.2319	.0158	–.0367	–.0263	–.0091	–.0003	.0012	.0005	1681.3756	1.5306
	.50	1.0000	.6493	.2175	.0114	–.0317	–.0192	–.0049	.0004	.0008	.0002	1757.0990	.0004
	.25	1.0000	.6375	.2031	.0078	–.0263	–.0130	–.0021	.0006	.0003	.0000	1838.7851	–1.0286
	.05	1.0000	.6277	.1915	.0055	–.0219	–.0087	–.0007	.0004	.0001	–.0000	1908.9047	.0364
18.0	1.00	1.0000	.6422	.2072	.0004	–.0428	–.0277	–.0090	–.0001	.0016	.0007	1921.6110	1.5165
	.75	1.0000	.6313	.1943	–.0025	–.0373	–.0209	–.0051	.0008	.0012	.0004	2000.5662	.4140
	.50	1.0000	.6201	.1814	–.0047	–.0314	–.0147	–.0023	.0009	.0006	.0001	2085.1380	–.6198
	.25	1.0000	.6084	.1686	–.0062	–.0255	–.0094	–.0006	.0006	.0002	–.0000	2175.9332	–.7249
	.05	1.0000	.5986	.1585	–.0069	–.0208	–.0060	.0000	.0003	.0000	–.0000	2253.5163	.2226
20.0	1.00	1.0000	.6145	.1754	–.0145	–.0417	–.0220	–.0050	.0012	.0016	.0005	2249.8065	.6001
	.75	1.0000	.6037	.1618	–.0157	–.0356	–.0160	–.0024	.0013	.0010	.0002	2337.3204	–.2579
	.50	1.0000	.5925	.1504	–.0161	–.0295	–.0108	–.0006	.0010	.0004	.0000	2430.7053	–.3598
	.25	1.0000	.5809	.1392	–.0159	–.0234	–.0065	.0002	.0006	.0001	–.0000	2530.5595	–.3598
	.05	1.0000	.5713	.1303	–.0154	–.0187	–.0039	.0004	.0002	–.0000	.0000	2615.5566	.1829
24.0	1.00	1.0000	.5633	.1184	–.0326	–.0356	–.0125	–.0004	.0018	.0010	.0002	2955.3358	–.3743
	.75	1.0000	.5527	.1094	–.0311	–.0294	–.0083	.0005	.0013	.0004	.0000	3059.9202	–.5110
	.50	1.0000	.5417	.1005	–.0291	–.0235	–.0050	.0009	0.007	.0001	–.0000	3170.8426	–.4262
	.25	1.0000	.5304	.0920	–.0266	–.0179	–.0025	.0008	.0003	–.0000	–.0000	3288.6874	.0492
	.05	1.0000	.5210	.0853	–.0243	–.0138	–.0012	.0005	.0001	–.0000	–.0000	3388.3866	.0248

Notes: M = Coefficient; D_2 or $D_4 = r^2/(Eh^b l_3)$.

Table 12.12 Hoop force variation; outward normal line load of intensity $F_3 = 1$ (force/unit length) at top edge; both edges free

η	h_t/h_b	x/l										
		0.0	*.1*	*.2*	*.3*	*.4*	*.5*	*.6*	*.7*	*.8*	*.9*	*1.0*
.4	1.00	−1.9459	−1.3689	−.7911	−.2110	.3738	.9655	1.5664	2.1778	2.7999	3.4313	4.0685
	.75	−2.2208	−1.4940	−.8007	−.1390	.4939	1.1010	1.6850	2.2477	2.7893	3.3072	3.7954
	.50	−2.6379	−1.6839	−.8154	−.0299	.6761	1.3066	1.8654	2.3548	2.7743	3.1189	3.3766
	.25	−3.3872	−2.0263	−.8434	.1647	1.0028	1.6768	2.1918	2.5510	2.7519	2.7811	2.6021
	.05	−4.7410	−2.6498	−.9011	.5099	1.5903	2.3486	2.7926	2.9255	2.7371	2.1789	1.0557
.8	1.00	−1.7982	−1.2847	−.7685	−.2433	.2991	.8681	1.4718	2.1159	2.8016	3.5238	4.2690
	.75	−2.0037	−1.3743	−.7715	−.1880	.3864	.9632	1.5529	2.1628	2.7941	3.4381	4.0719
	.50	−2.2921	−1.5003	−.7762	−.1108	.5085	1.0964	1.6672	2.2306	2.7869	3.3190	3.7810
	.25	−2.7587	−1.7068	−.7873	.0108	.7042	1.3131	1.8575	2.3510	2.7897	3.1299	3.2399
	.05	−3.5442	−2.0693	−.8273	.1966	1.0249	1.6862	2.2091	2.6117	2.8734	2.8470	1.8390
1.2	1.00	−1.5808	−1.1600	−.7338	−.2897	.1897	.7235	1.3297	2.0211	2.8019	3.6621	4.5732
	.75	−1.6952	−1.2030	−.7277	−.2554	.2345	.7654	1.3602	2.0357	2.7969	3.6278	4.4808
	.50	−1.8264	−1.2506	−.7190	−.2154	.2844	.8099	1.3909	2.0508	2.7954	3.5954	4.3577
	.25	−1.9826	−1.3065	−.7084	−.1694	.3399	.8572	1.4236	2.0737	2.8142	3.5766	4.1049
	.05	−2.2022	−1.4036	−.7200	−.1296	.4039	.9290	1.5030	2.1784	2.9594	3.6388	2.9052
1.6	1.00	−1.3255	−1.0122	−.6906	−.3417	.0620	.5516	1.1574	1.9030	2.7979	3.8285	4.9475
	.75	−1.3497	−1.0087	−.6746	−.3268	.0659	.5406	1.1355	1.8817	2.7924	3.8468	4.9682
	.50	−1.5402	−.9853	−.6517	−.3171	.0536	.5050	1.0861	1.8407	2.7889	3.8961	5.0167
	.25	−1.2568	−.9227	−.6187	−.3217	.0068	.4202	.9839	1.7642	2.7992	4.0200	5.0446
	.05	−1.0990	−.8344	−.5952	−.3591	−.0858	.2871	.8461	1.6969	2.9209	4.3496	4.0886

continued

Table 12.12 (continued) Hoop force variation; outward normal line load of intensity F_3 = 1 (force/unit length) at top edge; both edges free

		x/l										
η	h_t/h_b	0.0	.1	.2	.3	.4	.5	.6	.7	.8	.9	1.0
2.0	1.00	−1.0620	−.8575	−.6423	−.3916	−.0686	.3709	.9715	1.7707	2.7873	4.0063	5.3600
	.75	−1.0121	−.8153	−.6167	−.3906	−.0966	.3160	.9028	1.7138	2.7766	4.0705	5.4883
	.50	−.9013	−.7398	−.5807	−.3983	−.1502	.2221	.7885	1.6195	2.7610	4.1843	5.6928
	.25	−.6779	−.6045	−.5267	−.4219	−.2483	.0590	.5895	1.4509	2.7349	4.4058	5.9723
	.05	−.3502	−.4219	−.4666	−.4680	−.3904	−.1677	.3113	1.2157	2.7494	4.8959	5.3026
3.0	1.00	−.4899	−.5087	−.5152	−.4779	−.3436	−.0397	.5195	1.4199	2.7228	4.4277	6.4165
	.75	−.3505	−.4172	−.4703	−.4821	−.4011	−.1494	.3751	1.2863	2.6779	4.5618	6.7618
	.50	−.1571	−.2933	−.4086	−.4826	−.4705	−.2927	.1721	1.0814	2.5921	4.7547	7.2734
	.25	.0996	−.1254	−.3152	−.4625	−.5383	−.4719	−.1253	.7287	2.3899	5.0535	8.0790
	.05	.3367	.0496	−.1901	−.3911	−.5497	−.6231	−.4809	.1789	1.9409	5.5530	8.2929
4.0	1.00	−.1193	−.2612	−.3953	−.4953	−.5055	−.3345	.1446	1.0802	2.6042	4.7606	7.3916
	.75	.0117	−.1692	−.3380	−.4776	−.5419	−.4402	−.0275	.8881	2.5038	4.9132	7.8952
	.50	.1549	−.0630	−.2616	−.4375	−.5609	−.5507	−.2472	.6033	2.3162	5.1074	8.6431
	.25	.2785	.0507	−.1525	−.3441	−.5211	−.6314	−.5209	.1381	1.8959	5.3443	9.9058
	.05	.2845	.1219	−.0233	−.1747	−.3576	−.5724	−.7312	−.5069	.9807	5.5030	11.1517
5.0	1.00	.0754	−.1074	−.2901	−.4595	−.5685	−.5179	−.1448	.7676	2.4434	5.0020	8.2616
	.75	.1616	−.0348	−.2268	−.4141	−.5661	−.5925	−.3170	.5320	2.2790	5.1452	8.8960
	.50	.2282	.0366	−.1468	−.3362	−.5241	−.6430	−.5157	.1954	1.9845	5.3028	9.8512
	.25	.2334	.0909	−.0447	−.1995	−.3947	−.6096	−.7084	−.3109	1.3651	5.4110	11.5368
	.05	.1243	.0870	.0462	−.0209	−.1513	−.3832	−.7019	−.8598	.1366	5.0590	13.8704

6.0	1.00	.1549	–.0193	–.2016	–.3943	–.5658	–.6186	–.3634	.4845	2.2550	5.1713	9.0465
	.75	.1944	.0304	–.1382	–.3270	–.5250	–.6537	–.5199	.2192	2.0270	5.2908	9.7988
	.50	.2004	.0696	–.0639	–.2272	–.4344	–.6421	–.6770	–.1423	1.6341	5.3920	10.9446
	.25	.1418	.0816	.0163	–.0837	–.2558	–.5137	–.7648	–.6292	.8570	5.3365	13.0250
	.05	.0211	.0441	.0582	.0491	–.0174	–.2003	–.5559	–.9700	–.5101	4.4105	16.4558
8.0	1.00	.1517	.0485	–.0736	–.2434	–.4653	–.6685	–.6441	.0055	1.8402	5.3609	10.4369
	.75	.1329	.0626	–.0231	–.1595	–.3729	–.6284	–.7436	–.2807	1.5008	5.4110	11.4008
	.50	.0874	.0618	.0238	–.0611	–.2369	–.5184	–.7917	–.6254	.9552	5.3687	12.8898
	.25	.0166	.0380	.0505	.0343	–.0553	–.2836	–.6721	–.9582	–.0038	4.9412	15.6875
	.05	–.0285	.0008	.0278	.0548	.0694	.0181	–.2261	–.8099	–1.2415	2.9078	21.2707
10.0	1.00	.0894	.0542	–.0013	–.1180	–.3312	–.6141	–.7798	–.3637	1.4193	5.4207	11.6614
	.75	.0561	.0500	.0288	–.0451	–.2219	–.5188	–.8093	–.6307	.9968	5.3863	12.8121
	.50	.0149	.0350	.0458	.0218	–.0905	–.3546	–.7483	–.9008	.3585	5.1824	14.6046
	.25	–.0191	.0100	.0366	.0557	.0356	–.1054	–.4815	–1.0183	–.6345	4.3763	18.0444
	.05	–.0134	–.0063	.0030	.0214	.0535	.0782	–.0137	–.5005	–1.4379	1.4620	25.6792
12.0	1.00	.0371	.0421	.0318	–.0340	–.2105	–.5181	–.8215	–.6369	1.0188	5.3972	12.7680
	.75	.0102	.0311	.0430	.0167	–.1069	–.3902	–.7831	–.8593	.5405	5.2717	14.0868
	.50	–.0122	.0149	.0395	.0490	–.0036	–.2092	–.6349	–1.0283	–.1395	4.9027	16.1533
	.25	–.0174	–.0014	.0170	.0423	.0604	.0015	–.2944	–.9367	–1.0624	3.7438	20.1782
	.05	–.0015	–.0039	–.0045	.0015	.0239	.0679	.0771	–.2309	–1.3406	.2338	29.7473
14.0	1.00	.0063	.0278	.0411	.0144	–.1160	–.4117	–.8034	–.8305	.6496	5.3169	13.7844
	.75	–.0091	.0161	.0392	.0427	–.0297	–.2702	–.7077	–.9947	.1390	5.0987	15.2571
	.50	–.0159	.0037	.0257	.0487	.0395	–.0986	–.5003	–1.0566	–.5425	4.5692	17.5753
	.25	–.0088	–.0041	.0040	.0237	.0551	.0531	–.1459	–.7918	–1.3287	3.0966	22.1396
	.05	.0017	–.0013	–.0040	–.0047	.0047	.0407	.0947	–.0474	–1.1125	–.7396	33.5250

continued

Table 12.12 (continued) Hoop force variation; outward normal line load of intensity $F_3 = 1$ (force/unit length) at top edge; both edges free

η	h_t/h_b	*x/l*										
		0.0	*.1*	*.2*	*.3*	*.4*	*.5*	*.6*	*.7*	*.8*	*.9*	*1.0*
16.0	1.00	–.0077	.0160	.0383	.0379	–.0477	–.3100	–.7492	–.9602	.3151	5.1965	14.7292
	.75	–.0136	.0063	.0291	.0480	.0172	–.1696	–.6099	–1.0609	–.2081	4.8867	16.3446
	.50	–.0116	–.0015	.0132	.0380	.0550	–.0223	–.3696	–1.0213	–.8605	4.2056	18.8965
	.25	–.0026	–.0036	–.0020	.0096	.0400	.0697	–.0411	–.6286	–1.4722	2.4638	23.9624
	.05	.0015	–.0001	–.0020	–.0045	–.0036	.0177	.0797	.0574	–.8485	–1.4708	37.0509
18.0	1.00	–.0117	.0077	.0304	.0457	–.0015	–.2203	–.6753	–1.0395	.0157	5.0474	15.6153
	.75	–.0116	.0008	.0185	.0432	.0421	–.0907	–.5058	–1.0768	–.5041	4.6491	17.3642
	.50	–.0064	–.0031	.0047	.0256	.0550	.0256	–.2546	–.9475	–1.1049	–3.8273	20.1349
	.25	.0004	–.0023	–.0036	.0013	.0246	.0670	.0257	–.4714	–1.5249	1.8613	25.6705
	.05	.0006	.0003	–.0005	–.0028	–.0054	.0034	.0553	.1051	–.5984	–1.9910	40.3549
20.0	1.00	–.0109	.0025	.0217	.0446	.0274	–.1451	–.5925	–1.0796	–.2499	4.8775	16.4521
	.75	–.0078	–.0018	.0100	.0344	.0524	–.0322	–.4051	–1.0568	–.7533	4.3947	18.3269
	.50	–.0025	–.0031	–.0002	.0149	.0475	.0521	–.1593	–.8524	–1.2866	3.4444	21.3037
	.25	.0012	–.0011	–.0033	–.0026	.0125	.0557	.0635	–.3318	–1.5126	1.2973	27.2815
	.05	.0001	.0003	.0001	–.0012	–.0044	–.0036	.0327	.1167	–.3844	–2.3361	43.4614
24.0	1.00	–.0054	–.0020	.0077	.0317	.0508	–.0372	–.4257	–1.0764	–.6877	4.4972	18.0053
	.75	–.0019	–.0027	.0003	.0168	.0491	.0361	–.2314	–.9507	–1.1302	3.8599	20.1127
	.50	.0007	–.0016	–.0031	.0019	.0270	.0656	–.0263	–.6418	–1.5018	2.6898	23.4702
	.25	.0007	.0000	–.0013	–.0036	–.0008	.0288	.0843	–.1197	–1.3676	.2985	30.2630
	.05	–.0001	.0001	.0002	.0002	–.0015	–.0056	.0043	.0906	–.0837	–2.6344	49.1573

Note: N = Coefficient × *r/l*.

Table 12.13 Bending moment variation and rotations D_2 and D_4 at bottom and top edges, respectively; outward normal line load of intensity $F_3 = l$ (force/unit length) at top edge; both edges free

η	h_t/h_b	x/l									D_2	D_4
		.1	.2	.3	.4	.5	.6	.7	.8	.9		
.4	1.00	–.008794	–.031277	–.061668	–.094166	–.122920	–.142012	–.145431	–.127064	–.080691	5.7695	6.3833
	.75	–.009912	–.034792	–.067703	–.102028	–.131434	–.149849	–.151431	–.130552	–.081800	6.8848	7.9574
	.50	–.011609	–.040126	–.076866	–.113969	–.144373	–.161768	–.160565	–.135871	–.083495	8.6527	10.9000
	.25	–.014658	–.049724	–.093365	–.135497	–.167734	–.183332	–.177137	–.145560	–.086602	11.9657	19.0002
	.05	–.020177	–.067130	–.123367	–.174773	–.210536	–.223068	–.207925	–.163786	–.092571	18.1288	77.6683
.8	1.00	–.008159	–.029163	–.057845	–.088947	–.117035	–.136414	–.141043	–.124478	–.079868	5.1328	7.4954
	.75	–.008987	–.031739	–.062221	–.094591	–.123094	–.141955	–.145270	–.126939	–.080656	5.9391	10.0279
	.50	–.010149	–.035357	–.068373	–.102536	–.131639	–.149792	–.151279	–.130465	–.081800	7.1250	15.5808
	.25	–.012034	–.041244	–.078425	–.115584	–.145769	–.162875	–.161447	–.136552	–.083836	9.1315	35.2102
	.05	–.015235	–.051352	–.095919	–.138678	–.171324	–.187220	–.181121	–.149015	–.088394	12.5726	237.6202
1.2	1.00	–.007225	–.026042	–.052184	–.081196	–.108266	–.128043	–.134453	–.120579	–.078621	4.2033	9.2077
	.75	–.007669	–.027383	–.054376	–.083909	–.111063	–.130512	–.136292	–.121645	–.078971	4.6091	13.1529
	.50	–.008179	–.028899	–.056833	–.086923	–.114149	–.133230	–.138339	–.122869	–.079399	5.0949	22.4607
	.25	–.008784	–.030696	–.059740	–.090503	–.117861	–.136606	–.141046	–.124673	–.080134	5.6955	58.3733
	.05	–.009668	–.033466	–.064540	–.096956	–.125341	–.144397	–.148340	–.130406	–.082931	6.5061	475.8108
1.6	1.00	–.006121	–.022358	–.045479	–.071973	–.097777	–.117972	–.126478	–.115832	–.077095	3.1270	11.3603
	.75	–.006190	–.022474	–.045493	–.071743	–.097268	–.117290	–.125834	–.115426	–.076976	3.1453	16.9911
	.50	–.006115	–.022100	–.044603	–.070248	–.095292	–.115180	–.124065	–.114385	–.076683	3.0231	30.6792
	.25	–.005728	–.020708	–.041881	–.066246	–.090475	–.110379	–.120269	–.112306	–.076190	2.5853	85.4422
	.05	–.005053	–.018471	–.037845	–.060781	–.084494	–.105187	–.117182	–.111898	–.077195	1.7640	770.7586

continued

Table 12.13 (continued) Bending moment variation and rotations D_2 and D_4 at bottom and top edges, respectively; outward normal line load of intensity $F_3 = I$ (force/unit length) at top edge; both edges free

		x/l										
η	h_t/h_b	.1	.2	.3	.4	.5	.6	.7	.8	.9	D_2	D_4
2.0	1.00	–.004980	–.018526	–.038467	–.062266	–.086657	–.107208	–.117883	–.110673	–.075424	2.0377	13.8008
	.75	–.004739	–.017630	–.036667	–.059556	–.083314	–.103771	–.115018	–.108920	–.074866	1.7512	21.2489
	.50	–.004241	–.015882	–.033313	–.054674	–.077438	–.097823	–.110108	–.105944	–.073934	1.2180	39.5964
	.25	–.003270	–.012582	–.027140	–.045863	–.066961	–.087290	–.101455	–.100765	–.072395	.2374	114.5556
	.05	–.001877	–.007951	–.018657	–.033982	–.053095	–.073680	–.090810	–.095269	–.071683	–1.1637	1108.7348
3.0	1.00	–.002481	–.010040	–.022716	–.040094	–.060771	–.081638	–.097032	–.097899	–.071207	–.1958	20.4657
	.75	–.001863	–.007887	–.018582	–.034023	–.053339	–.073928	–.090452	–.093721	–.069805	–.7803	32.5879
	.50	–.001013	–.004943	–.012926	–.025667	–.042981	–.062991	–.080920	–.087544	–.067706	–1.5197	62.9472
	.25	.000119	.000987	–.005212	–.014005	–.028069	–.046629	–.066035	–.077497	–.064221	–2.3509	191.6634
	.05	.001197	.002929	.002790	–.001226	–.010674	–.026185	–.046098	–.063342	–.059629	–2.8135	2093.7119
4.0	1.00	–.000827	–.004260	–.011619	–.023860	–.041011	–.061257	–.079687	–.086835	–.067420	–1.4233	27.3104
	.75	–.000238	–.002160	–.007438	–.017434	–.032717	–.052152	–.071455	–.081306	–.065464	–1.8537	44.1051
	.50	.000413	.000212	–.002589	–.009723	–.022363	–.040275	–.060219	–.073429	–.062583	–2.2079	86.7442
	.25	.001011	.002549	.002571	–.000838	–.009408	–.024120	–.043607	.060829	–.057689	–2.2291	272.5678
	.05	.001149	.003531	.005676	.006046	.002812	–.006110	–.022055	–.042089	–.049857	–1.4890	3225.0867
5.0	1.00	.000081	–.000912	–.004797	–.013231	–.027223	–.046135	–.066059	–.077684	–.064148	–1.8260	34.1244
	.75	.000437	.000629	.001494	–.007733	–.019534	–.037020	–.057213	–.071356	–.061785	–1.9660	55.6211
	.50	.000824	.002021	.001745	–.001894	–.010723	–.025790	–.045548	.062479	.058300	–1.8859	110.8207
	.25	.000929	.002774	.004156	.003509	–.001105	–.011731	–.029057	–.048474	–.052290	–1.3402	356.1602
	.05	.000560	.001987	.003856	.005465	.005477	.001579	–.009235	–.027787	.041886	–.2753	4459.6655

6.0	1.00	.000493	.000786	−.000947	−.006608	−.017836	−.035004	−.055329	−.070058	−.061292	−1.7340	40.9080
	.75	.000704	.001709	.001315	−.002359	−.011231	−.026434	−.046351	−.063209	−.058595	−1.6199	67.1594
	.50	.000787	.002268	.003086	.001594	−.004248	−.016380	−.034834	−.053711	−.054611	−1.2576	135.1479
	.25	.000610	.002032	.003590	.004251	.002281	−.004832	−.019304	−.039009	−.047702	−.5261	441.6748
	.05	.000145	.000725	.001869	.003458	.004779	.003952	−.002509	−.018064	−.035328	.2786	5769.6935
8.0	1.00	.000593	.001656	.001944	−.000248	−.007083	−.020427	−.039712	−.057982	−.056459	−1.0160	54.4528
	.75	.000551	.001717	.002611	.001846	−.002689	−.013406	−.031104	−.050548	−.053230	−.6719	90.3303
	.50	.000396	.001401	.002607	.003127	.001188	−.005942	−.020655	−.040503	−.048474	−.2129	184.3237
	.25	.000119	.000612	.001588	.002849	.003442	.001060	−.007994	−.025701	−.040251	.2479	616.5177
	.05	−.000093	−.000180	.000011	.000743	.002119	.003523	.002375	−.006877	−.025355	.2911	8545.5730
10.0	1.00	.000392	.001311	.002168	.001764	−.002015	−.011351	−.029028	−.048741	−.052435	−.3364	67.9905
	.75	.000272	.001033	.002039	.002510	.000660	−.006387	−.021170	−.041106	−.048806	−.0380	113.5740
	.50	.000108	.000560	.001442	.002471	.002470	−.001195	−.012181	−.031081	−.043496	.2232	233.9076
	.25	−.000047	.000005	.000417	.001357	.002557	.002508	−.002523	−.017104	−.034416	.2957	794.6092
	.05	−.000055	−.000172	−.000251	−.000104	.000575	.001949	.002870	−.001692	−.018331	.0619	11453.6713
12.0	1.00	.000195	.000800	.001678	.002126	.000356	−.006606	−.021411	−.041408	−.048980	.0605	81.5103
	.75	.000086	.000476	.001267	.002147	.001824	−.002503	−.014438	−.033806	−.045045	.2205	136.8392
	.50	−.000016	.000115	.000631	.001589	.002386	.000903	−.006944	−.024113	−.039333	.2768	283.7155
	.25	−.000061	−.000133	−.000029	.000494	.001563	.002456	.000094	−.011360	−.029705	.1545	974.6729
	.05	−.000012	−.000061	−.000150	−.000212	−.000015	.000842	.002231	.000594	−.013291	−.0282	14435.1369
14.0	1.00	.000067	.000406	.001125	.001905	.001386	−.003343	−.015840	−.035451	−.045954	.2188	95.0054
	.75	−.000005	.000151	.000683	.001583	.002049	−.000362	−.009767	−.028025	−.041783	.2547	160.1029
	.50	−.000048	−.000056	.000193	.000907	.001914	.001714	−.003656	−.018832	−.035777	.1925	333.6487
	.25	−.000036	−.000111	−.000136	.000087	.000839	.001969	.001259	−.007441	−.025818	.0397	1155.9308
	.05	.000004	−.000006	−.000054	−.000141	−.000159	.000251	.001476	.001466	−.009625	−.0313	17451.5490

continued

Table 12.13 (continued) Bending moment variation and rotations D_2 and D_4 at bottom and top edges, respectively; outward normal line load of intensity $F_3 = I$ (force/unit length) at top edge; both edges free

η	h_t/h_b	x/l										
		.1	.2	.3	.4	.5	.6	.7	.8	.9	D_2	D_4
16.0	1.00	–.000000	.000159	.000685	.001522	.001737	–.001308	–.011697	–.030528	–.043265	.2359	108.4746
	.75	–.000036	–.000006	.000314	.001075	.001884	.000779	–.006477	.023365	–.038910	.1958	183.3533
	.50	–.000042	–.000095	–.000007	.000458	.001401	.001900	–.001585	–.014757	–.032693	.0952	383.6428
	.25	–.000015	–.000063	–.000123	–.000071	.000385	.001435	.001681	–.004732	–.022562	–.0146	1337.8650
	.05	.000005	.000009	–.000008	–.000067	–.000147	–.000013	.000877	.001663	–.006932	–.0146	20476.2999
18.0	1.00	–.000028	.000025	.000378	.001140	.001747	.000054	–.008577	.026408	–.040850	.1896	121.9180
	.75	–.000038	–.000064	.000102	.000683	.001582	.001336	–.004140	–.019559	–.036351	.1184	206.5824
	.50	–.000027	–.000082	–.000080	.000188	.000964	.001795	–.000291	–.011572	–.029985	.0267	433.6525
	.25	–.000003	–.000027	–.000083	–.000111	.000127	.000979	.001727	.002847	–.019801	–.0280	1520.1110
	.05	.000003	.000008	.000007	.000022	–.000097	–.000103	.000467	.001555	.004942	–.0028	23490.2327
20.0	1.00	–.000034	–.000038	.000180	.000813	.001597	.000696	–.006209	–.022923	–.038661	.1268	135.3359
	.75	–.000030	–.000074	–.000006	.000403	.001259	.001554	–.002473	–.016417	–.034051	.0543	229.7842
	.50	–.000014	–.000057	–.000092	.000038	.000626	.001568	.000499	.009057	–.027587	–.0096	483.6446
	.25	.000002	–.000007	.000047	–.000102	–.000007	.000630	.001593	–.001536	–.017437	–.0233	1702.4010
	.05	.000001	.000004	.000008	–.000002	–.000054	–.000114	.000209	.001329	–.003466	.0019	26479.1964
24.0	1.00	–.000022	–.000060	–.000008	.000360	.001155	.001329	–.003015	–.017398	.034829	.0282	162.0952
	.75	–.000011	–.000047	.000069	.000091	.000713	.001494	–.000443	–.011610	–.030073	–.0114	276.0884
	.50	–.000000	–.000017	.000060	–.000068	.000211	.001053	.001200	–.005458	–.023524	–.0235	583.4818
	.25	.000003	.000005	–.000007	–.000052	–.000083	.000204	.001160	–.000007	–.013627	–.0061	2066.3446
	.05	–.000000	.000000	.000003	.000006	–.000007	.000068	–.000020	.000840	–.001553	.0016	32342.2988

Notes: M = Coefficient × l; D_2 or D_4 = Coefficient × $r^2/(Eh_b l^2)$.

Table 12.14 Hoop force variation; uniformly distributed moment of intensity $F_4 = 1$ (force × length/length) at top edge; both edges free

η	h_t/h_b	x/l 0.0	.1	.2	.3	.4	.5	.6	.7	.8	.9	1.0
.4	1.00	−5.7695	−4.6642	−3.5566	−2.4412	−1.3101	−.1532	1.0404	2.2815	3.5805	4.9458	6.3833
	.75	−6.2348	−4.8708	−3.5662	−2.3147	−1.1068	.0696	1.2287	2.3849	3.5525	4.7439	5.9680
	.50	−6.8349	−5.1356	−3.5763	−2.1498	−.8445	.3551	1.4672	2.5125	3.5122	4.4862	5.4500
	.25	−7.6918	−5.5110	−3.5869	−1.9109	−.4687	.7598	1.8003	2.6835	3.4449	4.1222	4.7501
	.05	−8.8169	−6.0008	−3.5961	−1.5926	.0275	1.2898	2.2292	2.8920	3.3383	3.6471	3.8834
.8	1.00	−5.1328	−4.2892	−3.4373	−2.5573	−1.6189	−.5846	.5882	1.9437	3.5242	5.3662	7.4954
	.75	−5.3107	−4.3436	−3.4105	−2.4892	−1.5450	−.5325	.6024	1.9187	3.4763	5.3300	7.5209
	.50	−5.3871	−4.3379	−3.3613	−2.4333	−1.5139	−.5474	.5366	1.8215	3.3994	5.3632	7.7904
	.25	−5.1085	−4.1406	−3.2581	−2.4359	−1.6289	−.7693	.2381	1.5207	3.2421	5.6006	8.8025
	.05	−3.8065	−3.4266	−3.0429	−2.6332	−2.1538	−1.5293	−.6374	.7187	2.8591	6.3202	11.8810
1.2	1.00	−4.2033	−3.7362	−3.2534	−2.7166	−2.0653	−1.2214	−.0926	1.4209	3.4194	5.9928	9.2077
	.75	−4.0135	−3.5938	−3.1748	−2.7163	−2.1516	−1.3888	−.3126	1.2113	3.3270	6.1745	9.8646
	.50	−3.4759	−3.2652	−3.0439	−2.7723	−2.3795	−1.7584	−.7615	.8013	3.1587	6.5603	11.2303
	.25	−2.0401	−2.4655	−2.7871	−2.9733	−2.9577	−2.6230	−1.7786	−.1317	2.7446	7.4301	14.5933
	.05	1.3110	−.6616	−2.2482	−3.4441	−4.2165	−4.4696	−3.9840	−2.3082	1.4520	9.0689	23.7905
1.6	1.00	−3.1270	−3.0853	−3.0221	−2.8820	−2.5735	−1.9713	−.9198	.7594	3.2544	6.7425	11.3603
	.75	−2.5904	−2.7532	−2.8850	−2.9327	−2.8011	−2.3481	−1.3820	.3368	3.0833	7.1400	12.7433
	.50	−1.5472	−2.1485	−2.6646	−3.0517	−3.2196	−3.0131	−2.1934	−.4211	2.7479	7.8351	15.3396
	.25	.6330	−.9278	−2.2428	−3.2963	−4.0285	−4.2935	−3.8006	−2.0321	1.8582	9.1319	21.3605
	.05	4.7725	1.4229	−1.3462	−3.5848	−5.5274	−6.5189	−6.8832	−5.6661	−1.0990	10.8111	38.5379

continued

Table 12.14 (continued) Hoop force variation; uniformly distributed moment of intensity $F_4 = 1$ (force × length/length) at top edge; both edges free

η	h_t/h_b	*x/l*										
		0.0	*.1*	*.2*	*.3*	*.4*	*.5*	*.6*	*.7*	*.8*	*.9*	*1.0*
2.0	1.00	−2.0377	−2.4109	−2.7604	−3.0204	−3.0740	−2.7486	−1.8153	.0042	3.0197	7.5369	13.8008
	.75	−1.2417	−1.9303	−2.5645	−3.0892	−3.3918	−3.2849	−2.4915	−.6386	2.7304	8.1111	15.9367
	.50	.1021	−1.1452	−2.2567	−3.2009	−3.8880	−4.1287	−3.5893	−1.7453	2.1462	9.0150	19.7982
	.25	2.5124	.2594	−1.6780	−3.3230	−4.6544	−5.5306	−5.5756	−4.0112	.5719	10.4478	28.6389
	.05	6.2432	2.6023	−.4592	−3.0684	−5.3683	−7.4100	−8.9402	−8.9503	−4.6430	11.1160	55.4367
3.0	1.00	.1958	−.9339	−2.0568	−3.1264	−4.0101	−4.4450	−4.0032	−2.0759	2.1038	9.3745	20.4657
	.75	1.1700	−.3125	−1.7343	−3.0912	−4.2953	−5.0983	−5.0111	−3.2356	1.3580	10.1442	24.4409
	.50	2.4451	.5255	−1.2483	−2.9397	−4.5489	−5.8918	−6.4399	−5.1235	−.1398	11.0932	31.4736
	.25	3.9626	1.6655	−.3908	−2.3745	−4.4565	−6.6504	−8.5270	−8.7065	−4.0051	11.7231	47.9159
	.05	4.5094	2.7560	1.1708	−.5170	−2.7292	−5.9470	−10.4177	−15.2448	−15.6155	5.8553	104.6856
4.0	1.00	1.4233	.0414	−1.3773	−2.8634	−4.3440	−5.5323	−5.8210	−4.2000	.7519	10.7294	27.3104
	.75	2.1296	.5740	−.9713	−2.6078	−4.3568	−6.0030	−6.9047	−5.7896	−.6074	11.3966	33.0788
	.50	2.7818	1.1669	−.3923	−2.0862	−4.0829	−6.3304	−8.2466	−8.2592	−3.2384	11.8842	43.3721
	.25	2.9356	1.7023	.5131	−.9100	−2.9467	−5.9067	−9.6167	−12.4828	−9.5745	10.5663	68.1419
	.05	1.1599	1.5061	1.6877	1.4758	.3680	−2.5139	−8.3517	−17.6912	−26.0213	−5.7416	161.2543
5.0	1.00	1.8260	.5678	−.7786	−2.3487	−4.1829	−6.0427	−7.2007	−6.2379	−.9350	11.5731	34.1244
	.75	2.1437	.9303	−.3394	−1.8796	−3.8629	−6.1946	−8.1886	−8.1501	−2.9695	11.9450	41.7158
	.50	2.1818	1.2310	.2366	−1.0885	−3.0882	−5.9352	−9.1808	−10.9502	−6.7515	11.6499	55.4103
	.25	1.4547	1.2633	.9777	.3038	−1.2692	−4.3849	−9.3980	−15.0597	−15.2277	7.6533	89.0400
	.05	−.7449	.4426	1.4136	2.1062	2.1531	.5303	−4.8295	−16.7040	−33.5742	−21.0076	222.9833

6.0	1.00	1.7340	.7901	–.2847	–1.7269	–3.7152	–6.1232	–8.1888	–8.1301	–2.8470	11.9848	40.9080
	.75	1.7190	.9830	.1385	–1.1145	–3.1033	–5.9202	–8.9641	–10.2423	–5.5518	11.9310	50.3695
	.50	1.3560	1.0458	.6297	–.2175	–1.9794	–5.0995	–9.4417	–13.1069	–10.3857	10.6287	67.5739
	.25	.3265	.7748	1.0875	1.0356	.0799	–2.7052	–8.3455	–16.4505	–20.4572	3.4713	110.4187
	.05	–1.2316	–.1343	.8450	1.8155	2.6511	2.4709	–1.2570	–13.5192	–37.7689	–37.9575	288.4847
8.0	1.00	1.0160	.7729	.3707	–.5481	–2.4272	–5.4997	–9.2159	–11.3512	–7.0038	11.8065	54.4528
	.75	.6723	.7288	.6546	.1054	–1.4772	–4.6550	–9.3511	–13.4908	–10.8954	10.6269	67.7477
	.50	.1185	.5384	.8563	.8257	–.1436	–3.0066	–8.5917	–15.7948	–17.3451	6.8509	92.1619
	.25	–.5643	.1187	.7513	1.3271	1.4873	.0855	–5.1317	–16.3906	–28.8313	–7.3978	154.1294
	.05	–.5840	–.3475	–.0486	.5513	1.7405	3.4113	3.5995	–5.0081	–37.6154	–71.7542	427.2786
10.0	1.00	.3364	.5481	.6441	.2746	–1.1762	–4.3706	–9.2896	–13.7426	–11.2352	10.6187	67.9905
	.75	.0002	.4045	.7392	.7521	–.1891	–3.0845	–8.6758	–15.4981	–16.0199	8.0609	85.1805
	.50	–.3455	.1710	.6719	1.0837	.9141	–1.0756	–6.7487	–16.6288	–23.3662	1.4448	116.9538
	.25	–.4813	–.1228	.2857	.9009	1.6663	1.6800	–1.8700	–13.9450	–34.1734	–20.1452	198.6523
	.05	–.0029	–.1751	–.2850	–.1903	.4964	2.3547	5.0201	2.1771	–30.2322	–100.6537	572.6836
12.0	1.00	–.0605	.3207	.6633	.7138	–.1958	–3.1090	–8.7197	–15.3365	–15.2915	8.6758	31.5103
	.75	–.2625	.1642	.6029	.9494	.6390	–1.6066	–7.4033	–16.4363	–20.6545	4.5856	102.6294
	.50	–.3648	–.0194	.3850	.9293	1.3331	.3666	–4.6077	–16.1184	–28.2260	–5.0267	141.8578
	.25	–.2137	–.1515	–.0121	.4058	1.2962	2.2516	.6534	–10.4580	–36.7887	–33.6151	243.6682
	.05	.1322	–.0354	–.1979	–.3398	–.2020	1.0108	4.3321	6.4198	–19.9733	–122.2510	721.7568
14.0	1.00	–.2188	.1485	.5510	.8664	.4807	–1.9101	–7.7534	–16.2352	–19.0472	6.1524	95.0054
	.75	–.2869	.0199	.3995	.8861	1.0754	–.3878	–5.8680	–16.5314	–24.6901	.4468	120.0772
	.50	–.2380	–.0908	.1451	.6403	1.3493	1.2843	–2.5691	–14.7277	–31.9139	–12.1760	166.8244
	.25	–.0324	–.1082	–.1307	.0635	.7986	2.1736	2.2774	–6.7853	–37.2005	–47.0803	288.9827
	.05	.0870	.0202	–.0714	–.2404	–.4067	.0640	2.8820	8.0077	–9.6643	–136.2204	872.5774

continued

Table 12.14 (continued) Hoop force variation; uniformly distributed moment of intensity F_4 = 1 (force × length/length) at top edge; both edges free

η	h_t/h_b	x/l										
		0.0	*.1*	*.2*	*.3*	*.4*	*.5*	*.6*	*.7*	*.8*	*.9*	*1.0*
16.0	1.00	–.2359	.0361	.3940	.8412	.3919	–.8698	–6.5772	–16.5587	–22.4413	3.1783	108.4746
	.75	–.2154	–.0515	.2115	.7066	1.2282	.5257	–4.2904	–15.9960	–28.0957	–4.1744	137.5150
	.50	–.1097	–.0987	–.0074	.3601	1.1551	1.7591	–.8274	–12.8191	–34.5100	–19.7272	191.8214
	.25	.0423	–.0572	–.1403	–.1118	.3682	1.7727	3.1150	–3.4066	–35.9459	–60.0796	334.4662
	.05	.0282	.0273	.0026	–.1090	–.3548	–.4043	1.4594	7.7854	–.7984	–143.2240	1023.8150
18.0	1.00	–.1896	–.0280	.2424	.7229	1.0960	–.0253	–5.3240	–16.4216	–25.4485	–.1464	121.9180
	.75	–.1275	–.0765	.0707	.4992	1.1970	1.1489	–2.8037	–15.0106	–30.8819	–9.1408	154.9368
	.50	–.0241	–.0795	–.0823	.1437	.8797	1.9054	.5486	–10.6589	–36.1330	–27.4787	216.8263
	.25	.0535	–.0202	–.1046	–.1692	.0657	1.2704	3.3636	–.5529	–33.5029	–72.3228	380.0277
	.05	–.0022	.0180	.0269	–.0202	–.2242	–.5294	.3873	6.5710	6.0335	–144.3323	1174.5116
20.0	1.00	–.1268	–.0580	.1188	.5682	1.1504	.6207	–4.0835	–15.9257	–28.0644	–3.7427	135.3359
	.75	–.0565	–.0755	–.0192	.3094	1.0599	1.5244	–1.4810	–13.7217	–33.0811	–14.3464	172.3382
	.50	.0191	–.0536	–.1042	–.0002	.5999	1.8323	1.5603	–8.4350	–36.9130	–35.2815	241.8223
	.25	.0396	.0007	–.0612	–.1609	–.1126	.7929	3.2187	1.6988	–30.2691	–83.6339	425.6002
	.05	–.0102	.0079	.0253	.0212	–.1039	–.4694	–.2818	4.9719	10.7752	–140.7202	1323.9598
24.0	1.00	–.0282	–.0618	–.0285	.2707	.9956	1.3898	–1.8479	–14.1922	–32.1648	–11.5087	162.0952
	.75	.0147	–.0457	–.0852	.0436	.6734	1.7360	.5664	–10.6714	–35.9004	–25.1585	207.0663
	.50	.0319	–.0129	–.0756	–.1187	.1618	1.3614	2.6535	–4.2603	–36.4515	–50.6220	291.7409
	.25	.0082	.0111	–.0040	–.0830	–.2132	.1060	2.3539	4.4917	–22.6291	–103.1162	516.5861
	.05	–.0044	–.0013	.0071	.0269	.0201	–.2069	–.7137	1.9748	14.9934	–123.7146	1617.1149

Note: N = Coefficient × r/l^2.

Table 12.15 Bending moment variation and rotations D_2 and D_4 at bottom and top edges, respectively; uniformly distributed moment of intensity $F_4 = 1$ (force × length/length) at top edge; both edges free

η	h_t/h_b	x/l											
		.1	*.2*	*.3*	*.4*	*.5*	*.6*	*.7*	*.8*	*.9*	*1.0*	D_2	D_4
.4	1.00	–.0271	–.1008	–.2100	–.3437	–.4904	–.6386	–.7764	–.8914	–.9704	–1.0000	11.0513	14.7464
	.75	–.0289	–.1066	–.2200	–.3566	–.5043	–.6513	–.7860	–.8968	–.9722	–1.0000	12.3892	19.2144
	.50	–.0314	–.1142	–.2329	–.3732	–.5220	–.6674	–.7981	–.9038	–.9743	–1.0000	14.2876	30.1354
	.25	–.0348	–.1250	–.2512	–.3967	–.5470	–.6900	–.8150	–.9133	–.9773	–1.0000	17.3372	81.4055
	.05	–.0394	–.1390	–.2750	–.4272	–.5795	.7191	–.8366	–.9254	–.9809	–1.0000	21.8590	1432.2390
.8	1.00	–.0243	–.0915	–.1930	–.3201	–.4633	–.6122	–.7550	–.8783	–.9661	–1.0000	8.4301	22.7773
	.75	–.0250	–.0934	–.1960	–.3234	–.4662	–.6142	–.7561	–.8786	–.9661	–1.0000	8.5502	34.9105
	.50	–.0252	–.0938	–.1961	–.3228	–.4645	–.6116	–.7532	–.8764	–.9652	–1.0000	8.2026	69.0911
	.25	–.0239	–.0893	–.1874	–.3098	–.4485	–.5947	–.7383	–.8664	–.9616	–1.0000	6.3150	252.5715
	.05	–.0184	–.0711	–.1542	–.2635	–.3943	–.5402	–.6921	–.8361	–.9505	–1.0000	.1970	5559.2202
1.2	1.00	–.0203	–.0779	–.1680	–.2852	–.4228	–.5725	–.7228	–.8584	.9594	–1.0000	4.6600	35.4863
	.75	–.0194	–.0747	–.1617	–.2759	–.4113	–.5604	–.7123	–.8516	–.9570	–1.0000	3.2641	59.5106
	.50	–.0170	–.0667	–.1468	–.2545	–.3858	–.5344	–.6902	–.8373	–.9520	–1.0000	.3777	129.5670
	.25	–.0109	–.0464	–.1097	–.2025	–.3247	–.4727	–.6378	–.8032	–.9398	–1.0000	–6.2603	518.3739
	.05	.0032	.0001	–.0252	–.0845	–.1857	–.3309	–.5150	–.7205	–.9085	–1.0000	–20.4170	12244.3871
1.6	1.00	–.0156	–.0620	–.1385	–.2438	–.3745	–.5246	–.6834	–.8340	–.9512	–1.0000	.4011	52.1024
	.75	–.0132	–.0540	–.1235	–.2222	–.3486	–.4981	–.6609	–.8194	–.9460	–1.0000	–2.3476	91.3602
	.50	–.0087	–.0389	–.0956	–.1826	–.3016	–.4502	–.6199	–.7927	–.9365	–1.0000	–7.1526	207.3130
	.25	.0005	–.0080	–.0388	–.1022	–.2056	–.3513	–.5340	–.7353	–.9153	–1.0000	–16.3583	862.5898
	.05	.0182	.0511	.0709	.0554	–.0131	–.1460	–.3466	–.6012	–.8610	–1.0000	–31.9784	21317.0452

continued

Table 12.15 (continued) Bending moment variation and rotations D_2 and D_4 at bottom and top edges, respectively; uniformly distributed moment of intensity F_4 = 1 (force × length/length) at top edge; both edges free

η	h_t/h_b	x/l											
		.1	.2	.3	.4	.5	.6	.7	.8	.9	1.0	D_2	D_4
2.0	1.00	–.0108	–.0457	–.1081	–.2006	–.3235	–.4733	–.6407	–.8070	–.9419	–1.0000	–3.7483	71.8799
	.75	–.0073	–.0339	–.0861	–.1690	–.2854	–.4342	–.6070	–.7850	–.9341	–1.0000	–7.3917	129.0012
	.50	–.0016	–.0145	–.0498	–.1170	–.2226	–.3690	–.5501	–.7471	–.9202	–1.0000	–13.0776	299.0186
	.25	.0088	.0204	.0155	–.0224	–.1065	–.2452	–.4384	–.6693	–.8903	–1.0000	–22.3023	1274.0975
	.05	.0251	.0766	.1240	.1409	.1043	–.0060	–.2046	–.4893	–.8115	–1.0000	–33.6270	32636.2494
3.0	1.00	–.0009	–.0110	–.0418	–.1036	–.2052	–.3505	–.5347	–.7378	–.9173	–1.0000	–11.2972	131.6096
	.75	.0034	.0038	–.0132	–.0609	–.1513	–.2920	–.4814	–.7009	–.9034	–1.0000	–14.8893	242.3572
	.50	.0090	.0235	.0255	–.0019	–.0745	–.2054	–.3993	–.6414	–.8799	–1.0000	–18.8805	577.3670
	.25	.0160	.0488	.0778	.0829	.0434	–.0624	–.2522	–.5252	–.8295	–1.0000	–21.5519	2558.2374
	.05	.0196	.0669	.1259	.1792	.2044	.1691	.0292	–.2599	–.6890	–1.0000	–14.5571	70029.9685
4.0	1.00	.0049	.0101	.0016	–.0356	–.1160	–.2511	–.4428	–.6738	–.8932	–1.0000	–13.7840	202.4396
	.75	.0081	.0220	.0260	.0039	–.0617	–.1868	–.3793	–.6264	–.8742	–1.0000	–15.3501	377.7489
	.50	.0112	.0342	.0531	.0509	.0077	–.0986	–.2859	–.5518	–.8420	–1.0000	–15.4529	915.4210
	.25	.0126	.0423	.0770	.1020	.0968	.0319	–.1287	–.4096	–.7728	–1.0000	–10.8584	4164.3561
	.05	.0064	.0277	.0657	.1176	.1718	.1985	.1387	–.0980	–.5749	–1.0000	5.0870	119268.3282
5.0	1.00	.0071	.0198	.0245	.0056	–.0553	–.1760	–.3671	–.6170	–.8705	–1.0000	–12.5062	282.4920
	.75	.0087	.0267	.0411	.0364	–.0074	–.1128	–.2985	–.5616	–.8466	–1.0000	–11.8000	532.2379
	.50	.0093	.0310	.0547	.0670	.0477	–.0313	–.2011	–.4757	–.8064	–1.0000	–8.7564	1306.1073
	.25	.0070	.0265	.0555	.0868	.1042	.0761	–.0466	–.3160	–.7198	–1.0000	–.8121	6054.0994
	.05	.0017	.0008	.0173	.0543	.1116	.1712	.1771	.0109	–.4725	–1.0000	12.3110	179168.4457

6.0	1.00	.0072	.0221	.0339	.0280	–.0154	–.1199	–.3045	–.5663	–.8488	–1.0000	–9.3322	370.9266
	.75	.0074	.0246	.0428	.0492	.0239	–.0608	–.2338	–.5045	–.8205	–1.0000	–6.9950	704.0802
	.50	.0063	.0229	.0456	.0653	.0640	.0108	–.1365	–.4100	–.7728	–1.0000	–2.4546	1744.1706
	.25	.0024	.0124	.0331	.0633	.0929	.0931	.0076	–.2397	–.6704	–1.0000	5.1450	8198.6499
	.05	–.0043	–.0101	–.0074	.0134	.0600	.1284	.1775	.0806	–.3823	–1.0000	10.7490	248766.7541
8.0	1.00	.0047	.0170	.0326	.0420	.0260	–.0455	–.2080	–.4789	–.8082	–1.0000	–2.3133	570.4752
	.75	.0035	.0141	.0310	.0480	.0490	.0021	–.1381	–.4079	–.7716	–1.0000	.8379	1094.2085
	.50	.0013	.0079	.0228	.0452	.0647	.0518	–.0486	–.3029	–.7105	–1.0000	4.5087	2746.3704
	.25	–.0017	–.0022	.0047	.0246	.0582	.0896	.0645	–.1265	–.5816	–1.0000	6.8664	13169.5397
	.05	–.0025	–.0085	–.0147	–.0149	.0027	.0535	.1336	.1438	–.2358	–1.0000	1.9220	413919.9594
10.0	1.00	.0020	.0095	.0230	.0384	.0405	–.0025	–.1382	–.4059	–.7706	–1.0000	2.1874	796.8925
	.75	.0007	.0053	.0171	.0357	.0508	.0328	–.0732	–.3296	–.7267	–1.0000	4.1615	1539.1315
	.50	–.0009	–.0001	.0074	.0253	.0509	.0627	.0034	–.2206	–.6542	–1.0000	5.2036	3897.6029
	.25	–.0018	–.0048	–.0048	.0044	.0297	.0691	.0827	–.0511	–.5043	–1.0000	3.3963	18950.4936
	.05	–.0003	–.0023	–.0070	–.0132	–.0134	.0108	.0815	.1510	–.1274	–1.0000	–1.9135	609389.6955
12.0	1.00	.0003	.0038	.0137	.0300	.0427	.0221	–.0868	–.3442	–.7354	–1.0000	3.8142	1047.1472
	.75	–.0006	.0004	.0074	.0234	.0443	.0461	–.0290	–.2653	–.6852	–1.0000	4.2559	2032.7423
	.50	–.0013	–.0027	–.0001	.0117	.0358	.0603	.0332	–.1568	–.6030	–1.0000	3.3544	5181.6050
	.25	–.0010	–.0034	–.0057	–.0036	.0116	.0475	.0823	–.0012	–.4369	–1.0000	.4359	25453.7587
	.05	.0004	.0004	–.0015	–.0067	–.0130	–.0074	.0414	.1339	–.0487	–1.0000	–1.6800	830914.1356
14.0	1.00	–.0005	.0005	.0070	.0216	.0394	.0352	–.0489	–.2915	–.7024	–1.0000	3.6157	1319.0667
	.75	–.0009	–.0016	.0018	.0139	.0354	.0498	.0010	–.2121	–.6466	–1.0000	2.9830	2570.6552
	.50	–.0010	–.0028	–.0029	.0036	.0230	.0523	.0490	–.1071	–.5562	–1.0000	1.3384	6586.4327
	.25	–.0003	–.0016	–.0041	–.0055	.0017	.0298	.0737	.0311	–.3778	–1.0000	–.8651	32613.1353
	.05	.0003	.0008	.0006	–.0021	–.0084	–.0122	.0155	.1086	.0072	–1.0000	–.6135	1074937.7930

continued

Table 12.15 (continued) Bending moment variation and rotations D_2 and D_4 at bottom and top edges, respectively; uniformly distributed moment of intensity F_4 = 1 (force × length/length) at top edge; both edges free

η	h_t/h_b	x/l										D_2	D_4
		.1	.2	.3	.4	.5	.6	.7	.8	.9	1.0		
16.0	1.00	–.0008	–.0011	.0026	.0145	.0338	.0412	–.0210	–.2462	–.6714	–1.0000	2.6285	1610.9972
	.75	–.0008	–.0021	–.0010	.0072	.0267	,0482	.0209	–.1679	–.6107	–1.0000	1.5306	3149.4818
	.50	–.0005	–.0020	–.0033	–.0007	.0135	.0428	.0562	–.0682	–.5134	–1.0000	.0004	8102.7567
	.25	.0000	–.0005	–.0023	–.0049	–.0030	.0168	.0620	.0512	–.3259	–1.0000	–1.0286	40376.1433
	.05	.0001	.0005	.0009	.0000	–.0043	–.0112	.0006	.0829	.0461	–1.0000	.0364	1338410.1235
18.0	1.00	–.0007	–.0016	.0001	.0090	.0277	.0428	–.0004	–.2072	–.6422	–1.0000	1.5165	1921.6110
	.75	–.0006	–.0018	–.0021	.0028	.0192	.0440	.0338	–.1309	–.5770	–1.0000	.4140	3766.4850
	.50	–.0002	–.0012	–.0028	–.0026	.0068	.0335	.0578	–.0378	–.4740	–1.0000	–.6198	9723.0180
	.25	.0001	.0001	–.0010	–.0036	–.0047	.0078	.0499	.0628	–.2802	–1.0000	–.7249	48699.6719
	.05	.0000	.0002	.0006	.0007	–.0015	–.0083	–.0067	.0601	.0721	–1.0000	.2226	1618664.8951
20.0	1.00	–.0005	–.0016	–.0012	.0050	.0220	.0417	.0145	–.1734	–.6145	–1.0000	.6001	2249.8065
	.75	–.0003	–.0013	–.0023	.0001	.0131	.0386	.0417	–.0999	–.5455	–1.0000	–.2579	4419.3920
	.50	–.0000	–.0006	–.0020	–.0031	.0024	.0253	.0562	–.0140	–.4377	–1.0000	–.7500	11440.9249
	.25	.0001	.0003	–.0003	–.0023	–.0048	.0021	.0389	.0686	–.2399	–1.0000	–.3598	57547.3381
	.05	–.0000	.0000	.0003	.0008	–.0001	–.0053	–.0095	.0413	.0885	–1.0000	.1329	1913342.3980
24.0	1.00	–.0002	–.0010	–.0018	.0004	.0125	.0356	.0326	–.1184	–.5633	–1.0000	–.3743	2955.3368
	.75	–.0000	–.0005	–.0017	–.0020	.0048	.0274	.0476	–.0521	–.4881	–1.0000	–.6110	5825.4773
	.50	.0001	.0000	–.0008	–.0025	–.0019	.0127	.0479	.0186	–.3731	–1.0000	–.4262	15148.9951
	.25	.0000	.0002	.0002	–.0006	–.0032	–.0032	.0212	.0693	–.1726	–1.0000	.0492	76693.5362
	.05	–.0000	–.0000	.0000	.0003	.0007	–.0014	–.0085	.0155	.1021	–1.0000	.0248	2537755.1077

Notes: M = Coefficient; D_2 or D_4 = $r^2/(Eh_b\beta^3)$.

Table 12.16 Flexibility coefficients corresponding to the coordinate system in Figure 12.1b

η	h_t/h_b	*(1, 1)*	*(1, 2)*	*(1, 3)*	*(1, 4)*	*(2, 2)*	*(2, 3)*	*(2, 4)*	*(3, 3)*	*(3, 4)*	*(4, 4)*
.4	1.00	6.3571	−9.9740	−3.0404	−9.0148	23.0413	9.0148	17.2677	6.3571	9.9740	23.0413
	.75	6.8081	−10.9974	−3.4700	−9.7419	26.6038	10.7574	19.3581	7.9071	12.4334	30.0224
	.50	7.3715	−12.4288	−4.1218	−10.6795	31.9838	13.5198	22.3243	10.5517	17.0312	47.0865
	.25	8.1500	−14.7368	−5.2926	−12.0184	41.4173	18.6965	27.0893	16.2633	29.6879	127.1960
	.05	9.1659	−18.3355	−7.4078	−13.7765	57.2446	28.3262	34.1547	32.9905	121.3567	2237.8733
.8	1.00	1.6676	−2.9279	−.7024	−2.0050	8.8974	2.0050	3.2930	1.6676	2.9279	8.8974
	.75	1.7955	−3.2671	−.7827	−2.0745	10.2911	2.3199	3.3399	2.1208	3.9171	13.6369
	.50	1.9554	−3.7320	−.8953	−2.1044	12.2904	2.7832	3.2041	2.9539	6.0862	26.9887
	.25	2.1705	−4.4342	−1.0776	−1.9955	15.4601	3.5670	2.4668	5.0624	13.7540	98.6608
	.05	2.4253	−5.3770	−1.3845	−1.4869	19.8949	4.9112	.0770	14.3673	92.8204	2171.5704
1.2	1.00	.7940	−1.5986	−.2744	−.7297	6.1608	.7297	.8090	.7940	1.5986	6.1608
	.75	.8595	−1.7985	−.2943	−.6968	7.0921	.8002	.5667	1.0372	2.2835	10.3317
	.50	.9404	−2.0636	−.3171	−.6034	8.3624	.8845	.0656	1.5131	3.8994	22.4943
	.25	1.0447	−2.4357	−.3442	−.3542	10.2043	.9888	−1.0869	2.8506	10.1343	89.9955
	.05	1.1561	−2.8701	−.3823	.2276	12.4230	1.1295	−3.5446	10.0874	82.6060	2125.7616
1.6	1.00	.4832	−1.1094	−.1294	−.3054	5.0881	.3054	.0392	.4832	1.1094	5.0881
	.75	.5247	−1.2502	−.1318	−.2530	5.8064	.3072	−.2293	.6469	1.6593	8.9219
	.50	.5748	−1.4298	−.1309	−.1511	6.7437	.2952	−.6985	.9798	2.9960	20.2454
	.25	.6364	−1.6658	−.1227	.0618	8.0125	.2525	−1.5975	1.9706	8.3440	34.2373
	.05	.6970	−1.9152	−.1073	.4661	9.3996	.1723	−3.1229	7.9856	75.2694	2081.7427

continued

Table 12.16 (continued) Flexibility coefficients corresponding to the coordinate system in Figure 12.1b

η	h_t/h_b	*(1, 1)*	*(1, 2)*	*(1, 3)*	*(1, 4)*	*(2, 2)*	*(2, 3)*	*(2, 4)*	*(3, 3)*	*(3, 4)*	*(4, 4)*
2.0	1.00	.3350	−.8625	−.0664	−.1274	4.4925	.1274	−.2343	.3350	.8625	4.4925
	.75	.3640	−.9688	−.0633	−.0776	5.0761	.1094	−.4620	.4574	1.3281	8.0626
	.50	.3979	−1.0993	−.0563	.0064	5.8105	.0761	−.8174	.7116	2.4748	18.6887
	.25	.4380	−1.2623	−.0424	.1570	6.7573	.0148	−1.3939	1.4931	7.1597	79.6311
	.05	.4752	−1.4242	−.0219	.3902	7.7353	−.0727	−2.1017	6.6282	69.2959	2039.7655
3.0	1.00	.1782	−.5685	−.0136	.0054	3.6558	−.0054	−.3138	.1782	.5685	3.6558
	.75	.1924	−.6284	−.0097	.0325	4.0423	−.0217	−.4136	.2504	.9052	6.7321
	.50	.2081	−.6975	−.0044	.0679	4.5026	−.0422	−.5245	.4041	1.7485	16.0380
	.25	.2254	−.7782	.0028	.1101	5.0610	−.0653	−.5987	.8977	5.3240	71.0621
	.05	.2406	−.8536	.0094	.1253	5.6040	−.0782	−.4044	4.6072	58.1587	1945.2769
4.0	1.00	.1155	−.4267	−.0019	.0222	3.1631	−.0222	−.2154	.1155	.4267	3.1631
	.75	.1236	−.4650	.0002	.0333	3.4477	−.0290	−.2398	.1645	.6891	5.9023
	.50	.1323	−.5082	.0024	.0435	3.7790	−.0345	−.2415	.2701	1.3554	14.3035
	.25	.1417	−.5575	.0044	.0459	4.1707	−.0348	−.1697	.6191	4.2589	65.0681
	.05	.1498	−.6026	.0044	.0181	4.5403	−.0233	.0795	3.4849	50.3920	1863.5676
5.0	1.00	.0826	−.3412	.0008	.0183	2.8249	−.0183	−.1251	.0826	.3412	2.8249
	.75	.0878	−.3683	.0016	.0214	3.0506	−.0197	−.1180	.1186	.5562	5.3224
	.50	.0932	−.3984	.0023	.0218	3.3096	−.0189	−.0876	.1970	1.1082	13.0611
	.25	.0991	−.4325	.0023	.0145	3.6101	−.0134	−.0081	.4615	3.5616	60.5410
	.05	.1042	−.4631	.0012	−.0074	3.8873	−.0028	.1231	2.7741	44.5967	1791.6844

6.0	1.00	.0628	-.2841	.0011	.0120	2.5759	-.0120	-.0648	.0628	.2841	2.5759
	.75	.0664	-.3045	.0013	.0119	2.7635	-.0112	-.0486	.0907	.4664	4.8894
	.50	.0701	-.3271	.0014	.0094	2.9758	-.0087	-.0170	.1520	.9385	12.1123
	.25	.0742	-.3524	.0010	.0023	3.2186	-.0037	.0357	.3618	3.0672	56.9351
	.05	.0776	-.3748	.0001	-.0086	3.4388	.0019	.0746	2.2855	40.0673	1727.5469
8.0	1.00	.0408	-.2127	.0006	.0040	2.2284	-.0040	-.0090	.0408	.2127	2.2284
	.75	.0428	-.2259	.0005	.0026	2.3683	-.0026	.0033	.0594	.3529	4.2743
	.50	.0449	-.2403	.0003	.0005	2.5239	-.0008	.0176	.1007	.7200	10.7280
	.25	.0471	-.2561	.0001	-.0022	2.6979	.0010	.0268	.2451	2.4083	51.4435
	.05	.0489	-.2699	-.0001	-.0023	2.8526	.0011	.0075	1.6618	33.3811	1616.8748
10.0	1.00	.0292	-.1700	.0002	.0008	1.9922	-.0008	.0055	.0292	.1700	1.9922
	.75	.0304	-.1794	.0001	.0000	2.1035	-.0001	.0104	.0427	.2839	3.8478
	.50	.0318	-.1895	.0000	-.0009	2.2258	.0006	.0130	.0730	.5848	9.7440
	.25	.0331	-.2005	-.0000	-.0012	2.3607	.0007	.0085	.1804	1.9865	47.3762
	.05	.0343	-.2100	-.0000	-.0000	2.4791	.0002	-.0048	1.2840	28.6342	1523.4742
12.0	1.00	.0222	-.1415	.0001	-.0001	1.8180	.0001	.0066	.0222	.1415	1.8180
	.75	.0230	-.1486	.0000	-.0005	1.9103	.0004	.0074	.0326	.2376	3.5291
	.50	.0240	-.1563	-.0000	-.0006	2.0108	.0005	.0058	.0561	.4926	8.9958
	.25	.0249	-.1645	-.0000	-.0004	2.1207	.0003	.0008	.1401	1.6921	44.1906
	.05	.0257	-.1715	-.0000	.0002	2.2162	-.0000	-.0029	1.0329	25.0610	1442.5592
14.0	1.00	.0176	-.1212	.0000	-.0003	1.6825	.0003	.0046	.0176	.1212	1.6825
	.75	.0182	-.1268	-.0000	-.0004	1.7613	.0003	.0038	.0259	.2042	3.2789
	.50	.0189	-.1328	-.0000	-.0003	1.8465	.0002	.0017	.0448	.4256	8.4011
	.25	.0196	-.1392	-.0000	-.0000	1.9390	.0001	-.0011	.1130	1.4744	41.5984
	.05	.0202	-.1447	.0000	.0001	2.0189	.0000	-.0008	.8552	22.2596	1371.0941

continued

Table 12.16 (continued) Flexibility coefficients corresponding to the coordinate system in Figure 12.1b

η	h_t/h_b	*(1, 1)*	*(1, 2)*	*(1, 3)*	*(1, 4)*	*(2, 2)*	*(2, 3)*	*(2, 4)*	*(3, 3)*	*(3, 4)*	*(4, 4)*
16.0	1.00	.0144	–.1059	–.0000	–.0002	1.5732	.0002	.0026	.0144	.1059	1.5732
	.75	.0149	–.1105	–.0000	–.0002	1.6420	.0002	.0015	.0213	.1791	3.0757
	.50	.0154	–.1154	–.0000	–.0001	1.7159	.0001	.0000	.0369	.3747	7.9128
	.25	.0159	–.1206	–.0000	.0000	1.7957	–.0000	–.0010	.0936	1.3065	39.4298
	.05	.0163	–.1249	.0000	.0000	1.8642	–.0000	.0000	.7236	19.9964	1307.0411
18.0	1.00	.0120	–.0941	–.0000	–.0001	1.4827	.0001	.0012	.0120	.0941	1.4827
	.75	.0124	–.0979	–.0000	–.0001	1.5436	.0001	.0003	.0179	.1594	2.9062
	.50	.0128	–.1019	–.0000	–.0000	1.6089	.0000	–.0005	.0311	.3346	7.5023
	.25	.0132	–.1062	.0000	.0000	1.6790	–.0000	–.0006	.0792	1.1729	37.5769
	.05	.0136	–.1099	.0000	–.0000	1.7388	–.0000	.0002	.6228	18.1252	1248.9698
20.0	1.00	.0103	–.0846	–.0000	–.0001	1.4061	.0001	.0004	.0103	.0846	1.4061
	.75	.0106	–.0878	–.0000	–.0000	1.4608	.0000	–.0002	.0153	.1436	2.7621
	.50	.0109	–.0913	–.0000	.0000	1.5192	–.0000	–.0005	.0266	.3023	7.1506
	.25	.0113	–.0949	.0000	.0000	1.5816	–.0000	–.0002	.0682	1.0640	35.9671
	.05	.0115	–.0980	.0000	–.0000	1.6347	.0000	.0001	.5433	16.5495	1195.8390
24.0	1.00	.0078	–.0704	–.0000	–.0000	1.2827	.0000	–.0002	.0078	.0704	1.2827
	.75	.0080	–.0728	–.0000	.0000	1.3281	–.0000	–.0003	.0116	.1198	2.5284
	.50	.0083	–.0754	.0000	.0000	1.3762	–.0000	–.0002	.0204	.2532	6.5751
	.25	.0085	–.0781	.0000	.0000	1.4274	–.0000	.0000	.0525	.8969	33.2871
	.05	.0087	–.0804	–.0000	–.0000	1.4707	.0000	.0000	.4267	14.0375	1101.4562

Notes: Displacements due to unit value of the forces applied at one of the coordinates. f_{11}, f_{13}, or f_{33} = Coefficient × $l^3/(Eh_b^3)$; f_{12}, f_{14}, f_{23}, or f_{34} = Coefficient × $l^2/(Eh_b^3)$; f_{22}, f_{24}, or f_{44} = Coefficient × $l/(Eh_b^3)$.

Table 12.17 Stiffness coefficients corresponding to the coordinate system in Figure 12.1b

η	h_t/h_b	*(1, 1)*	*(1, 2)*	*(1, 3)*	*(1, 4)*	(2, 2)	(2, 3)	(2, 4)	(3, 3)	(3, 4)	(4, 4)
.4	1.00	1.2639	.5471	–.9472	.4945	.3487	–.4945	.1668	1.2639	–.5471	.3487
	.75	.9186	.4252	–.6105	.2767	.2846	–.3698	.1076	.8550	–.3137	.1836
	.50	.6467	.3110	–.3475	.1249	.2178	–.2517	.0583	.5194	–.1473	.0752
	.25	.4436	.2047	–.1531	.0341	.1464	–.1395	.0207	.2513	–.0434	.0168
	.05	.3269	.1240	–.0418	.0024	.0802	–.0491	.0022	.0765	–.0037	.0006
.8	1.00	1.9523	.6424	–.7225	.4399	.3659	–.4399	.1541	1.9523	–.6424	.3659
	.75	1.6088	.5291	–.4211	.2361	.3053	–.3180	.0970	1.3600	–.3768	.1937
	.50	1.3400	.4262	–.1987	.0987	.2439	–.2045	.0504	.8459	–.1820	.0798
	.25	1.1414	.3364	–.0546	.0223	.1818	–.1008	.0163	.4074	–.0554	.0179
	.05	1.0298	.2785	–.0006	.0007	.1322	–.0270	.0013	.1088	–.0047	.0007
1.2	1.00	3.0415	.7892	–.3994	.3603	.3920	–.3603	.1355	3.0415	–.7892	.3920
	.75	2.6882	.6855	–.1606	.1792	.3360	–.2452	.0820	2.1463	–.4718	.2086
	.50	2.4050	.5939	–.0091	.0644	.2809	–.1422	.0398	1.3406	–.2322	.0863
	.25	2.1825	.5168	.0526	.0089	.2280	–.0562	.0111	.6309	–.0715	.0193
	.05	2.0347	.4672	.0296	–.0006	.1894	–.0083	.0006	.1486	–.0058	.0007
1.6	1.00	4.4657	.9737	–.0343	.2680	.4240	–.2680	.1136	4.4657	–.9737	.4240
	.75	4.0816	.8774	.1155	.1168	.3724	–.1653	.0652	3.1557	–.5879	.2264
	.50	3.7575	.7923	.1714	.0300	.3227	–.0795	.0288	1.9578	–.2912	.0937
	.25	3.4779	.7184	.1338	–.0022	.2760	–.0182	.0065	.8957	–.0892	.0208
	.05	3.2687	.6648	.0411	–.0012	.2418	.0026	.0001	.1908	–.0069	.0007

continued

Table 12.17 (continued) Stiffness coefficients corresponding to the coordinate system in Figure 12.1b

η	h_t/h_b	*(1, 1)*	*(1, 2)*	*(1, 3)*	*(1, 4)*	*(2, 2)*	*(2, 3)*	*(2, 4)*	*(3, 3)*	*(3, 4)*	*(4, 4)*
2.0	1.00	6.1607	1.1828	.3213	.1746	.4594	−.1746	.0910	6.1607	−1.1828	.4594
	.75	5.7233	1.0897	.3633	.0577	.4113	−.0894	.0488	4.3373	−.7161	.2453
	.50	5.3340	1.0052	.3117	.0009	.3653	−.0257	.0190	2.6637	−.3540	.1012
	.25	4.9763	.9272	.1779	−.0096	.3218	.0086	.0030	1.1878	−.1070	.0223
	.05	4.7011	.8655	.0389	−.0013	.2887	.0072	−.0001	.2343	−.0080	.0008
3.0	1.00	11.2794	1.7540	.9683	−.0168	.5499	.0168	.0420	11.2794	−1.7540	.5499
	.75	10.6414	1.6532	.7302	−.0480	.5063	.0477	.0167	7.8364	−1.0543	.2916
	.50	10.0388	1.5546	.4446	−.0401	.4638	.0532	.0028	4.7059	−.5132	.1186
	.25	9.4687	1.4563	.1637	−.0147	.4218	.0333	−.0012	2.0080	−.1504	.0254
	.05	9.0426	1.3776	.0140	−.0007	.3884	.0063	−.0002	.3487	−.0104	.0008
4.0	1.00	17.3486	2.3404	1.1813	−.1220	.6334	.1220	.0102	17.3486	−2.3404	.6334
	.75	16.4824	2.2232	.7516	−.0903	.5907	.1047	−.0008	11.9363	−1.3936	.3326
	.50	15.6625	2.1066	.3606	−.0462	.5483	.0702	−.0038	7.0690	−.6698	.1335
	.25	14.8940	1.9911	.0797	−.0105	.5061	.0274	−.0019	2.9397	−.1924	.0280
	.05	14.3200	1.9007	−.0039	−.0001	.4725	.0021	−.0001	.4712	−.0127	.0009
5.0	1.00	24.2071	2.9242	1.0716	−.1565	.7079	.1565	−.0065	24.2071	−2.9242	.7079
	.75	23.1214	2.7914	.5769	−.0915	.6652	.1094	−.0079	16.5516	−1.7296	.3688
	.50	22.0959	2.6602	.2029	−.0363	.6226	.0580	−.0052	9.7148	−.8242	.1465
	.25	21.1324	2.5315	.0058	−.0051	.5803	.0149	−.0014	3.9699	−.2335	.0303
	.05	20.4033	2.4306	−.0090	.0001	.5468	−.0002	−.0000	.6010	−.0150	.0009

6.0	1.00	31.7827	3.5052	.7994	–.1486	.7752	.1486	–.0133	31.7827	–3.5052	.7752
	.75	30.4859	3.3595	.3415	–.0736	.7323	.0894	–.0095	21.6362	–2.0637	.4015
	.50	29.2574	3.2162	.0565	–.0226	.6897	.0376	–.0044	12.6155	–.9775	.1583
	.25	28.0942	3.0757	–.0360	–.0011	.6474	.0049	–.0008	5.0875	–.2741	.0323
	.05	27.2059	2.9649	–.0074	.0002	.6139	–.0008	.0000	.7373	–.0171	.0010
8.0	1.00	48.8729	4.6650	.1978	–.0870	.8941	.0870	–.0130	48.8729	–4.6650	.8941
	.75	47.1555	4.4979	–.0411	–.0290	.8513	.0365	–.0064	33.0619	–2.7294	.4593
	.50	45.5127	4.3333	–.1032	–.0022	.8088	.0054	–.0019	19.0929	–1.2814	.1792
	.25	43.9397	4.1710	–.0467	.0019	.7666	–.0036	–.0001	7.5542	–.3536	.0360
	.05	42.7294	4.0425	–.0013	.0001	.7330	–.0005	.0000	1.0282	–.0212	.0011
10.0	1.00	68.2579	5.8237	–.1877	–.0288	.9989	.0288	–.0077	68.2579	–5.8237	.9989
	.75	66.1126	5.6375	–.2029	–.0003	.9561	.0016	–.0027	45.9671	–3.3920	.5102
	.50	64.0479	5.4536	–.1183	.0055	.9136	–.0074	–.0003	26.3653	–1.5823	.1976
	.25	62.0612	5.2717	–.0226	.0016	.8714	–.0036	.0001	10.2936	–.4316	.0392
	.05	60.5267	5.1276	.0010	–.0000	.8378	–.0000	.0000	1.3409	–.0252	.0011
12.0	1.00	89.6766	6.9805	–.3269	.0052	1.0934	–.0052	–.0032	89.6766	–6.9805	1.0934
	.75	87.1002	6.7766	–.2071	.0110	1.0507	–.0125	–.0005	60.1840	–4.0514	.5561
	.50	84.6124	6.5749	–.0758	.0059	1.0082	–.0087	.0003	34.3432	–1.8804	.2141
	.25	82.2109	6.3754	–.0028	.0007	.9660	–.0017	.0001	13.2748	–.5083	.0421
	.05	80.3504	6.2173	.0009	–.0000	.9323	.0001	–.0000	1.6736	–.0291	.0012
14.0	1.00	112.9414	8.1346	–.3096	.0187	1.1803	–.0187	–.0005	112.9414	–8.1346	1.1803
	.75	109.9340	7.9144	–.1450	.0121	1.1375	–.0142	.0004	75.5922	–4.7080	.5982
	.50	107.0229	7.6965	–.0301	.0038	1.0950	–.0059	.0004	42.9613	–2.1763	.2293
	.25	104.2058	7.4808	.0056	.0001	1.0528	–.0004	.0000	16.4748	–.5839	.0447
	.05	102.0181	7.3097	.0003	–.0000	1.0191	.0001	–.0000	2.0249	–.0329	.0013

continued

Table 12.17 (continued) Stiffness coefficients corresponding to the coordinate system in Figure 12.1b

η	h_t/h_b	(1, 1)	(1, 2)	(1, 3)	(1, 4)	(2, 2)	(2, 3)	(2, 4)	(3, 3)	(3, 4)	(4, 4)
16.0	1.00	137.9094	9.2860	–.2249	.0202	1.2609	–.0202	.0007	137.9094	–9.2860	1.2609
	.75	134.4714	9.0506	–.0742	.0091	1.2182	–.0108	.0007	92.0987	–5.3617	.6373
	.50	131.1371	8.8176	.0001	.0018	1.1757	–.0028	.0003	52.1689	–2.4701	.2433
	.25	127.9037	8.5868	.0066	–.0001	1.1334	.0002	.0000	19.8760	–.6586	.0472
	.05	125.3880	8.4038	–.0000	–.0000	1.0997	.0000	–.0000	2.3939	–.0366	.0013
18.0	1.00	164.4656	10.4346	–.1296	.0162	1.3365	–.0162	.0010	164.4656	–10.4346	1.3365
	.75	160.5976	10.1851	–.0200	.0054	1.2938	–.0065	.0006	109.6286	–6.0129	.6739
	.50	156.8398	9.9379	.0139	.0004	1.2512	–.0007	.0001	61.9249	–2.7619	.2565
	.25	153.1898	9.6930	.0046	–.0002	1.2089	.0003	–.0000	23.4641	–.7324	.0495
	.05	150.3455	9.4987	–.0001	.0000	1.1752	.0000	–.0000	2.7799	–.0403	.0014
20.0	1.00	192.5149	11.5806	–.0511	.0108	1.4078	–.0108	.0009	192.5149	–11.5806	1.4078
	.75	188.2170	11.3176	.0126	.0024	1.3651	–.0030	.0004	128.1196	–6.6615	.7084
	.50	184.0358	11.0570	.0168	–.0003	1.3226	.0004	.0001	72.1953	–3.0519	.2689
	.25	179.9687	10.7988	.0023	–.0001	1.2802	.0003	–.0000	27.2273	–.8055	.0516
	.05	176.7955	10.5939	–.0001	.0000	1.2465	–.0000	–.0000	3.1823	–.0440	.0014
24.0	1.00	252.7801	13.8646	.0322	.0024	1.5401	–.0024	.0005	252.7801	–13.8646	1.5401
	.75	247.6233	13.5765	.0297	–.0006	1.4973	.0007	.0001	167.7813	–7.9517	.7724
	.50	242.5955	13.2910	.0095	–.0005	1.4548	.0007	–.0000	94.1682	–3.6270	.2918
	.25	237.6939	13.0080	–.0003	–.0000	1.4125	.0001	–.0000	35.2409	–.9495	.0556
	.05	233.8618	12.7833	–.0000	.0000	1.3787	–.0000	.0000	4.0353	–.0514	.0016

Notes: Forces corresponding to unit displacement introduced at one of the coordinates. S_{11}, S_{13}, or S_{33} = Coefficient × Eh_b^3/l^3; S_{12}, S_{14}, S_{23}, or S_{34} = Coefficient × Eh_b^3/l^2; S_{22}, S_{24}, or S_{44} = Coefficient × Eh_b^3/l.

Table 12.18 Hoop force variation and shear at edges; triangular load of intensity q per unit area at bottom edge and zero at top; both edges encastré

η	h_t/h_b	*x/l*											Shear	Shear
		0.0	*.1*	*.2*	*.3*	*.4*	*.5*	*.6*	*.7*	*.8*	*.9*	*1.0*	*at base*	*at top*
.4	1.00	0.0000	.0015	.0044	.0073	.0092	.0096	.0085	.0062	.0035	.0010	0.0000	.3473	–.1476
	.75	0.0000	.0017	.0053	.0089	.0115	.0122	.0110	.0083	.0047	.0015	0.0000	.3604	–.1331
	.50	0.0000	.0020	.0065	.0114	.0150	.0164	.0153	.0119	.0070	.0022	0.0000	.3772	–.1141
	.25	0.0000	.0026	.0086	.0156	.0214	.0245	.0239	.0195	.0122	.0042	0.0000	.4006	–.0861
	.05	0.0000	.0056	.0125	.0235	.0339	.0414	.0437	.0396	.0285	.0123	0.0000	.4328	–.0433
.8	1.00	0.0000	.0056	.0170	.0282	.0353	.0367	.0325	.0238	.0132	.0040	0.0000	.3398	–.1406
	.75	0.0000	.0065	.0201	.0339	.0434	.0462	.0417	.0312	.0177	.0055	0.0000	.3504	–.1249
	.50	0.0000	.0077	.0244	.0423	.0556	.0608	.0565	.0436	.0256	.0082	0.0000	.3631	–.1045
	.25	0.0000	.0095	.0311	.0558	.0760	.0866	.0842	.0685	.0427	.0147	0.0000	.3783	–.0748
	.05	0.0000	.0121	.0410	.0764	.1090	.1313	.1371	.1226	.0873	.0374	0.0000	.3919	–.0326
1.2	1.00	0.0000	.0120	.0361	.0595	.0744	.0772	.0680	.0498	.0276	.0083	0.0000	.3284	–.1303
	.75	0.0000	.0136	.0419	.0705	.0899	.0952	.0856	.0639	.0362	.0112	0.0000	.3359	–.1133
	.50	0.0000	.0157	.0496	.0855	.1118	.1215	.1124	.0865	.0505	.0162	0.0000	.3436	–.0914
	.25	0.0000	.0186	.0605	.1073	.1449	.1635	.1577	.1274	.0790	.0272	0.0000	.3503	–.0610
	.05	0.0000	.0219	.0729	.1334	.1872	.2214	.2270	.1995	.1398	.0592	0.0000	.3510	–.0226
1.6	1.00	0.0000	.0198	.0595	.0976	.1214	.1255	.1102	.0805	.0445	.0135	0.0000	.3148	–.1180
	.75	0.0000	.0222	.0678	.1133	.1436	.1513	.1355	.1009	.0570	.0176	0.0000	.3191	–.1000
	.50	0.0000	.0250	.0782	.1335	.1731	.1869	.1717	.1315	.0766	.0244	0.0000	.3224	–.0775
	.25	0.0000	.0284	.0911	.1597	.2131	.2375	.2267	.1816	.1119	.0384	0.0000	.3230	–.0482
	.05	0.0000	.0314	.1029	.1849	.2541	.2940	.2949	.2537	.1746	.0729	0.0000	.3180	–.0155

continued

Table 12.18 (continued) Hoop force variation and shear at edges; triangular load of intensity q per unit area at bottom edge and zero at top; both edges encastré

η	h_t/h_b	*x/l*											*Shear*	*Shear*
		0.0	*.1*	*.2*	*.3*	*.4*	*.5*	*.6*	*.7*	*.8*	*.9*	*1.0*	*at base*	*at top*
2.0	1.00	0.0000	.0286	.0852	.1387	.1716	.1765	.1545	.1125	.0622	.0188	0.0000	.3002	–.1050
	.75	0.0000	.0314	.0953	.1580	.1988	.2081	.1854	.1375	.0775	.0239	0.0000	.3018	–.0866
	.50	0.0000	.0346	.1072	.1811	.2326	.2488	.2269	.1727	.1002	.0319	0.0000	.3019	–.0645
	.25	0.0000	.0381	.1205	.2081	.2737	.3010	.2836	.2249	.1375	.0470	0.0000	.2989	–.0376
	.05	0.0000	.0407	.1307	.2299	.3091	.3495	.3424	.2880	.1943	.0800	0.0000	.2925	–.0108
3.0	1.00	0.0000	.0514	.1501	.2397	.2911	.2948	.2549	.1839	.1011	.0304	0.0000	.2650	–.0752
	.75	0.0000	.0547	.1618	.2617	.3218	.3302	.2894	.2121	.1186	.0364	0.0000	.2629	–.0584
	.50	0.0000	.0579	.1738	.2848	.3551	.3697	.3295	.2465	.1414	.0448	0.0000	.2595	–.0401
	.25	0.0000	.0609	.1852	.3074	.3883	.4106	.3735	.2877	.1725	.0584	0.0000	.2546	–.0209
	.05	0.0000	.0627	.1926	.3223	.4105	.4384	.4056	.3234	.2091	.0843	0.0000	.2496	–.0053
4.0	1.00	0.0000	.0733	.2088	.3253	.3863	.3840	.3270	.2335	.1276	.0383	0.0000	.2364	–.0532
	.75	0.0000	.0764	.2198	.3454	.4135	.4144	.3563	.2576	.1429	.0437	0.0000	.2334	–.0396
	.50	0.0000	.0793	.2304	.3651	.4403	.4447	.3864	.2837	.1611	.0509	0.0000	.2297	–.0260
	.25	0.0000	.0819	.2402	.3832	.4645	.4719	.4143	.3105	.1834	.0621	0.0000	.2256	–.0131
	.05	0.0000	.0838	.2472	.3954	.4796	.4869	.4289	.3282	.2072	.0842	0.0000	.2224	–.0033
5.0	1.00	0.0000	.0938	.2598	.3942	.4569	.4447	.3727	.2633	.1430	.0429	0.0000	.2144	–.0385
	.75	0.0000	.0967	.2699	.4117	.4790	.4681	.3946	.2815	.1552	.0474	0.0000	.2115	–.0281
	.50	0.0000	.0994	.2796	.4283	.4995	.4893	.4148	.2995	.1689	.0535	0.0000	.2084	–.0182
	.25	0.0000	.1021	.2888	.4436	.5172	.5062	.4306	.3160	.1855	.0634	0.0000	.2053	–.0093
	.05	0.0000	.1042	.2961	.4549	.5284	.5137	.4352	.3239	.2034	.0849	3.0000	.2029	–.0024

6.0	1.00	0.0000	.1130	.3046	.4499	.5085	.4846	.3996	.2795	.1513	.0454	0.0000	.1974	–.0289
	.75	0.0000	.1158	.3140	.4650	.5259	.5014	.4148	.2926	.1608	.0493	0.0000	.1949	–.0211
	.50	0.0000	.1186	.3233	.4795	.5416	.5155	.4274	.3048	.1716	.0547	0.0000	.1924	–.0139
	.25	0.0000	.1214	.3326	.4931	.5548	.5252	.4351	.3150	.1855	.0647	0.0000	.1899	–.0073
	.05	0.0000	.1238	.3403	.5040	.5633	.5279	.4339	.3178	.2006	.0866	0.0000	.1880	–.0019
8.0	1.00	0.0000	.1488	.3807	.5336	.5747	.5260	.4218	.2910	.1574	.0476	0.0000	.1730	–.0187
	.75	0.0000	.1518	.3897	.5456	.5852	.5335	.4279	.2978	.1643	.0512	0.0000	.1712	–.0140
	.50	0.0000	.1550	.3989	.5573	.5941	.5381	.4313	.3040	.1731	.0570	0.0000	.1694	–.0096
	.25	0.0000	.1585	.4086	.5689	.6012	.5389	.4303	.3087	.1860	.0683	0.0000	.1676	–.0053
	.05	0.0000	.1611	.4170	.5787	.6062	.5365	.4243	.3076	.1987	.0906	0.0000	.1662	–.0014
10.0	1.00	0.0000	.1823	.4441	.5930	.6111	.5405	.4247	.2918	.1594	.0492	0.0000	.1562	–.0141
	.75	0.0000	.1856	.4531	.6028	.6170	.5423	.4262	.2963	.1661	.0533	0.0000	.1547	–.0108
	.50	0.0000	.1891	.4624	.6125	.6215	.5416	.4254	.3006	.1755	.0600	0.0000	.1533	–.0077
	.25	0.0000	.1928	.4723	.6224	.6247	.5380	.4211	.3037	.1886	.0726	0.0000	.1519	–.0043
	.05	0.0000	.1959	.4807	.6309	.6271	.5335	.4146	.3020	.1989	.0938	0.0000	.1508	–.0010
12.0	1.00	0.0000	.2138	.4980	.6362	.6308	.5427	.4213	.2906	.1616	.0511	0.0000	.1436	–.0116
	.75	0.0000	.2174	.5069	.6443	.6334	.5412	.4206	.2946	.1690	.0559	0.0000	.1424	–.0091
	.50	0.0000	.2212	.5162	.6523	.6349	.5376	.4180	.2985	.1790	.0635	0.0000	.1412	–.0064
	.25	0.0000	.2252	.5261	.6606	.6355	.5321	.4130	.3010	.1916	.0767	0.0000	.1401	–.0035
	.05	0.0000	.2286	.5344	.6675	.6358	.5270	.4075	.2996	.1995	.0959	0.0000	.1391	–.0008
14.0	1.00	0.0000	.2436	.5444	.6680	.6405	.5394	.4163	.2900	.1647	.0535	0.0000	.1337	–.0101
	.75	0.0000	.2475	.5531	.6746	.6407	.5359	.4147	.2939	.1727	.0588	0.0000	.1327	–.0079
	.50	0.0000	.2515	.5623	.6811	.6402	.5310	.4117	.2977	.1828	.0669	0.0000	.1317	–.0056
	.25	0.0000	.2558	.5718	.6878	.6391	.5250	.4070	.2998	.1944	.0800	0.0000	.1306	–.0030
	.05	0.0000	.2592	.5798	.6931	.6379	.5202	.4030	.2989	.2000	.0972	0.0000	.1298	–.0007

continued

Table 12.18 (continued) Hoop force variation and shear at edges; triangular load of intensity q per unit area at bottom edge and zero at top; both edges encastré

η	h_t/h_b	*x/l*											*Shear at base*	*Shear at top*
		0.0	*.1*	*.2*	*.3*	*.4*	*.5*	*.6*	*.7*	*.8*	*.9*	*1.0*		
16.0	1.00	0.0000	.2718	.5845	.6914	.6441	.5338	.4117	.2903	.1682	.0560	0.0000	.1256	–.0089
	.75	0.0000	.2759	.5929	.6966	.6428	.5295	.4098	.2942	.1765	.0617	0.0000	.1247	–.0070
	.50	0.0000	.2802	.6018	.7018	.6409	.5242	.4069	.2977	.1864	.0700	0.0000	.1238	–.0049
	.25	0.0000	.2846	.6109	.7069	.6385	.5185	.4031	.2995	.1965	.0828	0.0000	.1229	–.0026
	.05	0.0000	.2882	.6184	.7109	.6363	.5142	.4005	.2989	.2002	.0981	0.0000	.1222	–.0006
18.0	1.00	0.0000	.2986	.6192	.7086	.6440	.5277	.4078	.2912	.1720	.0585	0.0000	.1189	–.0080
	.75	0.0000	.3029	.6274	.7126	.6417	.5232	.4061	.2950	.1802	.0644	0.0000	.1181	–.0062
	.50	0.0000	.3073	.6356	.7165	.6389	.5181	.4036	.2982	.1894	.0727	0.0000	.1173	–.0043
	.25	0.0000	.3118	.6444	.7202	.6357	.5130	.4007	.2995	.1980	.0850	0.0000	.1165	–.0023
	.05	0.0000	.3156	.6514	.7229	.6329	.5094	.3992	.2991	.2003	.0987	0.0000	.1158	–.0005
20.0	1.00	0.0000	.3241	.6495	.7209	.6418	.5220	.4049	.2924	.1756	.0608	0.0000	.1131	–.0073
	.75	0.0000	.3285	.6572	.7239	.6389	.5175	.4034	.2961	.1835	.0668	0.0000	.1124	–.0056
	.50	0.0000	.3330	.6651	.7266	.6355	.5129	.4014	.2988	.1919	.0750	0.0000	.1117	–.0039
	.25	0.0000	.3377	.6732	.7291	.6319	.5087	.3994	.2997	.1990	.0868	0.0000	.1109	–.0020
	.05	0.0000	.3415	.6797	.7307	.6288	.5058	.3986	.2994	.2003	.0992	0.0000	.1103	–.0004
24.0	1.00	0.0000	.3715	.6989	.7356	.6346	.5125	.4014	.2952	.1817	.0650	0.0000	.1038	–.0061
	.75	0.0000	.3761	.7057	.7368	.6311	.5089	.4004	.2981	.1886	.0709	0.0000	.1031	–.0047
	.50	0.0000	.3807	.7125	.7377	.6274	.5056	.3993	.2999	.1954	.0788	0.0000	.1025	–.0032
	.25	0.0000	.3855	.7194	.7381	.6236	.5029	.3985	.3001	.2001	.0898	0.0000	.1019	–.0017
	.05	0.0000	.3894	.7250	.7381	.6203	.5013	.3987	.2998	.2002	.0997	0.0000	.1014	–.0003

Notes: N = Coefficient × qr; V = Coefficient × ql.

Table 12.19 Bending moment variation; triangular load of intensity q per unit area at bottom edge and zero at top; both edges encastré

η	h_t/h_b	x/l										
		0.0	*.1*	*.2*	*.3*	*.4*	*.5*	*.6*	*.7*	*.8*	*.9*	*1.0*
.4	1.00	–.049411	–.019512	.001402	.014360	.020391	.020513	.015730	.007030	–.004607	–.018210	–.032802
	.75	–.056480	–.025269	–.003040	.011242	.018613	.020096	.016701	.009414	–.000791	–.012948	–.026089
	.50	–.066569	–.033684	–.009777	.006196	.015281	.018515	.016911	.011458	.003122	–.007144	–.018386
	.25	–.083049	–.047819	–.021560	–.003214	.008287	.014000	.014954	.012144	.006527	–.000969	–.009419
	.05	–.111569	–.073121	–.043633	–.022018	–.007169	.002017	.006613	.007641	.006059	.002757	–.001420
.8	1.00	–.047820	–.018673	.001535	.013913	.019570	.019574	.014942	.006631	–.004445	–.017387	–.031285
	.75	–.054127	–.023911	–.002623	.010865	.017688	.018941	.015649	.008770	–.000801	–.012193	–.024526
	.50	–.062753	–.031271	–.008706	.006105	.014334	.017113	.015492	.010429	.002798	–.006578	–.016864
	.25	–.075698	–.042696	–.018590	–.002170	.007804	.012531	.013112	.010525	.005613	–.000873	–.008201
	.05	–.093751	–.059383	–.033880	–.015963	–.004284	.002477	.005538	.005953	.004576	.002061	–.001074
1.2	1.00	–.045442	–.017420	.001731	.013243	.018342	.018175	.013770	.006038	–.004200	–.016160	–.029028
	.75	–.050719	–.021947	–.002027	.010313	.016349	.017273	.014136	.007845	–.000812	–.011105	–.022276
	.50	–.057505	–.027962	–.007247	.005966	.013027	.015191	.013555	.009030	.002360	–.005803	–.014790
	.25	–.066581	–.036365	–.014945	–.000916	.007178	.010705	.010847	.008546	.004504	–.000751	–.006715
	.05	–.076331	–.046043	–.024513	–.010247	–.001652	.002799	.004441	.004326	.003179	.001412	–.000750
1.6	1.00	–.042583	–.015918	.001960	.012432	.016868	.016501	.012373	.005334	–.003904	–.014694	–.026334
	.75	–.046784	–.019687	–.001349	.009667	.014803	.015357	.012404	.006792	–.000820	–.009858	–.019703
	.50	–.051818	–.024390	–.005689	.005795	.011602	.013119	.011481	.007539	.001900	–.004971	–.012575
	.25	–.057773	–.030285	–.011485	.000230	.006530	.008937	.008689	.006681	.003468	–.000630	–.005317
	.05	–.062783	–.035787	–.017444	–.006063	.000155	.002890	.003534	.003092	.002156	.000947	–.000516

continued

Table 12.19 (continued) Bending moment variation; triangular load of intensity q per unit area at bottom edge and zero at top; both edges encastré

η	h_t/h_b	x/l										
		0.0	*.1*	*.2*	*.3*	*.4*	*.5*	*.6*	*.7*	*.8*	*.9*	*1.0*
2.0	1.00	–.039529	–.014320	.002196	.011561	.015297	.014725	.010896	.004597	–.003585	–.013140	–.023487
	.75	–.042762	–.017387	–.000672	.008994	.013223	.013413	.010658	.005737	–.000820	–.008597	–.017111
	.50	–.046371	–.020988	–.004229	.005603	.010227	.011149	.009526	.006146	.001479	–.004184	–.010499
	.25	–.050184	–.025091	–.008581	.001138	.005920	.007407	.006868	.005130	.002618	–.000522	–.004159
	.05	–.052834	–.028376	–.012472	–.003255	.001246	.002805	.002820	.002220	.001469	.000644	–.000361
3.0	1.00	–.032284	–.010567	.002701	.009461	.011587	.010583	.007493	.002925	–.002813	–.009530	–.016910
	.75	–.033873	–.012364	.000733	.007440	.009732	.009199	.006928	.003521	–.000777	–.005881	–.011585
	.50	–.035417	–.014247	–.001453	.005074	.007415	.007262	.005760	.003517	.000721	–.002654	–.006541
	.25	–.036787	–.016110	–.003774	.002411	.004635	.004695	.003812	.002625	.001292	–.000317	–.002299
	.05	–.037617	–.017429	–.005561	.000232	.002214	.002251	.001623	.001011	.000606	.000283	–.000171
4.0	1.00	–.026517	–.007644	.003010	.007735	.008667	.007409	.004947	.001725	–.002173	–.006781	–.011965
	.75	–.027340	–.008766	.001624	.006196	.007174	.006232	.004386	.002069	–.000685	–.003998	–.007829
	.50	–.028099	–.009886	.000176	.004527	.005480	.004777	.003471	.001992	.000333	–.001705	–.004187
	.25	–.028772	–.010966	–.001282	.002791	.003643	.003080	.002182	.001390	.000683	000185	–.001387
	.05	–.029253	–.011774	–.002395	.001442	.002179	.001649	.000935	.000476	.000282	.000158	–.000100
5.0	1.00	–.022233	–.005540	.003144	.006399	.006535	.005179	.003222	.000961	–.001676	–.004866	–.008582
	.75	–.022722	–.006314	.002119	.005222	.005380	.004263	.002776	.001202	–.000568	–.002771	–.005452
	.50	–.023182	–.007081	.001075	.004000	.004142	.003213	.002125	.001152	.000161	–.001129	–.002835
	.25	–.023621	–.007835	.000038	.002770	.002870	.002072	.001285	.000772	.000406	–.000099	–.000925
	.05	–.023981	–.008431	–.000772	.001819	.001888	.001168	.000533	.000224	.000148	.000106	–.000068

6.0	1.00	–.019057	–.004044	.003157	.005362	.004995	.003645	.002091	.000505	–.001293	–.003560	–.006325
	.75	–.019393	–.004626	.002361	.004445	.004106	.002955	.001769	.000702	–.000447	–.001973	–.003956
	.50	–.019722	–.005207	.001558	.003514	.003186	.002201	.001322	.000689	.000095	–.000770	–.002040
	.25	–.020056	–.005790	.000758	.002587	.002265	.001416	.000769	.000450	.000272	–.000044	–.000673
	.05	–.020345	–.006266	.000121	.001865	.001562	.000815	.000298	.000104	.000087	.000079	–.000051
8.0	1.00	–.014788	–.002174	.002986	.003884	.003026	.001841	.000873	.000100	–.000765	–.002037	–.003784
	.75	–.014997	–.002563	.002450	.003289	.002487	.001462	.000729	.000256	–.000243	–.001083	–.002362
	.50	–.015213	–.002957	.001911	.002694	.001948	.001067	.000527	.000283	.000077	–.000387	–.001236
	.25	–.015442	–.003361	.001368	.002104	.001422	.000674	.000279	.000178	.000163	.000011	–.000424
	.05	–.015641	–.003698	.000924	.001633	.001018	.000384	.000082	.000020	.000041	.000049	–.000033
10.0	1.00	–.012090	–.001125	.002712	.002892	.001894	.000937	.000348	–.000006	–.000442	–.001267	–.002552
	.75	–.012243	–.001418	.002316	.002478	.001555	.000731	.000297	.000115	–.000105	–.000650	–.001616
	.50	–.012405	–.001717	.001916	.002065	.001222	.000522	.000210	.000146	.000088	–.000210	–.000866
	.25	–.012575	–.002025	.001509	.001653	.000899	.000320	.000098	.000088	.000117	.000029	–.000304
	.05	–.012719	–.002281	.001174	.001320	.000646	.000170	.000011	.000004	.000025	.000033	–.000023
12.0	1.00	–.010235	–.000494	.002422	.002192	.001207	.000463	.000120	–.000017	–.000246	–.000845	–.001885
	.75	–.010356	–.000728	.002113	.001890	.000989	.000356	.000113	.000072	–.000023	–.000419	–.001212
	.50	–.010483	–.000969	.001798	.001589	.000776	.000249	.000080	.000093	.000091	–.000120	–.000658
	.25	–.010615	–.001216	.001477	.001286	.000570	.000146	.000032	.000054	.000089	.000034	–.000232
	.05	–.010724	–.001420	.001213	.001039	.000404	.000066	–.000008	.000003	.000017	.000023	–.000017
14.0	1.00	–.008881	–.000094	.002149	.001683	.000773	.000209	.000023	–.000002	–.000127	–.000594	–.001482
	.75	–.008979	–.000289	.001897	.001455	.000632	.000160	.000035	.000059	.000021	–.000285	–.000962
	.50	009082	–.000489	.001642	.001228	.000495	.000109	.000028	.000069	.000086	–.000071	–.000525
	.25	–.009187	–.000694	.001380	.000998	.000359	.000060	.000010	.000039	.000068	.000033	–.000184
	.05	–.009274	–.000863	.001165	.000810	.000248	.000016	–.000010	.000004	.000012	.000017	–.000013

continued

Table 12.19 (continued) Bending moment variation; triangular load of intensity q per unit area at bottom edge and zero at top; both edges encastré

η	h_t/h_b	x/l										
		0.0	.1	.2	.3	.4	.5	.6	.7	.8	.9	1.0
16.0	1.00	–.007846	.000168	.001902	.001303	.000492	.000073	–.000016	.000013	–.000056	–.000437	–.001214
	.75	–.007929	.000001	.001692	.001128	.000402	.000057	.000004	.000053	.000043	–.000203	–.000791
	.50	–.008013	–.000169	.001479	.000953	.000313	.000039	.000009	.000056	.000077	–.000043	–.000431
	.25	–.008100	–.000343	.001261	.000775	.000223	.000018	.000004	.000031	.000053	.000030	–.000151
	.05	–.008170	–.000486	.001083	.000629	.000148	–.000006	–.000008	.000005	.000009	.000013	–.000011
18.0	1.00	–.007030	.000343	.001682	.001014	.000308	.000002	–.000028	.000024	–.000015	–.000331	–.001021
	.75	–.007100	.000198	.001505	.000878	.000251	.000005	–.000006	.000049	.000051	–.000148	–.000665
	.50	–.007171	.000051	.001324	.000741	.000195	.000005	.000004	.000046	.000067	–.000025	–.000362
	.25	–.007244	–.000100	.001140	.000601	.000135	–.000001	.000004	.000025	.000041	.000027	–.000126
	.05	–.007303	–.000223	.000990	.000487	.000084	–.000014	–.000004	.000005	.000007	.000010	–.000009
20.0	1.00	–.006368	.000462	.001489	.000793	.000186	–.000033	–.000028	.000030	.000008	–.000258	–.000875
	.75	–.006428	.000334	.001337	.000686	.000152	–.000019	–.000007	.000044	.000052	–.000111	–.000570
	.50	–.006489	.000204	.001182	.000577	.000117	–.000010	.000003	.000038	.000058	–.000014	–.000309
	.25	–.006552	.000072	.001024	.000467	.000078	–.000008	.000006	.000021	.000033	.000024	–.000107
	.05	–.006602	–.000036	.000896	.000377	.000043	–.000016	–.000002	.000005	.000005	.000008	–.000007
24.0	1.00	–.005360	.000596	.001171	.000487	.000051	–.000052	–.000019	.000031	.000027	–.000165	–.000669
	.75	–.005406	.000493	.001055	.000419	.000044	–.000030	–.000003	.000034	.000046	–.000065	–.000434
	.50	–.005453	.000389	.000938	.000350	.000033	–.000015	.000006	.000027	.000041	–.000002	–.000235
	.25	–.005501	.000284	.000820	.000280	.000018	–.000009	.000008	.000015	.000021	.000020	–.000081
	.05	–.005540	.000198	.000724	.000223	.000002	–.000012	.000002	.000004	.000003	.000005	–.000005

Note: M = Coefficient × ql^2.

Table 12.20 Hoop force variation and shear at edges; uniform load of intensity q per unit area; both edges encastré

η	h_t/h_b	x/l											Shear	Shear
		0.0	*.1*	*.2*	*.3*	*.4*	*.5*	*.6*	*.7*	*.8*	*.9*	*1.0*	*at base*	*at top*
.4	1.00	0.0000	.0025	.0079	.0136	.0177	.0192	.0177	.0136	.0079	.0025	0.0000	.4949	–.4949
	.75	0.0000	.0030	.0097	.0170	.0227	.0252	.0237	.0186	.0111	.0036	0.0000	.5217	–.4649
	.50	0.0000	.0038	.0124	.0225	.0309	.0353	.0344	.0280	.0173	.0059	0.0000	.5580	–.4230
	.25	0.0000	.0052	.0176	.0330	.0472	.0564	.0579	.0500	.0332	.0124	0.0000	.6143	–.3544
	.05	0.0000	.0082	.0293	.0575	.0869	.1113	.1244	.1200	.0934	.0447	0.0000	.7102	–.2221
.8	1.00	0.0000	.0096	.0302	.0519	.0678	.0735	.0678	.0519	.0302	.0096	0.0000	.4804	–.4804
	.75	0.0000	.0140	.0366	.0643	.0857	.0951	.0897	.0705	.0420	.0137	0.0000	.5018	–.4473
	.50	0.0000	.0184	.0461	.0831	.1140	.1304	.1271	.1034	.0642	.0219	0.0000	.5286	–.4010
	.25	0.0000	.0184	.0625	.1166	.1663	.1988	.2041	.1765	.1178	.0441	0.0000	.5637	–.3258
	.05	0.0000	.0261	.0922	.1800	.2709	.3460	.3866	.3742	.2935	.1426	0.0000	.6001	–.1883
1.2	1.00	0.0000	.0203	.0637	.1093	.1424	.1543	.1424	.1093	.0637	.0203	0.0000	.4587	–.4587
	.75	0.0000	.0236	.0757	.1326	.1766	.1958	.1848	.1454	.0870	.0285	0.0000	.4731	–.4219
	.50	0.0000	.0282	.0925	.1663	.2276	.2601	.2537	.2069	.1288	.0441	0.0000	.4884	–.3708
	.25	0.0000	.0350	.1183	.2199	.3124	.3728	.3829	.3321	.2231	.0843	0.0000	.5014	–.2905
	.05	0.0000	.0439	.1538	.2978	.4450	.5659	.6319	.6146	.4880	.2428	0.0000	.4954	–.1555
1.6	1.00	0.0000	.0333	.1040	.1780	.2316	.2509	.2316	.1780	.1040	.0333	0.0000	.4328	–.4328
	.75	0.0000	.0379	.1210	.2113	.2807	.3110	.2938	.2316	.1390	.0457	0.0000	.4402	–.3926
	.50	0.0000	.0438	.1431	.2562	.3495	.3987	.3892	.3183	.1992	.0688	0.0000	.4450	–.3382
	.25	0.0000	.0515	.1728	.3189	.4509	.5367	.5515	.4805	.3255	.1246	0.0000	.4420	–.2566
	.05	0.0000	.0586	.2026	.3876	.5737	.7250	.8084	.7907	.6377	.3268	0.0000	.4174	–.1305

continued

Table 12.20 (continued) Hoop force variation and shear at edges; uniform load of intensity q per unit area; both edges encastré

η	h_t/h_b	*x/l*											*Shear*	*Shear*
		0.0	*.1*	*.2*	*.3*	*.4*	*.5*	*.6*	*.7*	*.8*	*.9*	*1.0*	*at base*	*at top*
2.0	1.00	0.0000	.0473	.1473	.2512	.3261	.3531	.3261	.2512	.1473	.0473	0.0000	.4051	–.4051
	.75	0.0000	.0528	.1676	.2914	.3861	.4273	.4040	.3193	.1925	.0637	0.0000	.4067	–.3628
	.50	0.0000	.0592	.1920	.3417	.4645	.5290	.5168	.4242	.2672	.0931	0.0000	.4039	–.3072
	.25	0.0000	.0663	.2205	.4036	.5672	.6729	.6918	.6057	.4147	.1613	0.0000	.3918	–.2276
	.05	0.0000	.0705	.2402	.4532	.6626	.8303	.9233	.9085	.7457	.3953	0.0000	.3633	–.1124
3.0	1.00	0.0000	.0818	.2512	.4236	.5459	.5897	.5459	.4236	.2512	.0818	0.0000	.3402	–.3402
	.75	0.0000	.0873	.2723	.4669	.6129	.6758	.6408	.5111	.3128	.1056	0.0000	.3336	–.2978
	.50	0.0000	.0924	.2931	.5122	.6870	.7777	.7619	.6329	.4072	.1460	0.0000	.3230	–.2459
	.25	0.0000	.0959	.3097	.5521	.7600	.8911	.9169	.8158	.5770	.2355	0.0000	.3067	–.1776
	.05	0.0000	.0949	.3098	.5601	.7879	.9587	1.0528	1.0496	.9037	.5245	0.0000	.2880	–.0856
4.0	1.00	0.0000	.1116	.3363	.5588	.7133	.7679	.7133	.5588	.3363	.1116	0.0000	.2896	–.2896
	.75	0.0000	.1156	.3528	.5940	.7699	.8448	.8038	.6488	.4048	.1401	0.0000	.2813	–.2511
	.50	0.0000	.1185	.3661	.6247	.8236	.9248	.9088	.7665	.5062	.1880	0.0000	.2711	–.2060
	.25	0.0000	.1194	.3726	.6432	.8622	.9951	1.0234	.9274	.6812	.2936	0.0000	.2592	–.1487
	.05	0.0000	.1176	.3677	.6355	.8558	1.0046	1.0823	1.0884	.9813	.6220	0.0000	.2502	–.0711
5.0	1.00	0.0000	.1366	.4028	.6575	.8296	.8895	.8296	.6575	.4028	.1366	0.0000	.2528	–.2528
	.75	0.0000	.1392	.4142	.6826	.8716	.9505	.9081	.7435	.4744	.1688	0.0000	.2455	–.2190
	.50	0.0000	.1407	.4219	.7009	.9053	1.0064	.9921	.8513	.5791	.2234	0.0000	.2377	–.1800
	.25	0.0000	.1407	.4240	.7071	.9202	1.0421	1.0696	.9879	.7550	.3439	0.0000	.2301	–.1304
	.05	0.0000	.1403	.4219	.6995	.9036	1.0230	1.0785	1.0893	1.0222	.7013	0.0000	.2260	–.0619

6.0	1.00	0.0000	.1583	.4558	.7294	.9081	.9691	.9081	.7294	.4558	.1583	0.0000	.2263	–.2263
	.75	0.0000	.1600	.4636	.7461	.9368	1.0143	.9735	.8097	.5296	.1942	0.0000	.2206	–.1965
	.50	0.0000	.1610	.4683	.7562	.9552	1.0497	1.0380	.9073	.6367	.2553	0.0000	.2149	–.1620
	.25	0.0000	.1614	.4701	.7575	.9562	1.0611	1.0856	1.0214	.8115	.3899	0.0000	.2102	–.1177
	.05	0.0000	.1627	.4727	.7554	.9405	1.0308	1.0647	1.0768	1.0437	.7668	0.0000	.2081	–.0554
8.0	1.00	0.0000	.1965	.5381	.8246	.9965	1.0520	.9965	.8246	.5381	.1965	0.0000	.1917	–.1917
	.75	0.0000	.1980	.5430	.8316	1.0061	1.0722	1.0391	.8942	.6158	.2398	0.0000	.1884	–.1673
	.50	0.0000	.1995	.5473	.8349	1.0068	1.0792	1.0728	.9732	.7264	.5138	0.0000	.1855	–.1387
	.25	0.0000	.2018	.5530	.8368	.9976	1.0641	1.0798	1.0491	.8941	.4727	0.0000	.1834	–.1008
	.05	0.0000	.2052	.5622	.8446	.9913	1.0345	1.0362	1.0457	1.0577	.8659	0.0000	.1821	–.0464
10.0	1.00	0.0000	.2315	.6035	.8848	1.0359	1.0811	1.0359	.8848	.6035	.2315	0.0000	.1702	–.1702
	.75	0.0000	.2337	.6090	.8882	1.0352	1.0848	1.0614	.9452	.6846	.2823	0.0000	.1682	–.1490
	.50	0.0000	.2366	.6155	.8911	1.0287	1.0761	1.0742	1.0081	.7961	.3679	0.0000	.1665	–.1235
	.25	0.0000	.2404	.6250	.8967	1.0194	1.0522	1.0591	1.0538	.9503	.5450	0.0000	.1650	–.0894
	.05	0.0000	.2443	.6360	.9078	1.0194	1.0317	1.0170	1.0224	1.0543	.9330	0.0000	.1637	–.0405
12.0	1.00	0.0000	.2649	.6597	.9269	1.0521	1.0855	1.0521	.9269	.6597	.2649	0.0000	.1552	–.1552
	.75	0.0000	.2681	.6665	.9296	1.0463	1.0791	1.0651	.9789	.7426	.3225	0.0000	.1538	–.1359
	.50	0.0000	.2720	.6752	.9336	1.0378	1.0629	1.0640	1.0274	.8519	.4179	0.0000	.1525	–.1125
	.25	0.0000	.2766	.6866	.9415	1.0308	1.0388	1.0389	1.0499	.9886	.6078	0.0000	.1512	–.0811
	.05	0.0000	.2805	.6970	.9518	1.0328	1.0264	1.0063	1.0085	1.0450	.9779	0.0000	.1500	–.0362
14.0	1.00	0.0000	.2970	.7090	.9580	1.0568	1.0788	1.0568	.9580	.7090	.2970	0.0000	.1437	–.1437
	.75	0.0000	.3010	.7172	.9610	1.0489	1.0668	1.0608	1.0020	.7922	.3606	0.0000	.1426	–.1257
	.50	0.0000	.3055	.7271	.9660	1.0406	1.0481	1.0511	1.0377	.8966	.4638	0.0000	.1415	–.1039
	.25	0.0000	.3104	.7388	.9743	1.0359	1.0274	1.0232	1.0430	1.0139	.6622	0.0000	.1403	–.0747
	.05	0.0000	.3142	.7482	.9823	1.0374	1.0206	1.0010	1.0012	1.0347	1.0076	0.0000	.1392	–.0330

continued

Table 12.20 (continued) Hoop force variation and shear at edges; uniform load of intensity q per unit area; both edges encastré

η	h_t/h_b	x/l											Shear	Shear
		0.0	.1	.2	.3	.4	.5	.6	.7	.8	.9	1.0	at base	at top
16.0	1.00	0.0000	.3278	.7527	.9817	1.0557	1.0677	1.0557	.9817	.7527	.3278	0.0000	.1345	–.1345
	.75	0.0000	.3323	.7617	.9850	1.0474	1.0531	1.0533	1.0180	.8344	.3964	0.0000	.1336	–.1176
	.50	0.0000	.3371	.7719	.9902	1.0401	1.0548	1.0387	1.0425	.9322	.5059	0.0000	.1325	–.0970
	.25	0.0000	.3421	.7830	.9976	1.0368	1.0186	1.0121	1.0355	1.0301	.7094	0.0000	.1314	–.0695
	.05	0.0000	.3459	.7915	1.0034	1.0371	1.0151	.9987	.9979	1.0252	1.0267	0.0000	.1305	–.0304
18.0	1.00	0.0000	.3571	.7912	.9997	1.0518	1.0555	1.0518	.9997	.7912	.3571	0.0000	.1269	–.1269
	.75	0.0000	.3619	.8004	1.0031	1.0440	1.0404	1.0451	1.0288	.8702	.4300	0.0000	.1260	–.1108
	.50	0.0000	.3668	.8104	1.0080	1.0379	1.0240	1.0283	1.0438	.9602	.5444	0.0000	.1250	–.0913
	.25	0.0000	.3718	.8207	1.0140	1.0353	1.0122	1.0049	1.0283	1.0396	.7505	0.0000	.1240	–.0652
	.05	0.0000	.3757	.8284	1.0178	1.0342	1.0104	.9979	.9968	1.0174	1.0386	0.0000	.1232	–.0283
20.0	1.00	0.0000	.3850	.8251	1.0134	1.0468	1.0439	1.0468	1.0134	.8251	.3850	0.0000	.1204	–.1204
	.75	0.0000	.3899	.8341	1.0165	1.0397	1.0294	1.0372	1.0357	.9005	.4614	0.0000	.1196	–.1050
	.50	0.0000	.3949	.8435	1.0207	1.0347	1.0156	1.0198	1.0427	.9822	.5797	0.0000	.1187	–.0864
	.25	0.0000	.3998	.8528	1.0251	1.0323	1.0076	1.0005	1.0220	1.0446	.7865	0.0000	.1178	–.0616
	.05	0.0000	.4038	.8600	1.0274	1.0302	1.0067	.9978	.9968	1.0114	1.0454	0.0000	.1170	–.0265
24.0	1.00	0.0000	.4365	.8806	1.0308	1.0360	1.0249	1.0360	1.0308	.8806	.4365	0.0000	.1098	–.1098
	.75	0.0000	.4414	.8886	1.0330	1.0307	1.0135	1.0240	1.0418	.9474	.5186	0.0000	.1091	–.0957
	.50	0.0000	.4464	.8566	1.0352	1.0272	1.0050	1.0084	1.0370	1.0124	.6422	0.0000	.1084	–.0786
	.25	0.0000	.4515	.9043	1.0367	1.0247	1.0021	.9969	1.0121	1.0462	.8460	0.0000	.1076	–.0558
	.05	0.0000	.4556	.9104	1.03/0	1.0217	1.0018	.9985	.9978	1.0035	1.0494	0.0000	.1070	–.0237

Notes: N = Coefficient × qr; V = Coefficient × ql.

Table 12.21 Bending moment variation; uniform load of intensity q per unit area; both edges encastré

η	h_t/h_b	*0.0*	*.1*	*.2*	*.3*	*.4*	*.5*	*.6*	*.7*	*.8*	*.9*	*1.0*
.4	1.00	–.082213	–.037723	–.003205	.021391	.036121	.041026	.036121	.021391	–.003205	–.037723	–.082213
	.75	–.096681	–.049513	–.012312	.014986	.032453	.040144	.038083	.026256	.004614	–.026917	–.068409
	.50	–.118661	–.067860	–.027017	.003951	.025142	.036639	.038484	.030669	.013130	–.014237	–.051540
	.25	–.158504	–.102070	–.055579	–.018909	.008089	.025556	.033580	.032177	.021266	.000683	–.029769
	.05	–.244918	–.178886	–.122762	–.076338	–.039338	–.011473	.007496	.017695	.019076	.011373	–.005886
.8	1.00	–.079105	–.036060	–.002909	.020544	.034512	.039149	.034512	.020544	–.002909	–.036060	–.079105
	.75	–.091920	–.046729	–.011413	.014271	.030593	.037761	.035868	.024860	.004550	–.025340	–.065082
	.50	–.110532	–.062660	–.024634	.003858	.023176	.033622	.035356	.028344	.012354	–.012998	–.048114
	.25	–.141403	–.090021	–.048433	–.016213	.007169	.022201	.029199	.028211	.018962	.000878	–.026744
	.05	–.194800	–.139764	–.094436	–.058167	–.030096	–.009328	.004871	.012894	.014602	.009187	–.004812
1.2	1.00	–.074470	–.033580	–.002469	.019280	.032112	.036350	.032112	.019280	–.002469	–.033580	–.074470
	.75	–.085045	–.042710	–.010116	.013239	.027908	.034323	.032671	.022844	.004455	–.023065	–.060277
	.50	–.099410	–.055548	–.021376	.003729	.020488	.029499	.031078	.025162	.011288	–.011304	–.043423
	.25	–.120389	–.075219	–.039660	–.012903	.006045	.018091	.023824	.023336	.016122	.001112	–.023017
	.05	–.147131	–.102556	–.067489	–.040855	–.021241	–.007199	.002458	.008366	.010331	.007077	–.003776
1.6	1.00	–.068917	–.030612	–.001944	.017767	.029241	.033002	.029241	.017767	–.001944	–.030612	–.068917
	.75	–.077147	–.038096	–.008630	.012051	.024828	.030380	.029003	.020528	.004343	–.020453	–.054754
	.50	–.087450	–.047907	–.017880	.003588	.017602	.025074	.026486	.021739	.010136	–.009486	–.038372
	.25	–.100393	–.061145	–.031325	–.009758	.004986	.014202	.018728	.018698	.013405	.001326	–.019454
	.05	–.111708	–.074920	–.047477	–.027975	–.014596	–.005508	.000775	.005060	.007142	.005469	–.002990

continued

Table 12.21 (continued) Bending moment variation; uniform load of intensity q per unit area; both edges encastré

η	h_t/h_b	*0.0*	*.1*	*.2*	*.3*	*.4*	*.5*	*.6*	*.7*	*.8*	*.9*	*1.0*
2.0	1.00	–.063016	–.027460	–.001389	.016158	.026193	.029450	.026193	.016158	–.001389	–.027460	–.063016
	.75	–.069130	–.033418	–.007128	.010845	.021707	.026386	.025286	.018176	.004224	–.017806	–.049146
	.50	–.076121	–.040676	–.014579	.003450	.014875	.020897	.022146	.018497	.009036	–.007768	–.033579
	.25	–.083533	–.049293	–.024319	–.007117	.004105	.010950	.014454	.014790	.011096	.001495	–.016430
	.05	–.087279	–.055892	–.033720	–.019110	–.009972	–.004242	–.000269	.002850	.004935	.004321	–.002429
3.0	1.00	–.049194	–.020097	–.000112	.012387	.019079	.021167	.019079	.012387	–.000112	–.020097	–.049194
	.75	–.051725	–.023289	–.003901	.008215	.014961	.017768	.017253	.013068	.003936	–.012077	–.036955
	.50	–.053955	–.026573	–.008181	.003155	.009577	.012799	.013712	.012152	.006840	–.004432	–.024157
	.25	–.055250	–.029493	–.012683	–.002754	.002666	.005624	.007408	.008257	.007152	.001727	–.011277
	.05	–.054282	–.030395	–.015462	–.007404	–.003764	–.002293	–.001296	.000150	.001979	.002655	–.001621
4.0	1.00	–.038483	–.014425	.000838	.009460	.013614	.014819	.013614	.009460	.000838	–.014425	–.038483
	.75	–.039405	–.016169	–.001681	.006337	.010239	.011759	.011632	.009450	.003682	–.008058	–.028301
	.50	–.039970	–.017754	–.004250	.002921	.006290	.007815	.008494	.008158	.005389	–.002370	–.018154
	.25	–.039935	–.018912	–.006590	–.000530	.001919	.002919	.003786	.004786	.004938	.001780	–.008396
	.05	–.039256	–.019129	–.007720	–.002623	–.001211	–.001300	–.001400	–.000733	.000730	.001807	–.001212
5.0	1.00	–.030815	–.010406	.001469	.007360	.009757	.010359	.009757	.007360	.001469	–.010406	–.030815
	.75	–.031124	–.011442	–.000262	.005053	.007126	.007826	.007932	.007020	.003455	–005399	–.022462
	.50	–.031223	–.012326	–.001912	.002716	.004290	.004833	.005347	.005679	.004416	–.001130	–.014350
	.25	–.031085	–.012952	–.003300	.000589	.001489	.001517	.001890	.002871	.003609	.001745	–.006669
	.05	–.030958	–.013227	–.003983	–.000526	.000137	–.000782	–.001248	–.000964	.000156	.001302	–.000968

6.0	1.00	−.025382	−.007604	.001864	.005867	.007086	.007290	.007086	.005867	.001864	−.007604	−.025382
	.75	−.025491	−.008288	.000627	.004156	.005070	.005260	.005498	.005370	.003244	−.003636	−.018471
	.50	−.025511	−.008870	−.000507	.002520	.003035	.003017	.003411	.004090	.003726	−.000369	−.011830
	.25	−.025497	−.009333	−.001440	.001132	.001205	.000762	.000866	.001757	.002745	.001667	−.005540
	.05	−.025652	−.009691	−.001989	.000414	.000289	−.000510	−.001047	−.000952	−.000118	.000970	−.000804
8.0	1.00	−.018572	−.004212	.002221	.003984	.003898	.003683	.003898	.003984	.002221	−.004212	−.018572
	.75	−.018638	−.004609	.001509	.003006	.002724	.002415	.002749	.005384	.002851	−.001611	−.013581
	.50	−.018724	−.004981	.000868	.002135	.001654	.001161	.001397	.002289	.002792	.000414	−.008777
	.25	−.018884	−.005356	.000302	.001428	.000819	.000109	.000000	.000655	.001705	.001471	−.004158
	.05	−.019129	−.005726	−.000157	.000966	.000422	.000294	−.000699	−.000733	−.000302	.000565	−.000598
10.0	1.00	−.014641	−.002393	.002270	.002886	.002243	.001873	.002243	.002886	.002270	−.002393	−.014641
	.75	−.014737	−.002687	.001799	.002289	.001552	.001089	.001416	.002284	.002486	−.000591	−.010772
	.50	−.014866	−.002987	.001359	.001767	.000971	.000390	.000533	.001368	.002166	.000739	−.007007
	.25	−.015049	−.003315	.000927	.001317	.000550	−.000095	.000237	.000206	.001113	.001274	−.003334
	.05	−.015227	−.003619	.000538	.000948	.000302	−.000230	−.000466	−.000515	−.000315	.000339	−.000471
12.0	1.00	−.012120	−.001339	.002177	.002176	.001327	.000925	.001327	.002176	.002177	−.001339	−.012120
	.75	−.012227	−.001582	.001829	.001785	.000917	.000445	.000729	.001608	.002154	−.000035	−.008953
	.50	−.012357	−.001838	.001486	.001435	.000592	.000064	.000149	.000847	.001709	.000868	−.005842
	.25	−.012510	−.002118	.001128	.001103	.000356	−.000148	−.000268	.000014	.000743	.001100	−.002780
	.05	−.012632	−.002360	.000808	.000803	.000172	−.000202	−.000320	−.000356	−.000278	.000203	−.000387
14.0	1.00	−.010363	−.000689	.002022	.001680	.000795	.000418	.000795	.001680	.002022	−.000689	−.010363
	.75	−.010464	−.000898	.001744	.001407	.000551	.000130	.000361	.001160	.001857	.000283	−.007671
	.50	−.010577	−.001120	.001460	.001150	.000365	−.000067	−.000018	.000532	.001361	.000907	−.005010
	.25	−.010697	−.001357	.001156	.000888	.000218	−.000148	−.000235	−.000067	.000500	.000951	−.002382
	.05	−.010786	−.001552	.000894	.000647	.000075	−.000179	−.000228	−.000250	−.000233	.000117	−.000326

continued

Table 12.21 (continued) Bending moment variation; uniform load of intensity q per unit area; both edges encastré

η	h_t/h_b	*0.0*	*.1*	*.2*	*.3*	*.4*	*.5*	*.6*	*.7*	*.8*	*.9*	*1.0*
16.0	1.00	–.009060	–.000269	.001845	.001316	.000476	.000146	.000476	.001316	.001845	–.000269	–.009060
	.75	–.009149	–.000453	.001612	.001113	.000332	–.000019	.000161	.000849	.001597	.000470	–.006712
	.50	–.009243	–.000645	.001369	.000914	.000222	–.000110	–.000084	.000334	.001090	.000901	–.004384
	.25	–.009337	–.000847	.001111	.000700	.000121	–.000132	–.000188	–.000096	.000336	.000826	–.002082
	.05	–.009407	–.001009	.000895	.000509	.000012	–.000157	–.000168	–.000180	–.000191	.000063	–.000280
18.0	1.00	–.008051	.000012	.001667	.001038	.000280	.000003	.000280	.001038	.001667	.000012	–.008051
	.75	–.008127	–.000151	.001466	.000881	.000196	–.000083	.000052	.000625	.001373	.000581	–.005965
	.50	–.008204	–.000319	.001256	.000721	.000129	–.000116	–.000103	.000204	.000877	.000874	–.003895
	.25	–.008280	–.000492	.001035	.000546	.000056	–.000113	–.000145	–.000102	.000223	.000720	–.001847
	.05	–.008339	–.000630	.000856	.000395	–.000027	–.000136	–.000129	–.000134	–.000156	.000027	–.000244
20.0	1.00	–.007243	.000204	.001497	.000823	.000157	–.000067	.000157	.000823	.001497	.000204	–.007243
	.75	–.007308	.000060	.001322	.000697	.000110	–.000104	–.000006	.000461	.001180	.000645	–.005366
	.50	–.007373	–.000089	.001139	.000566	.000069	–.000105	–.000102	.000119	.000708	.000836	–.003503
	.25	–.007436	–.000239	.000950	.000422	.000013	–.000094	–.000110	–.000098	.000145	.000630	–.001658
	.05	–.007486	–.000360	.000798	.000303	–.000049	–.000116	–.000102	–.000103	–.000127	.000004	–.000216
24.0	1.00	–.006029	.000430	.001198	.000518	.000032	–.000103	.000032	.000518	.001198	.000430	–.006029
	.75	–.006077	.000315	.001062	.000434	.000020	–.000094	–.000049	.000246	.000875	.000695	–.004465
	.50	–.006125	.000197	.000922	.000344	.000003	–.000073	–.000077	.000025	.000465	.000751	–.002912
	.25	–.006173	.000079	.000780	.000248	–.000030	–.000065	–.000063	–.000078	.000051	.000488	–.001374
	.05	–.006212	–.000017	.000667	.000171	–.000065	–.000083	–.000068	–.000067	–.000086	–.000021	–.000173

Note: M = Coefficient $\times\, ql^2$.

Table 12.22 Hoop force variation and shear at bottom edge; triangular load of intensity q per unit area at bottom edge and zero at top; bottom edge encastré, top edge free

η	h_t/h_b	x/l											Shear
		0.0	*.1*	*.2*	*.3*	*.4*	*.5*	*.6*	*.7*	*.8*	*.9*	*1.0*	*at base*
.4	1.00	0.0000	.0040	.0141	.0282	.0446	.0621	.0801	.0981	.1161	.1339	.1517	.4344
	.75	0.0000	.0041	.0145	.0287	.0449	.0617	.0781	.0936	.1080	.1211	.1332	.4380
	.50	0.0000	.0042	.0149	.0294	.0455	.0614	.0759	.0881	.0976	.1043	.1082	.4425
	.25	0.0000	.0044	.0156	.0305	.0466	.0616	.0736	.0813	.0836	.0802	.0710	.4486
	.05	0.0000	.0046	.0164	.0320	.0483	.0626	.0721	.0745	.0676	.0498	.0203	.4559
.8	1.00	0.0000	.0101	.0341	.0652	.0984	.1310	.1615	.1897	.2160	.2412	.2661	.3722
	.75	0.0000	.0106	.0359	.0683	.1022	.1342	.1622	.1858	.2054	.2219	.2365	.3756
	.50	0.0000	.0112	.0382	.0726	.1077	.1391	.1641	.1316	.1919	.1962	.1960	.3799
	.25	0.0000	.0122	.0416	.0789	.1162	.1477	.1690	.1779	.1739	.1585	.1340	.3855
	.05	0.0000	.0133	.0457	.0871	.1281	.1610	.1793	.1782	.1550	.1089	.0421	.3916
1.2	1.00	0.0000	.0167	.0545	.0997	.1436	.1817	.2122	.2359	.2544	.2701	.2848	.3390
	.75	0.0000	.0177	.0579	.1059	.1516	.1893	.2167	.2342	.2439	.2488	.2517	.3408
	.50	0.0000	.0189	.0623	.1140	.1624	.2001	.2237	.2332	.2310	.2211	.2076	.3428
	.25	0.0000	.0205	.0682	.1252	.1780	.2168	.2362	.2347	.2147	.1816	.1423	.3450
	.05	0.0000	.0223	.0749	.1387	.1977	.2398	.2566	.2435	.2005	.1317	.0470	.3464
1.6	1.00	0.0000	.0241	.0765	.1355	.1378	.2272	.2522	.2646	.2682	.2668	.2640	.3165
	.75	0.0000	.0255	.0815	.1444	.1994	.2385	.2596	.2646	.2579	.2449	.2300	.3167
	.50	0.0000	.0272	.0875	.1556	.2142	.2535	.2702	.2657	.2456	.2170	.1868	.3168
	.25	0.0000	.0293	.0951	.1701	.2343	.2750	.2867	.2699	.2307	.1790	.1266	.3163
	.05	0.0000	.0314	.1030	.1858	.2572	.3017	.3106	.2817	.2194	.1341	.0433	.3147

continued

Table 12.22 (continued) Hoop force variation and shear at bottom edge; triangular load of intensity *q* per unit area at bottom edge and zero at top; bottom edge encastré, top edge free

η	h_t/h_b	*x/l*											*Shear*
		0.0	*.1*	*.2*	*.3*	*.4*	*.5*	*.6*	*.7*	*.8*	*.9*	*1.0*	*at base*
2.0	1.00	0.0000	.0322	.0998	.1723	.2317	.2704	.2873	.2860	.2720	.2513	.2286	.2981
	.75	0.0000	.0339	.1060	.1833	.2458	.2841	.2965	.2869	.2620	.2295	.1957	.2972
	.50	0.0000	.0359	.1131	.1965	.2632	.3015	.3087	.2888	.2500	.2026	.1560	.2958
	.25	0.0000	.0383	.1217	.2127	.2854	.3249	.3264	.2934	.2357	.1676	.1046	.2937
	.05	0.0000	.0404	.1299	.2289	.3086	.3513	.3494	.3044	.2256	.1288	.0373	.2908
3.0	1.00	0.0000	.0532	.1584	.2607	.3322	.3636	.3574	.3217	.2668	.2022	.1347	.2611
	.75	0.0000	.0554	.1660	.2740	.3485	.3786	.3666	.3219	.2569	.1836	.1106	.2587
	.50	0.0000	.0578	.1745	.2890	.3672	.3959	.3773	.3220	.2452	.1623	.0856	.2560
	.25	0.0000	.0604	.1839	.3061	.3889	.4167	.3906	.3228	.2315	.1371	.0589	.2527
	.05	0.0000	.0626	.1919	.3211	.4087	.4366	.4051	.3270	.2220	.1126	.0258	.2494
4.0	1.00	0.0000	.0738	.2122	.3364	.4117	.4312	.4027	.3396	.2558	.1623	.0660	.2333
	.75	0.0000	.0762	.2204	.3499	.4269	.4435	.4084	.3372	.2461	.1486	.0535	.2308
	.50	0.0000	.0788	.2293	.3648	.4438	.4570	.4143	.3337	.2347	.1339	.0433	.2281
	.25	0.0000	.0816	.2389	.3811	.4625	.4722	.4209	.3297	.2215	.1174	.0353	.2251
	.05	0.0000	.0838	.2469	.3948	.4785	.4854	.4274	.3283	.2130	.1029	.0207	.2224
5.0	1.00	0.0000	.0935	.2603	.3989	.4714	.4763	.4282	.3457	.2445	.1356	.0248	.2124
	.75	0.0000	.0961	.2688	.4121	.4348	.4852	.4300	.3405	.2351	.1265	.0221	.2101
	.50	0.0000	.0989	.2781	.4264	.4993	.4947	.4314	.3341	.2242	.1171	.0227	.2077
	.25	0.0000	.1018	.2879	.4419	.5151	.5048	.4327	.3271	.2123	.1068	.0252	.2051
	.05	0.0000	.1042	.2961	.4547	.5279	.5128	.4338	.3230	.2060	.0986	.0184	.2029

6.0	1.00	0.0000	.1124	.3036	.4509	.5161	.5053	.4406	.3447	.2343	.1190	.0029	.1962
	.75	0.0000	.1152	.3125	.4636	.5276	.5112	.4392	.3376	.2253	.1135	.0071	.1942
	.50	0.0000	.1182	.3220	.4774	.5400	.5172	.4373	.3295	.2154	.1078	.0139	.1921
	.25	0.0000	.1213	.3320	.4920	.5530	.5233	.4351	.3214	.2055	.1014	.0212	.1899
	.05	0.0000	.1239	.3403	.5040	.5632	.5274	.4329	.3166	.2016	.0970	.0171	.1880
8.0	1.00	0.0000	.1482	.3790	.5318	.5754	.5344	.4446	.3340	.2183	.1033	–.0107	.1726
	.75	0.0000	.1514	.3884	.5435	.5836	.5358	.4396	.3257	.2110	.1019	–.0004	.1710
	.50	0.0000	.1548	.3982	.5558	.5921	.5369	.4343	.3175	.2038	.1002	.0103	.1694
	.25	0.0000	.1582	.4085	.5685	.6003	.5374	.4289	.3105	.1982	.0978	.0184	.1676
	.05	0.0000	.1611	.4170	.5788	.6062	.5364	.4239	.3067	.1982	.0969	.0151	.1662
10.0	1.00	0.0000	.1819	.4428	.5909	.6096	.5431	.4375	.3217	.2080	.0984	–.0094	.1561
	.75	0.0000	.1854	.4523	.6012	.6150	.5417	.4314	.3145	.2028	.0987	.0013	.1547
	.50	0.0000	.1890	.4621	.6118	.6200	.5399	.4255	.3081	.1984	.0984	.0107	.1534
	.25	0.0000	.1928	.4723	.6224	.6244	.5371	.4196	.3036	.1962	.0974	.0166	.1519
	.05	0.0000	.1959	.4808	.6309	.6271	.5335	.4144	.3015	.1981	.0976	.0133	.1508
12.0	1.00	0.0000	.2136	.4972	.6346	.6288	.5426	.4282	.3122	.2023	.0975	–.0054	.1436
	.75	0.0000	.2173	.5066	.6433	.6317	.5396	.4224	.3067	.1989	.0983	.0036	.1424
	.50	0.0000	.2212	.5162	.6520	.6340	.5360	.4170	.3025	.1967	.0984	.0107	.1413
	.25	0.0000	.2252	.5261	.6606	.6354	.5315	.4118	.3001	.1964	.0977	.0148	.1401
	.05	0.0000	.2286	.5344	.6675	.6359	.5271	.4075	.2993	.1987	.0982	.0120	.1391
14.0	1.00	0.0000	.2435	.5440	.6669	.6386	.5381	.4197	.3057	.1994	.0979	–.0021	.1337
	.75	0.0000	.2475	.5530	.6740	.6395	.5342	.4148	.3020	.1975	.0986	.0048	.1327
	.50	0.0000	.2516	.5623	.6810	.6397	.5298	.4102	.2995	.1966	.0987	.0101	.1317
	.25	0.0000	.2558	.5719	.6878	.6391	.5248	.4062	.2987	.1972	.0980	.0131	.1306
	.05	0.0000	.2592	.5798	.6931	.6379	.5202	.4031	.2987	.1994	.0986	.0110	.1298

continued

Table 12.22 (continued) Hoop force variation and shear at bottom edge; triangular load of intensity q per unit area at bottom edge and zero at top; bottom edge encastré, top edge free

η	h_t/h_b	x/l											Shear at base
		0.0	*.1*	*.2*	*.3*	*.4*	*.5*	*.6*	*.7*	*.8*	*.9*	*1.0*	
16.0	1.00	0.0000	.2718	.5843	.6907	.6426	.5321	.4130	.3018	.1983	.0985	–.0002	.1256
	.75	0.0000	.2760	.5930	.6964	.6419	.5279	.4089	.2995	.1973	.0991	.0052	.1248
	.50	0.0000	.2802	.6018	.7018	.6406	.5234	.4055	.2983	.1971	.0990	.0092	.1239
	.25	0.0000	.2846	.6109	.7070	.6385	.5184	.4025	.2984	.1981	.0982	.0117	.1229
	.05	0.0000	.2882	.6184	.7109	.6363	.5143	.4005	.2988	.1998	.0988	.0102	.1222
18.0	1.00	0.0000	.2987	.6192	.7082	.6429	.5260	.4079	.2996	.1981	.0991	.0008	.1189
	.75	0.0000	.3029	.6274	.7125	.6411	.5218	.4048	.2983	.1976	.0994	.0050	.1181
	.50	0.0000	.3073	.6358	.7165	.6388	.5175	.4023	.2980	.1978	.0991	.0084	.1173
	.25	0.0000	.3118	.6444	.7202	.6358	.5130	.4003	.2985	.1987	.0983	.0107	.1165
	.05	0.0000	.3156	.6514	.7229	.6329	.5095	.3992	.2991	.2000	.0990	.0096	.1158
20.0	1.00	0.0000	.3242	.6495	.7208	.6410	.5204	.4042	.2985	.1983	.0996	.0012	.1131
	.75	0.0000	.3285	.6573	.7239	.6385	.5165	.4020	.2980	.1981	.0997	.0047	.1124
	.50	0.0000	.3331	.6652	.7267	.6355	.5126	.4003	.2981	.1983	.0992	.0076	.1117
	.25	0.0000	.3377	.6732	.7291	.6320	.5087	.3992	.2989	.1992	.0984	.0098	.1109
	.05	0.0000	.3415	.6797	.7307	.6288	.5058	.3987	.2994	.2001	.0992	.0090	.1103
24.0	1.00	0.0000	.3715	.6990	.7357	.6343	.5113	.4000	.2981	.1990	.1001	.0012	.1038
	.75	0.0000	.3761	.7057	.7369	.6310	.5083	.3991	.2983	.1990	.0999	.0040	.1031
	.50	0.0000	.3807	.7125	.7377	.6274	.5055	.3986	.2988	.1991	.0993	.0065	.1025
	.25	0.0000	.3855	.7194	.7381	.6236	.5030	.3985	.2995	.1997	.0986	.0086	.1019
	.05	0.0000	.3894	.7250	.7381	.6203	.5013	.3987	.2998	.2001	.0994	.0083	.1014

Notes: N = Coefficient × qr; V = Coefficient × ql.

Table 12.23 Bending moment variation and rotation D_4 at top edge; triangular load of intensity q per unit area at bottom edge and zero at top; bottom edge encastré, top edge free

η	h_t/h_b	x/l										D_4
		0.0	.1	.2	.3	.4	.5	.6	.7	.8	.9	
.4	1.00	–.120280	–.081666	–.052008	–.030205	–.015118	–.005585	–.000429	.001527	.001465	.000563	.177765
	.75	–.123356	–.084390	–.054377	–.032217	–.016768	–.006870	–.001355	.000940	.001171	.000480	.212442
	.50	–.127331	–.087914	–.057449	–.034832	–.018920	–.008553	–.002574	.000163	.000779	.000369	.267290
	.25	–.132854	–.092S21	–.061738	–.038447	–.021950	–.010939	–.004314	–.000957	.000209	.000206	.373270
	.05	–.139552	–.098792	–.066979	–.042999	–.025699	–.013917	–.006513	–.002393	–.000536	–.000012	.622200
.8	1.00	–.078600	–.046206	–.022700	–.006848	.002658	.007147	.007944	.006354	.003659	.001124	.248347
	.75	–.082335	–.049595	–.023738	–.009516	.000390	.005317	.006582	.005466	.003204	.000993	.289274
	.50	–.087124	–.055956	–.029663	–.012982	–.002575	.002906	.004774	.004276	.002589	.000815	.352545
	.25	–.093723	–.059997	–.035135	–.017852	–.006780	–.000551	.002147	.002525	.001670	.000546	.477979
	.05	–.101715	–.067375	–.041887	–.023934	–.012110	–.005013	–.001317	.000155	.000392	.000160	.866818
1.2	1.00	–.059802	–.030720	–.010454	.002362	.009174	.011417	.010471	.007642	.004167	.001234	.146906
	.75	–.062960	–.033696	–.013236	–.000192	.006910	.009521	.009017	.006672	.003662	.001088	.143725
	.50	–.066861	–.037395	–.016720	–.003415	.004026	.007083	.007129	.005401	.002996	.000894	.129021
	.25	–.071933	–.042252	–.021345	–.007748	.000098	.003712	.004478	.003588	.002031	.000610	.091071
	.05	–.077524	–.047703	–.026634	–.012807	–.004597	–.000423	.001127	.001220	.000723	.000209	.190062
1.6	1.00	–.049823	–.022988	–.004890	.005979	.011197	.012282	.010628	.007486	.003933	.001158	–.029735
	.75	–.052339	–.025476	–.007334	.003628	.009028	.010408	.009158	.006492	.003463	.001008	–.094957
	.50	–.055321	–.028453	–.010287	.000760	.006354	.008075	.007312	.005234	.002800	.000816	–.214939
	.25	–.058956	–.032139	.013999	–.002902	.002888	.005001	.004840	.003524	.001888	.000549	–.466512
	.05	–.062553	–.035893	–.017887	–.006842	–.000949	.001494	.001924	.001430	.000725	.000194	–.790519

continued

Table 12.23 (continued) Bending moment variation and rotation D_4 at top edge; triangular load of intensity q per unit area at bottom edge and zero at top; bottom edge encastré, top edge free

η	h_t/h_b	x/l										D_4
		0.0	.1	.2	.3	.4	.5	.6	.7	.8	.9	
2.0	1.00	–.043330	–.018326	–.001973	.007382	.011448	.011815	.009868	.006778	.003538	.001013	–.229468
	.75	–.045300	–.020388	–.004106	.005239	.009404	.010008	.008430	.005800	.003027	.000866	–.354475
	.50	–.047549	–.022770	–.006599	.002708	.006966	.007832	.006687	.004608	.002401	.000687	–.570415
	.25	–.050147	–.025577	–.009588	–.000376	.003948	.005098	.004466	.003070	.001587	.000453	–1.003808
	.05	–.052499	–.028221	–.012500	–.003473	.000828	.002184	.002016	.001306	.000612	.000160	–1.675431
3.0	1.00	–.053011	–.011687	.001210	.007688	.009747	.009094	.007047	.004549	.002253	.000616	–.677259
	.75	–.034094	–.013003	–.000315	.006030	.008087	.007592	.005849	.003745	.001843	.000503	–.897290
	.50	–.035260	–.014441	–.001997	.004189	.006235	.005911	.004508	.002848	.001391	.000381	–1.247505
	.25	–.036507	–.016013	–.003866	.002119	.004131	.005986	.002964	.001814	.000873	.000244	–1.896896
	.05	–.037522	–.017357	–.005513	.000250	.002188	.002164	.001456	.000762	.000316	.000087	–2.944824
4.0	1.00	–.026546	–.007978	.002378	.006842	.007630	.006488	.004620	.002750	.001259	.000319	–.964993
	.75	–.027209	–.008889	.001245	.005570	.006350	.005350	.003743	.002191	.000994	.000253	–1.207884
	.50	–.027911	–.009861	.000034	.004207	.004980	.004137	.002819	.001613	.000729	.000191	–1.575945
	.25	–.028647	–.010895	–.001269	.002732	.003493	.002820	.001819	.000995	.000453	.000131	–2.235028
	.05	–.029237	–.011756	–.002376	.001460	.002190	.001642	.000895	.000389	.000153	.000051	–3.390384
5.0	1.00	–.022122	–.005630	.002852	.005912	.005906	.004559	.002932	.001559	.000628	.000137	–1.109622
	.75	–.022583	–.006315	.001972	.004921	.004932	.003732	.002339	.001218	.000490	.000111	–1.339936
	.50	–.023069	–.007039	.001042	.003876	.003910	.002875	.001738	.000888	.000369	.000093	–1.682622
	.25	–.023574	–.007800	.000057	.002765	.002S24	.001965	.001104	.000543	.000246	.000079	–2.311630
	.05	–.023981	–.008428	–.000765	.001328	.001898	.001174	.000525	.000190	.000080	.000037	–3.654541

6.0	1.00	–.018947	–.004056	.003017	.005087	.004597	.003207	.001826	.000826	.000261	.000036	–1.161532
	.75	–.019297	–.004603	.002304	.004295	.003850	.002615	.001447	.000648	.000216	.000039	–1.370427
	.50	–.019663	–.005176	.001557	.003467	.003073	.002009	.001071	.000484	.000186	.000049	–1.686192
	.25	–.020041	–.005774	.000773	.002595	.002255	.001370	.000671	.000303	.000148	.000057	–2.311030
	.05	–.020347	–.006266	.000123	.001869	.001568	.000821	.000299	.000090	.000048	.000030	–3.954863
8.0	1.00	–.014734	–.002159	.002956	.003793	.002857	.001606	.000660	.000143	–.000038	–.000036	–1.138847
	.75	–.014964	–.002545	.002448	.003253	.002398	.001307	.000536	.000146	.000013	–.000006	–1.319481
	.50	–.015201	–.002946	.001920	.002692	.001922	.000998	.000405	.000144	.000061	.000024	–1.621146
	.25	–.015443	–.003359	.001373	.002111	.001427	.000668	.000247	.000108	.000079	.000042	–2.325942
	.05	–.015642	–.003698	.000924	.001634	.001019	.000387	.000085	.000018	.000027	.000023	–4.658739
10.0	1.00	–.012070	–.001114	.002710	.002864	.001821	.000804	.000187	–.000062	–.000093	–.000039	–1.075775
	.75	–.012235	–.001410	.002322	.002473	.001525	.000656	.000175	.000004	–.000018	–.000005	–1.255732
	.50	–.012404	–.001713	.001922	.002070	.001219	.000497	.000147	.000050	.000040	.000023	–1.583804
	.25	–.012577	–.002025	.001511	.001657	.000903	.000322	.000088	.000052	.000060	.000036	–2.406345
	.05	–.012719	–.002282	.001174	.001320	.000646	.000171	.000013	.000005	.000020	.000017	–5.305748
12.0	1.00	–.010229	–.000488	.002427	.002187	.001175	.000386	.000002	–.000102	–.000078	–.000027	–1.026549
	.75	–.010355	–.000725	.002118	.001893	.000980	.000319	.000034	–.000024	–.000008	.000002	–1.220532
	.50	–.010484	–.000968	.001801	.001593	.000779	.000240	.000047	.000027	.000039	.000023	–1.585382
	.25	–.010616	–.001217	.001477	.001288	.000572	.000149	.000029	.000035	.000049	.000030	–2.505536
	.05	–.010724	–.001420	.001213	.001039	.000404	.000066	–.000007	.000004	.000015	.000013	–5.824918
14.0	1.00	–.008880	–.000091	.002153	.001684	.000760	.000164	–.000063	–.000089	–.000051	–.000016	–.998441
	.75	–.008980	–.000288	.001901	.001459	.000631	.000142	–.000016	–.000019	.000004	.000007	–1.209431
	.50	–.009082	–.000489	.001643	.001231	.000498	.000108	.000010	.000023	.000038	.000022	–1.605619
	.25	–.009188	–.000695	.001380	.000999	.000361	.000062	.000010	.000029	.000040	.000025	–2.593720
	.05	–.009274	–.000863	.001165	.000810	.000248	.000016	–.000010	.000005	.000012	.000010	–6.240837

continued

Table 12.23 (continued) Bending moment variation and rotation D_4 at top edge; triangular load of intensity q per unit area at bottom edge and zero at top; bottom edge encastré, top edge free

η	h_t/h_b	x/l										D_4
		0.0	.1	.2	.3	.4	.5	.6	.7	.8	.9	
16.0	1.00	–.007847	.000169	.001905	.001306	.000488	.000047	–.000078	–.000066	–.000030	–.000008	–.985952
	.75	–.007929	.000002	.001694	.001131	.000403	.000049	–.000028	–.000008	.000012	.000009	–1.211372
	.50	–.008014	–.000169	.001479	.000954	.000315	.000040	–.000000	.000024	.000035	.000019	–1.629805
	.25	–.008100	–.000344	.001261	.000775	.000224	.000020	.000005	.000026	.000033	.000021	–2.664176
	.05	–.008170	–.000486	.001083	.000629	.000148	–.000006	–.000008	.000006	.000009	.000008	–6.584965
18.0	1.00	–.007031	.000343	.001684	.001017	.000308	–.000013	–.000072	–.000044	–.000015	–.000003	–.982466
	.75	–.007100	.000198	.001505	.000880	.000254	.000002	–.000027	.000001	.000016	.000009	–1.218327
	.50	–.007171	.000051	.001324	.000742	.000196	.000007	–.000001	.000024	.000031	.000017	–1.651414
	.25	–.007244	–.000100	.001140	.000601	.000135	.000000	.000005	.000023	.000027	.000017	–2.719480
	.05	–.007303	.000223	.000990	.000487	.000084	.000014	.000004	.000006	.000007	.000007	–6.877138
20.0	1.00	–.006369	.000462	.001490	.000795	.000187	–.000041	–.000060	–.000028	–.000007	–.000001	–.983211
	.75	–.006429	.000334	.001337	.000687	.000154	–.000019	.000021	.000007	.000017	.000009	–1.225992
	.50	–.006489	.000204	.001182	.000577	.000118	–.000008	.000002	.000023	.000027	.000014	–1.668742
	.25	–.006552	.000072	.001024	.000467	.000078	–.000008	.000007	.000020	.000022	.000015	–2.763854
	.05	–.006602	–.000036	.000896	.000377	.000043	–.000016	.000002	.000005	.000006	.000005	–7.128685
24.0	1.00	–.005360	.000595	.001171	.000488	.000053	–.000053	–.000035	.000009	.000000	–.000000	–.987562
	.75	–.005407	.000493	.001055	.000419	.000045	–.000029	–.000008	.000011	.000014	.000006	–1.237591
	.50	–.005453	.000389	.000938	.000350	.000033	–.000014	.000007	.000019	.000020	.000011	–1.692527
	.25	–.005501	.000284	.000820	.000280	.000018	–.000009	.000009	.000015	.000015	.000011	–2.832097
	.05	–.005540	.000198	.000724	.000223	.000002	–.000012	.000002	.000004	.000004	.000004	–7.536554

Notes: M = Coefficient × ql^2; D_4 = Coefficient × $qr^2/(Eh_b l)$.

Table 12.24 Hoop force variation and shear at bottom edge; uniform load of intensity q per unit area; bottom edge encastré, top edge free

η	h_t/h_b	x/l											Shear
		0.0	*.1*	*.2*	*.3*	*.4*	*.5*	*.6*	*.7*	*.8*	*.9*	*1.0*	*at base*
.4	1.00	0.0000	.0115	.0424	.0883	.1450	.2096	.2792	.3521	.4266	.5019	.5774	.7661
	.75	0.0000	.0118	.0434	.0898	.1465	.2095	.2756	.3419	.4063	.4672	.5237	.7751
	.50	0.0000	.0121	.0446	.0920	.1489	.2104	.2720	.3295	.3795	.4191	.4464	.7870
	.25	0.0000	.0127	.0467	.0957	.1535	.2135	.2693	.3142	.3415	.3450	.3191	.8045
	.05	0.0000	.0135	.0495	.1013	.1611	.2209	.2711	.3007	.2963	.2405	.1089	.8276
.8	1.00	0.0000	.0247	.0891	.1810	.2908	.4116	.5385	.6681	.7988	.9297	1.0605	.5548
	.75	0.0000	.0258	.0931	.1885	.3009	.4217	.5443	.6640	.7778	.8841	.9821	.5616
	.50	0.0000	.0273	.0986	.1991	.3159	.4377	.5554	.6618	.7511	.8197	.8651	.5703
	.25	0.0000	.0296	.1072	.2161	.3408	.4666	.5796	.6670	.7166	.7172	.6589	.5823
	.05	0.0000	.0329	.1195	.2416	.3805	.5169	.6304	.6973	.6889	.5644	.2598	.5963
1.2	1.00	0.0000	.0352	.1233	.2434	.3802	.5236	.6678	.8099	.9495	1.0873	1.2244	.4579
	.75	0.0000	.0370	.1299	.2559	.3975	.5423	.6823	.8131	.9332	1.0427	1.1426	.4602
	.50	0.0000	.0394	.1386	.2729	.4218	.5699	.7057	.8220	.9144	.9811	1.0217	.4626
	.25	0.0000	.0427	.1510	.2980	.4597	.6159	.7497	.8471	.8962	.8861	.8070	.4641
	.05	0.0000	.0468	.1671	.3321	.5146	.6894	.8319	.9159	.9077	.7542	.3559	.4616
1.6	1.00	0.0000	.0454	.1547	.2967	.4503	.6027	.7478	.8839	1.0128	1.1374	1.2606	.4048
	.75	0.0000	.0477	.1630	.3126	.4724	.6268	.7674	.8914	.9996	1.0949	1.1804	.4041
	.50	0.0000	.0505	.1735	.3330	.5016	.6603	.7970	.9060	.9859	1.0381	1.0648	.4023
	.25	0.0000	.0540	.1870	.3606	.5434	.7116	.8476	.9394	.9781	.9557	.8631	.3980
	.05	0.0000	.0577	.2016	.3923	.5957	.7838	.9329	1.0199	1.0142	.8573	.4169	.3883

continued

Table 12.24 (continued) Hoop force variation and shear at bottom edge; uniform load of intensity q per unit area; bottom edge encastré, top edge free

η	h_t/h_b	x/l											Shear
		0.0	*.1*	*.2*	*.3*	*.4*	*.5*	*.6*	*.7*	*.8*	*.9*	*1.0*	*at base*
2.0	1.00	0.0000	.0557	.1855	.3469	.5129	.6684	.8075	.9305	1.0414	1.1456	1.2475	.3690
	.75	0.0000	.0583	.1949	.3649	.5377	.6953	.8296	.9400	1.0300	1.1054	1.1714	.3663
	.50	0.0000	.0613	.2061	.3867	.5688	.7306	.8609	.9565	1.0193	1.0539	1.0650	.3623
	.25	0.0000	.0647	.2194	.4138	.6096	.7806	.9105	.9910	1.0171	.9840	.8830	.3556
	.05	0.0000	.0675	.2311	.4395	.6526	.8411	.9849	1.0675	1.0648	.9173	.4608	.3446
3.0	1.00	0.0000	.0816	.2592	.4611	.6468	.7985	.9134	.9974	1.0604	1.1124	1.1609	.3092
	.75	0.0000	.0845	.2695	.4802	.6722	.8249	.9342	1.0061	1.0511	1.0801	1.1016	.3048
	.50	0.0000	.0874	.2806	.5013	.7010	.8559	.9603	1.0194	1.0434	1.0425	1.0238	.2994
	.25	0.0000	.0903	.2919	.5236	.7329	.8927	.9952	1.0442	1.0457	.9994	.8933	.2926
	.05	0.0000	.0920	.2989	.5381	.7553	.9223	1.0327	1.0912	1.0952	.9898	.5397	.2851
4.0	1.00	0.0000	.1062	.3248	.5557	.7491	.8889	.9783	1.0294	1.0565	1.0720	1.0840	.2698
	.75	0.0000	.1091	.3350	.5736	.7713	.9100	.9931	1.0339	1.0480	1.0484	1.0435	.2655
	.50	0.0000	.1119	.3455	.5924	.7949	.9329	1.0100	1.0407	1.0411	1.0230	.9918	.2608
	.25	0.0000	.1147	.3559	.6113	.8189	.9568	1.0295	1.0536	1.0433	.9989	.9000	.2556
	.05	0.0000	.1165	.3627	.6232	.8327	.9692	1.0417	1.0743	1.0826	1.0203	.5977	.2511
5.0	1.00	0.0000	.1293	.3824	.6324	.8248	.9487	1.0153	1.0425	1.0474	1.0425	1.0351	.2416
	.75	0.0000	.1322	.3924	.6489	.8434	.9642	1.0239	1.0427	1.0393	1.0263	1.0103	.2380
	.50	0.0000	.1351	.4028	.6661	.8626	.9799	1.0327	1.0439	1.0329	1.0102	.9776	.2342
	.25	0.0000	.1380	.4132	.6833	.8813	.9946	1.0412	1.0482	1.0345	.9983	.9105	.2301
	.05	0.0000	.1404	.4212	.6957	.8924	.9992	1.0400	1.0524	1.0632	1.0349	.6431	.2270
6.0	1.00	0.0000	.1510	.4334	.6954	.8809	.9875	1.0344	1.0450	1.0375	1.0235	1.0080	.2205
	.75	0.0000	.1541	.4435	.7108	.8964	.9981	1.0380	1.0420	1.0300	1.0129	.9944	.2175

	.50	0.0000	.1572	.4540	.7268	.9121	1.0083	1.0410	1.0394	1.0242	1.0031	.9740	.2144
	.25	0.0000	.1604	.4648	.7430	.9272	1.0167	1.0424	1.0386	1.0257	.9989	.9217	.2111
	.05	0.0000	.1631	.4738	.7559	.9371	1.0178	1.0350	1.0340	1.0459	1.0416	.6792	.2087
8.0	1.00	0.0000	.1917	.5210	.7920	.9550	1.0274	1.0446	1.0365	1.0208	1.0048	.9896	.1909
	.75	0.0000	.1951	.5314	.8057	.9657	1.0313	1.0421	1.0305	1.0149	1.0009	.9876	.1887
	.50	0.0000	.1987	.5422	.8198	.9762	1.0343	1.0388	1.0252	1.0111	.9981	.9784	.1863
	.25	0.0000	.2023	.5535	.8343	.9859	1.0353	1.0335	1.0207	1.0136	1.0014	.9399	.1840
	.05	0.0000	.2055	.5633	.8466	.9933	1.0332	1.0235	1.0111	1.0226	1.0439	.7326	.1821
10.0	1.00	0.0000	.2294	.5941	.8619	.9978	1.0411	1.0394	1.0241	1.0096	.9987	.9896	.1707
	.75	0.0000	.2331	.6045	.8738	1.0048	1.0411	1.0345	1.0182	1.0057	.9978	.9910	.1689
	.50	0.0000	.2370	.6153	.8860	1.0113	1.0400	1.0291	1.0133	1.0041	.9979	.9846	.1671
	.25	0.0000	.2410	.6267	.8983	1.0169	1.0372	1.0223	1.0092	1.0075	1.0036	.9517	.1652
	.05	0.0000	.2444	.6363	.9087	1.0213	1.0337	1.0141	1.0011	1.0103	1.0397	.7702	.1637
12.0	1.00	0.0000	.2646	.6558	.9132	1.0222	1.0430	1.0304	1.0140	1.0031	.9973	.9935	.1558
	.75	0.0000	.2686	.6661	.9233	1.0262	1.0406	1.0251	1.0093	1.0010	.9976	.9949	.1543
	.50	0.0000	.2727	.6767	.9334	1.0296	1.0373	1.0195	1.0058	1.0011	.9988	.9887	.1527
	.25	0.0000	.2769	.6878	.9437	1.0321	1.0326	1.0132	1.0030	1.0047	1.0050	.9593	.1512
	.05	0.0000	.2805	.6970	.9520	1.0337	1.0285	1.0074	.9976	1.0042	1.0336	.7985	.1499
14.0	1.00	0.0000	.2976	.7085	.9511	1.0353	1.0395	1.0217	1.0069	.9998	.9975	.9968	.1441
	.75	0.0000	.3017	.7185	.9594	1.0370	1.0358	1.0168	1.0037	.9990	.9983	.9973	.1429
	.50	0.0000	.3061	.7288	.9677	1.0379	1.0314	1.0119	1.0017	1.0001	.9997	.9910	.1416
	.25	0.0000	.3105	.7394	.9758	1.0381	1.0262	1.0068	1.0000	1.0034	1.0056	.9646	.1403
	.05	0.0000	.3142	.7481	.9822	1.0377	1.0219	1.0032	.9970	1.0012	1.0275	.8208	.1392

continued

Table 12.24 (continued) Hoop force variation and shear at bottom edge; uniform load of intensity q per unit area; bottom edge encastré, top edge free

η	h_t/h_b	*x/l*											*Shear at base*
		0.0	*.1*	*.2*	*.3*	*.4*	*.5*	*.6*	*.7*	*.8*	*.9*	*1.0*	
16.0	1.00	0.0000	.3286	.7538	.9790	1.0412	1.0339	1.0145	1.0025	.9984	.9982	.9989	.1348
	.75	0.0000	.3329	.7633	.9857	1.0412	1.0297	1.0104	1.0006	.9985	.9989	.9986	.1337
	.50	0.0000	.3374	.7731	.9922	1.0404	1.0249	1.0065	.9996	1.0000	1.0004	.9923	.1325
	.25	0.0000	.3420	.7832	.9985	1.0389	1.0198	1.0028	.9989	1.0026	1.0056	.9686	.1314
	.05	0.0000	.3459	.7914	1.0033	1.0371	1.0157	1.0007	.9975	.9997	1.0221	.8388	.1305
18.0	1.00	0.0000	.3578	.7928	.9996	1.0428	1.0278	1.0090	1.0000	.9981	.9988	1.0000	.1270
	.75	0.0000	.3623	.8019	1.0048	1.0415	1.0235	1.0058	.9991	.9986	.9995	.9991	.1260
	.50	0.0000	.3670	.8111	1.0098	1.0395	1.0189	1.0029	.9988	1.0001	1.0007	.9931	.1250
	.25	0.0000	.3717	.8206	1.0144	1.0368	1.0142	1.0004	.9986	1.0021	1.0054	.9717	.1240
	.05	0.0000	.3757	.8284	1.0178	1.0341	1.0106	.9994	.9982	.9991	1.0176	.8537	.1232
20.0	1.00	0.0000	.3855	.8266	1.0145	1.0417	1.0220	1.0050	.9988	.9983	.9993	1.0004	.1204
	.75	0.0000	.3901	.8352	1.0184	1.0395	1.0179	1.0026	.9984	.9990	.9998	.9993	.1195
	.50	0.0000	.3949	.8439	1.0220	1.0367	1.0137	1.0006	.9986	1.0003	1.0009	.9937	.1187
	.25	0.0000	.3998	.8528	1.0252	1.0333	1.0097	.9991	.9987	1.0016	1.0051	.9743	.1178
	.05	0.0000	.4038	.8600	1.0274	1.0301	1.0067	.9988	.9988	.9988	1.0138	.8663	.1170
24.0	1.00	0.0000	.4367	.8816	1.0326	1.0356	1.0125	1.0004	.9982	.9990	.9998	1.0004	.1098
	.75	0.0000	.4415	.8890	1.0344	1.0324	1.0092	.9993	.9985	.9996	1.0001	.9992	.1091
	.50	0.0000	.4464	.8966	1.0357	1.0288	1.0062	.9986	.9990	1.0005	1.0010	.9946	.1084
	.25	0.0000	.4514	.9042	1.0367	1.0250	1.0035	.9984	.9992	1.0010	1.0044	.9783	.1076
	.05	0.0000	.4556	.9104	1.0370	1.0216	1.0017	.9987	.9997	.9990	1.0083	.8864	.1070

Notes: N = Coefficient × qr; V = Coefficient × ql.

Table 12.25 Bending moment variation and rotation D_4 at top edge; uniform load of intensity q per unit area; bottom edge encastré, and top edge free

η	h_t/h_b	x/l										D_4
		0.0	*.1*	*.2*	*.3*	*.4*	*.5*	*.6*	*.7*	*.8*	*.9*	
.4	1.00	–.331614	–.259997	–.198249	–.146065	–.102990	–.068458	–.041826	–.022400	–.009451	–.002235	.755081
	.75	–.339640	–.267124	–.204475	–.151380	–.107379	–.071909	–.044341	–.024016	–.010274	–.002471	.954649
	.50	–.350500	–.276787	–.212936	–.158626	–.113389	–.076659	–.047824	–.026273	–.011433	–.002807	1.310009
	.25	–.366706	–.291244	–.225639	–.169554	–.122506	–.083921	–.053204	–.029803	–.013274	–.003350	2.156684
	.05	–.388881	–.311108	–.243182	–.184748	–.135295	–.094231	–.060965	–.035005	–.016066	–.004205	5.465168
.8	1.00	–.182576	–.132074	–.091293	–.059599	–.036081	–.019646	–.009090	–.003147	–.000522	.000091	1.308620
	.75	–.190628	–.139443	–.097966	–.065536	–.041207	–.023863	–.012299	–.005295	–.001656	–.000245	1.686859
	.50	–.201472	–.149414	–.107047	–.073671	–.048289	–.029745	–.016828	–.008365	–.003299	–.000738	2.400378
	.25	–.217640	–.164390	–.120804	–.086121	–.059264	–.038999	–.024078	–.013382	–.006047	–.001586	4.304012
	.05	–.240236	–.185573	–.140539	–.104280	–.075593	–.053102	–.035460	–.021552	–.010732	–.003116	13.908194
1.2	1.00	–.118807	–.077990	–.046778	–.024307	–.009389	–.000663	.003299	.003937	.002672	.000900	1.370945
	.75	–.124558	–.083502	–.052030	–.029233	–.013865	–.004519	.000246	.001826	.001529	.000555	1.780427
	.50	–.131924	–.090634	–.058902	–.035756	–.019870	–.009766	–.003973	–.001139	–.000104	.000053	2.596530
	.25	–.142079	–.100635	–.068712	–.045247	–.028790	–.017741	–.010552	–.005896	–.002809	–.000807	5.033630
	.05	–.154554	–.113352	–.081623	–.058187	–.041415	–.029505	–.020726	–.013676	–.007540	–.002444	20.802635
1.6	1.00	–.088261	–.052743	–.026720	–.009124	.001447	.006521	.007616	.006181	.003580	.001103	1.230349
	.75	–.092302	–.056855	–.030877	–.013241	–.002471	.003017	.004762	.004168	.002475	.000767	1.609741
	.50	–.097168	–.061895	–.036058	–.018458	–.007521	–.001576	.000953	.001430	.000942	.000291	2.419655
	.25	–.103209	–.068359	–.042905	–.025547	–.014577	–.008184	–.004703	–.002781	–.001509	–.000504	5.140445
	.05	–.109171	–.075290	–.050762	–.034181	–.023667	–.017208	–.012943	–.009400	–.005731	–.002041	26.805743

continued

Table 12.25 (continued) Bending moment variation and rotation D_4 at top edge; uniform load of intensity q per unit area; bottom edge encastré, and top edge free

η	h_t/h_b	x/l										D_4
		0.0	.1	.2	.3	.4	.5	.6	.7	.8	.9	
2.0	1.00	–.070846	–.038899	–.016336	–.001893	.006023	.009059	.008766	.006535	.003599	.001073	1.017417
	.75	–.073727	–.042045	–.019717	–.005413	.002542	.005860	.006113	.004641	.002553	.000753	1.352929
	.50	–.076998	–.045712	–.023746	–.009691	–.001767	.001827	.002702	.002158	.001152	.000316	2.133807
	.25	–.080662	–.050042	–.028702	–.015136	–.007431	–.003651	–.002098	–.001481	–.000998	–.000392	5.073898
	.05	–.083446	–.053928	–.033656	–.021038	–.014022	–.010500	–.008604	–.006909	–.004607	–.001769	32.510531
3.0	1.00	–.048090	–.022093	–.005205	.004294	.008390	.008926	.007417	.005017	.002574	.000726	.482720
	.75	–.049432	–.023879	–.007400	.001793	.005771	.006439	.005321	.003513	.001745	.000475	.747298
	.50	–.050809	–.025790	–.009814	–.001010	.002788	.003559	.002847	.001702	.000722	.000156	1.517274
	.25	–.052116	–.027779	–.012449	–.004177	–.000688	.000089	–.000255	–.000690	–.000721	–.000333	5.036660
	.05	–.052762	–.029168	–.014563	–.006944	–.003963	–.003470	–.003801	–.003848	–.003025	–.001331	46.279144
4.0	1.00	–.036398	–.014324	–.001100	.005381	.007387	.006840	.005140	.003192	.001519	.000401	.117729
	.75	–.037132	–.015482	–.002649	.003541	.005433	.004990	.003600	.002108	.000933	.000227	.383189
	.50	–.037866	–.016684	–.004285	.001576	.003325	.002969	.001893	.000880	.000249	.000015	1.236187
	.25	–.038556	–.017895	–.005987	–.000511	.001038	.000719	–.000087	–.000636	–.000677	–.000311	5.361717
	.05	–.038987	–.018773	–.007290	–.002171	–.000862	–.001287	–.002072	–.002472	–.002146	–.001043	59.036955
5.0	1.00	–.029190	–.009910	.000759	.005246	.006009	.004964	.003359	.001876	.000799	.000188	–.076118
	.75	–.029668	–.010748	–.000408	.003851	.004546	.003615	.002276	.001142	.000417	.000078	.226088
	.50	–.030151	–.011612	–.001618	.002397	.003016	.002196	.001122	.000343	–.000016	–.000055	1.189596
	.25	–.030627	–.012487	–.002859	.000894	.001420	.000687	–.000153	–.000614	–.000608	–.000275	5.836507
	.05	–.031004	–.013172	–.003826	–.000276	.000165	–.000544	–.001315	–.001707	–.001576	–.000834	70.409507

6.0	1.00	−.024327	−.007132	.001675	.004795	.004804	.003558	.002139	.001034	.000364	.000065	−.155403
	.75	−.024680	−.007786	.000752	.003702	.003692	.002577	.001393	.000559	.000133	.000002	.190297
	.50	−.025041	−.008460	−.000199	.002577	.002548	.001567	.000618	.000050	.000133	−.000079	1.250990
	.25	−.025408	−.009147	−.001171	.001428	.001378	.000521	−.000220	−.000560	−.000521	−.000235	6.299577
	.05	−.025725	−.009709	−.001947	.000526	.000473	−.000292	−.000930	−.001233	−.001184	−.000679	80.442327
8.0	1.00	−.018228	−.003953	.002340	.003790	.003071	.001827	.000812	.000223	−.000006	−.000028	−.151805
	.75	−.018457	−.004406	.001701	.003066	.002392	.001297	.000471	.000049	−.000071	−.000040	.256398
	.50	−.018693	−.004870	.001047	.002326	.001702	.000760	.000118	−.000151	−.000167	−.000071	1.434695
	.25	−.018939	−.005347	.000379	.001576	.001008	.000215	−.000265	−.000419	−.000359	−.000167	7.003055
	.05	−.019149	−.005743	−.000167	.000975	.000465	−.000197	−.000567	−.000703	−.000708	−.000471	97.376207
10.0	1.00	−.014568	−.002276	.002406	.002942	.001987	.000935	.000258	−.000035	−.000086	−.000057	−.089289
	.75	−.014734	−.002618	.001927	.002425	.001547	.000640	.000110	−.000082	−.000088	−.000034	.341996
	.50	−.014904	−.002968	.001438	.001901	.001103	.000340	−.000055	−.000161	−.000127	−.000050	1.563263
	.25	−.015081	−.003327	.000939	.001371	.000658	.000035	−.000243	−.000296	−.000246	−.000123	7.449026
	.05	−.015227	−.003623	.000530	.000940	.000304	−.000200	−.000392	−.000437	−.000453	−.000344	111.384600
12.0	1.00	−.012129	−.001291	.002275	.002281	.001297	.000463	.000033	−.000095	−.000078	−.000027	−.036835
	.75	−.012256	−.001564	.001898	.001897	.001001	.000297	−.000025	−.000094	−.000064	−.000021	.395070
	.50	−.012386	−.001842	.001515	.001509	.000704	.000124	−.000101	−.000128	−.000088	−.000035	1.629078
	.25	−.012520	−.002127	.001124	.001117	.000407	−.000053	−.000200	−.000208	−.000175	−.000095	7.754803
	.05	−.012630	−.002360	.000806	.000798	.000166	−.000197	−.000287	−.000292	−.000308	−.000260	123.365365
14.0	1.00	−.010387	−.000673	.002082	.001775	.000847	.000209	−.000050	−.000090	−.000053	−.000015	−.006016
	.75	−.010488	−.000898	.001777	.001481	.000645	.000117	−.000066	−.000075	−.000040	−.000011	.419074
	.50	−.010592	−.001127	.001466	.001185	.000442	.000017	−.000102	−.000093	−.000060	−.000025	1.659778
	.25	−.010698	−.001361	.001151	.000887	.000237	−.000090	−.000159	−.000148	−.000129	−.000076	7.992065
	.05	−.010785	−.001552	.000894	.000645	.000071	−.000182	−.000216	−.000207	−.000220	−.000202	133.814758

continued

Table 12.25 (continued) Bending moment variation and rotation D_4 at top edge; uniform load of intensity q per unit area; bottom edge encastré, and top edge free

η	h_t/h_b	*x/l*										D_4
		0.0	*.1*	*.2*	*.3*	*.4*	*.5*	*.6*	*.7*	*.8*	*.9*	
16.0	1.00	–.009080	–.000266	.001878	.001386	.000550	.000073	–.000074	–.000068	–.000031	–.000007	.007927
	.75	–.009163	–.000457	.001625	.001157	.000411	.000024	–.000071	–.000053	–.000023	–.000006	.426294
	.50	–.009243	–.000650	.001368	.000927	.000270	–.000033	–.000088	–.000066	–.000043	–.000020	1.675683
	.25	–.009336	–.000848	.001107	.000696	.000127	–.000099	–.000123	–.000109	–.000099	–.000063	8.191772
	.05	–.009407	–.001009	.000895	.000509	.000010	–.000161	–.000166	–.000154	–.000164	–.000161	143.050770
18.0	1.00	–.008063	.000010	.001683	.001086	.000352	.000002	–.000072	–.000047	–.000016	–.000002	.011852
	.75	–.008134	–.000155	.001469	.000906	.000255	–.000022	–.000062	–.000035	–.000013	–.000003	.426042
	.50	–.008205	–.000322	.001253	.000724	.000157	–.000054	–.000070	–.000047	–.000032	–.000017	1.686856
	.25	–.008279	–.000492	.001033	.000542	.000056	–.000096	–.000096	–.000082	–.000079	–.000053	8.365982
	.05	–.008339	–.000630	.000856	.000395	–.000028	–.000138	–.000130	–.000120	–.000127	–.000130	151.300735
20.0	1.00	–.007250	.000200	.001503	.000853	.000219	–.000033	–.000061	–.000030	–.000007	.000000	.010914
	.75	–.007310	.000056	.001321	.000709	.000152	–.000042	–.000050	–.000023	–.000007	–.000002	.423521
	.50	–.007372	–.000090	.001136	.000565	.000083	–.000059	–.000055	–.000035	–.000026	–.000014	1.697132
	.25	–.007435	–.000239	.000949	.000420	.000011	–.000087	–.000075	–.000064	–.000064	.000046	8.519909
	.05	–.007486	–.000360	.000798	.000303	–.000050	–.000117	–.000103	–.000096	–.000101	–.000107	158.735073
24.0	1.00	–.006030	.000428	.001197	.000528	.000069	–.000052	–.000037	–.000010	.000001	.000001	.005354
	.75	–.006077	.000313	.001059	.000435	.000038	–.000047	.000028	.000009	–.000003	–.000002	.419502
	.50	–.006124	.000197	.000920	.000341	.000005	–.000051	–.000032	–.000020	–.000018	–.000011	1.717754
	.25	–.006172	.000079	.000780	.000247	–.000032	–.000065	–.000048	–.000042	–.000044	–.000035	8.778773
	.05	–.006212	–.000017	.000667	.000171	–.000065	–.000083	–.000069	.000066	–.000068	.000075	171.649878

Notes: M = Coefficient × ql^2; D_4 = Coefficient × $qr^2/(Eh_b l)$.

Table 12.26 Hoop force variation and shear at bottom edge; triangular load of intensity q per unit area at bottom edge and zero at top; bottom edge hinged, top edge free

η	h_t/h_b	x/l											Shear
		0.0	*.1*	*.2*	*.3*	*.4*	*.5*	*.6*	*.7*	*.8*	*.9*	*1.0*	*at base*
.4	1.00	0.0000	.0566	.1118	.1648	.2151	.2628	.3082	.3520	.3946	.4366	.4784	.2457
	.75	0.0000	.0678	.1306	.1877	.2387	.2836	.3229	.3572	.3874	.4142	.4382	.2387
	.50	0.0000	.0851	.1599	.2234	.2754	.3160	.3456	.3654	.3762	.3794	.3759	.2278
	.25	0.0000	.1159	.2118	.2867	.3405	.3734	.3862	.3800	.3565	.3175	.2650	.2084
	.05	0.0000	.1660	.2961	.3897	.4467	.4676	.4532	.4051	.3251	.2159	.0806	.1767
.8	1.00	0.0000	.0746	.1438	.2044	.2553	.2967	.3300	.3571	.3801	.4011	.4214	.2342
	.75	0.0000	.0883	.1668	.2322	.2834	.3209	.3461	.3618	.3706	.3753	.3781	.2258
	.50	0.0000	.1085	.2005	.2728	.3244	.3559	.3695	.3684	.3565	.3379	.3159	.2136
	.25	0.0000	.1412	.2548	.3380	.3902	.4124	.4073	.3793	.3339	.2774	.2159	.1940
	.05	0.0000	.1871	.3311	.4298	.4832	.4931	.4629	.3976	.3039	.1901	.0661	.1661
1.2	1.00	0.0000	.0992	.1873	.2581	.3093	.3418	.3586	.3634	.3606	.3538	.3458	.2186
	.75	0.0000	.1156	.2145	.2903	.3412	.3684	.3753	.3669	.3486	.3257	.3018	.2091
	.50	0.0000	.1383	.2517	.3343	.3846	.4042	.3976	.3711	.3320	.2877	.2439	.1961
	.25	0.0000	.1717	.3061	.3983	.4473	.4560	.4298	.3770	.3076	.2324	.1616	.1773
	.05	0.0000	.2126	.3725	.4763	.5243	.5204	.4717	.3877	.2800	.1624	.0510	.1539
1.6	1.00	0.0000	.1262	.2345	.3154	.3661	.3886	.3875	.3691	.3397	.3048	.2681	.2021
	.75	0.0000	.1444	.2641	.3499	.3994	.4152	.4032	.3708	.3262	.2766	.2273	.1920
	.50	0.0000	.1684	.3027	.3944	.4420	.4489	.4224	.3720	.3082	.2410	.1782	.1792
	.25	0.0000	.2011	.3548	.4541	.4985	.4932	.4473	.3729	.2836	.1937	.1162	.1619
	.05	0.0000	.2373	.4121	.5193	.5604	.5425	.4766	.3769	.2590	.1401	.0393	.1428

continued

Table 12.26 (continued) Hoop force variation and shear at bottom edge; triangular load of intensity q per unit area at bottom edge and zero at top; bottom edge hinged, top edge free

η	h_t/h_b	*x/l*											*Shear*
		0.0	*.1*	*.2*	*.3*	*.4*	*.5*	*.6*	*.7*	*.8*	*.9*	*1.0*	*at base*
2.0	1.00	0.0000	.1526	.2800	.3696	.4188	.4308	.4128	.3731	.3200	.2602	.1984	.1865
	.75	0.0000	.1718	.3106	.4044	.4513	.4556	.4261	.3727	.3057	.2340	.1637	.1766
	.50	0.0000	.1960	.3487	.4472	.4906	.4851	.4409	.3708	.2875	.2029	.1258	.1644
	.25	0.0000	.2275	.3975	.5012	.5397	.5212	.4585	.3672	.2638	.1644	.0830	.1492
	.05	0.0000	.2600	.4474	.5560	.5892	.5579	.4776	.3662	.2422	.1241	.0315	.1334
3.0	1.00	0.0000	.2099	.3756	.4787	.5190	.5062	.4536	.3745	.2804	.1795	.0766	.1558
	.75	0.0000	.2296	.4050	.5095	.5448	.5226	.4586	.3688	.2666	.1621	.0614	.1474
	.50	0.0000	.2529	.4394	.5450	.5737	.5399	.4626	.3606	.2500	.1435	.0495	.1378
	.25	0.0000	.2812	.4805	.5864	.6064	.5586	.4657	.3498	.2303	.1230	.0406	.1267
	.05	0.0000	.3080	.5187	.6242	.6356	.5750	.4689	.3424	.2157	.1040	.0228	.1163
4.0	1.00	0.0000	.2559	.4477	.5558	.5804	.5452	.4684	.3671	.2536	.1356	.0165	.1352
	.75	0.0000	.2752	.4749	.5800	.5992	.5537	.4666	.3577	.2413	.1261	.0165	.1284
	.50	0.0000	.2977	.5060	.6092	.6195	.5619	.4631	.3461	.2271	.1167	.0208	.1209
	.25	0.0000	.3238	.5416	.6418	.6413	.5697	.4584	.3331	.2118	.1065	.0272	.1124
	.05	0.0000	.3477	.5734	.6699	.6590	.5753	.4546	.3249	.2032	.0977	.0202	.1048
5.0	1.00	0.0000	.2945	.5042	.6071	.6175	.5627	.4688	.3557	.2352	.1132	−.0087	.1210
	.75	0.0000	.3136	.5296	.6294	.6310	.5656	.4627	.3446	.2246	.1091	.0003	.1153
	.50	0.0000	.3353	.5581	.6538	.6451	.5678	.4553	.3322	.2129	.1051	.0122	.1091
	.25	0.0000	.3599	.5897	.6801	.6594	.5692	.4472	.3199	.2017	.1002	.0233	.1023
	.05	0.0000	.3820	.6173	.7018	.6699	.5688	.4406	.3131	.1978	.0963	.0188	.0963

6.0	1.00	0.0000	.3284	.5507	.6464	.6401	.5684	.4625	.3436	.2224	.1024	−.0167	.1105
	.75	0.0000	.3471	.5746	.6656	.6497	.5676	.4542	.3323	.2134	.1014	−.0034	.1056
	.50	0.0000	.3681	.6007	.6861	.6592	.5661	.4452	.3207	.2043	.1002	.0107	.1004
	.25	0.0000	.3916	.6291	.7074	.6681	.5636	.4362	.3106	.1966	.0980	.0220	.0946
	.05	0.0000	.4123	.6535	.7245	.6757	.5598	.4289	.3058	.1960	.0963	.0175	.0897
8.0	1.00	0.0000	.3863	.6237	.6995	.6617	.5640	.4445	.3233	.2071	.0956	−.0140	.0958
	.75	0.0000	.4044	.6445	.7135	.6656	.5595	.4355	.3141	.2014	.0971	−.0001	.0920
	.50	0.0000	.4242	.6666	.7277	.6688	.5543	.4268	.3060	.1964	.0980	.0122	.0879
	.25	0.0000	.4457	.6900	.7416	.6706	.5480	.4138	.3006	.1939	.0974	.0198	.0836
	.05	0.0000	.4645	.7097	.7522	.6704	.5415	.4127	.2991	.1965	.0973	.0152	.0800
10.0	1.00	0.0000	.4352	.6783	.7312	.6669	.5526	.4284	.3101	.2002	.0956	−.0071	.0857
	.75	0.0000	.4524	.6963	.7410	.6671	.5467	.4207	.3038	.1969	.0974	.0040	.0826
	.50	0.0000	.4710	.7151	.7505	.6664	.5403	.4139	.2992	.1948	.0983	.0126	.0793
	.25	0.0000	.4911	.7346	.7592	.6642	.5332	.4081	.2972	.1949	.0978	.0172	.0759
	.05	0.0000	.5084	.7509	.7656	.6611	.5266	.4038	.2976	.1980	.0979	.0133	.0729
12.0	1.00	0.0000	.4773	.7202	.7499	.6643	.5406	.4164	.3026	.1976	.0970	−.0023	.0783
	.75	0.0000	.4938	.7358	.7564	.6622	.5345	.4106	.2989	.1960	.0984	.0058	.0757
	.50	0.0000	.5114	.7517	.7622	.6591	.5282	.4058	.2968	.1954	.0989	.0117	.0729
	.25	0.0000	.5302	.7682	.7673	.6548	.5214	.4020	.2967	.1964	.0981	.0149	.0700
	.05	0.0000	.5463	.7817	.7706	.6503	.5158	.3996	.2978	.1991	.0983	.0120	.0676
14.0	1.00	0.0000	.5143	.7529	.7602	.6583	.5298	.4084	.2989	.1970	.0982	.0002	.0725
	.75	0.0000	.5300	.7663	.7642	.6547	.5242	.4043	.2970	.1964	.0992	.0061	.0702
	.50	0.0000	.5468	.7799	.7673	.6503	.5185	.4011	.2964	.1965	.0992	.0105	.0678
	.25	0.0000	.5645	.7937	.7696	.6450	.5128	.3990	.2973	.1977	.0983	.0130	.0654
	.05	0.0000	.5796	.8050	.7708	.6401	.5084	.3978	.2985	.1997	.0986	.0110	.0633

continued

Table 12.26 (continued) Hoop force variation and shear at bottom edge; triangular load of intensity q per unit area at bottom edge and zero at top; bottom edge hinged, top edge free

η	h_t/h_b	x/l											Shear
		0.0	.1	.2	.3	.4	.5	.6	.7	.8	.9	1.0	at base
16.0	1.00	0.0000	.5472	.7786	.7652	.6509	.5210	.4032	.2974	.1974	.0992	.0012	.0678
	.75	0.0000	.5623	.7901	.7671	.6465	.5161	.4005	.2966	.1972	.0996	.0057	.0658
	.50	0.0000	.5782	.8017	.7682	.6416	.5113	.3987	.2969	.1975	.0993	.0093	.0637
	.25	0.0000	.5950	.8134	.7686	.6360	.5068	.3977	.2981	.1985	.0983	.0117	.0616
	.05	0.0000	.6092	.8227	.7682	.6311	.5035	.3974	.2991	.2000	.0988	.0102	.0597
18.0	1.00	0.0000	.5768	.7990	.7666	.6432	.5141	.4001	.2970	.1981	.0997	.0015	.0640
	.75	0.0000	.5912	.8089	.7669	.6386	.5100	.3984	.2969	.1980	.0999	.0052	.0622
	.50	0.0000	.6064	.8187	.7666	.6335	.5062	.3976	.2976	.1982	.0994	.0083	.0603
	.25	0.0000	.6224	.8285	.7655	.6280	.5028	.3974	.2988	.1991	.0984	.0106	.0584
	.05	0.0000	.6358	.8363	.7640	.6235	.5005	.3976	.2995	.2001	.0990	.0096	.0568
20.0	1.00	0.0000	.6035	.8152	.7656	.6360	.5088	.3983	.2972	.1987	.1001	.0014	.0607
	.75	0.0000	.6174	.8236	.7648	.6314	.5055	.3975	.2975	.1986	.1000	.0047	.0591
	.50	0.0000	.6319	.8319	.7633	.6264	.5026	.3973	.2983	.1988	.0993	.0075	.0574
	.25	0.0000	.6471	.8401	.7612	.6213	.5002	.3976	.2993	.1994	.0984	.0098	.0557
	.05	0.0000	.6598	.8465	.7539	.6172	.4987	.3980	.2998	.2001	.0992	.0090	.0542
24.0	1.00	0.0000	.6502	.8382	.7598	.6236	.5021	.3972	.2982	.1995	.1003	.0011	.0555
	.75	0.0000	.6630	.8442	.7574	.6195	.5002	.3973	.2987	.1994	.1000	.0039	.0541
	.50	0.0000	.6763	.8500	.7546	.6153	.4987	.3977	.2993	.1993	.0993	.0065	.0527
	.25	0.0000	.6901	.8555	.7512	.6113	.4978	.3984	.2999	.1997	.0986	.0086	.0513
	.05	0.0000	.7016	.8597	.7481	.6082	.4974	.3989	.3001	.2001	.0994	.0083	.0500

Notes: N = Coefficient $\times\ qr$; V = Coefficient $\times\ ql$.

Table 12.27 Bending moment variation and rotations D_2 and D_4 at bottom and top edges, respectively; triangular load of intensity q per unit area at bottom edge and zero at top; bottom edge hinged, top edge free

η	h_t/h_b	x/l									D_2	D_4
		.1	.2	.3	.4	.5	.6	.7	.8	.9		
.4	1.00	.019831	.031227	.035739	.034897	.030204	.023136	.015149	.007681	.002158	.5691	.4182
	.75	.019148	.029970	.034094	.033090	.028468	.021677	.014110	.007113	.001987	.6978	.4984
	.50	.018087	.028017	.031536	.030280	.025769	.019409	.012497	.006231	.001721	.8986	.6192
	.25	.016203	.024548	.026994	.025290	.020975	.015378	.009627	.004662	.001249	1.2558	.8288
	.05	.013124	.018879	.019566	.017121	.013114	.008754	.004899	.002069	.000465	1.8367	1.2113
.8	1.00	.018707	.029155	.033033	.031948	.027407	.020827	.013542	.006825	.001907	.7559	.2026
	.75	.017895	.027666	.031093	.029832	.025393	.019153	.012366	.006192	.001720	.9160	.1976
	.50	.016712	.025495	.028268	.026752	.022463	.016720	.010659	.005274	.001449	1.1524	.1714
	.25	.014807	.022004	.023724	.021798	.017750	.012803	.007911	.003797	.001014	1.5355	.0917
	.05	.012098	.017032	.017240	.014710	.010976	.007141	.003907	.001625	.000366	2.0759	.0285
1.2	1.00	.017190	.026363	.029396	.027993	.023667	.017747	.011404	.005688	.001576	1.0134	−.0805
	.75	.016267	.024676	.027211	.025629	.021441	.015919	.010138	.005018	.001381	1.2069	−.1795
	.50	.015010	.022383	.024248	.022430	.018432	.013456	.008439	.004123	.001123	1.4762	−.3557
	.25	.013183	.019053	.019949	.017792	.014077	.009892	.005984	.002833	.000753	1.8746	−.7014
	.05	.010921	.014925	.014609	.012010	.008612	.005381	.002839	.001155	.000263	2.3650	−1.1404
1.6	1.00	.015582	.023410	.025560	.023840	.019757	.014541	.009186	.004513	.001234	1.2962	−.3675
	.75	.014610	.021644	.023291	.021409	.017493	.012707	.007936	.003863	.001049	1.5151	−.5409
	.50	.013366	.019388	.020402	.018324	.014632	.010402	.006377	.003060	.000823	1.8049	−.8235
	.25	.011702	.016376	.016553	.014227	.010845	.007362	.004329	.002013	.000533	2.2040	−1.3333
	.05	.009852	.013027	.012268	.009647	.006583	.003904	.001964	.000778	.000181	2.6490	−1.9706

continued

Table 12.27 (continued) Bending moment variation and rotations D_2 and D_4 at bottom and top edges, respectively; triangular load of intensity q per unit area at bottom edge and zero at top; bottom edge hinged, top edge free

η	h_t/h_b	x/l										
		.1	.2	.3	.4	.5	.6	.7	.8	.9	D_2	D_4
2.0	1.00	.014071	.020646	.021990	.019998	.016163	.011611	.007171	.003451	.000925	1.5747	−.6192
	.75	.013112	.018915	.019786	.017664	.014019	.009903	.006029	.002870	.000764	1.8098	−.8390
	.50	.011943	.016810	.017120	.014856	.011460	.007882	.004693	.002201	.000583	2.1093	−1.1788
	.25	.010471	.014170	.013790	.011369	.008298	.005403	.003070	.001400	.000371	2.5015	−1.7591
	.05	.008956	.011453	.010359	.007764	.005008	.002793	.001327	.000513	.000126	2.9110	−2.4585
3.0	1.00	.011097	.015257	.015121	.012721	.009468	.006245	.003536	.001560	.000383	2.1895	−1.0296
	.75	.010292	.013835	.013370	.010943	.007917	.005084	.002816	.001226	.000301	2.4419	−1.2747
	.50	.009378	.012231	.011412	.008980	.006234	.003853	.002077	.000900	.000227	2.7463	−1.6242
	.25	.008319	.010383	.009175	.006761	.004357	.002502	.001286	.000565	.000157	3.1187	−2.1843
	.05	.007332	.008664	.007097	.004695	.002591	.001200	.000481	.000187	.000061	3.4769	−2.8908
4.0	1.00	.009121	.011748	.010780	.008286	.005546	.003225	.001568	.000572	.000109	2.6947	−1.1913
	.75	.008476	.010641	.009478	.007044	.004550	.002559	.001218	.000447	.000091	2.9550	−1.4075
	.50	.007763	.009429	.008068	.005723	.003518	.001898	.000895	.000351	.000086	3.2605	−1.7089
	.25	.006964	.008078	.006512	.004281	.002406	.001196	.000560	.000257	.000085	3.6217	−2.2152
	.05	.006250	.006876	.005129	.002995	.001391	.000511	.000170	.000084	.000043	3.9564	−3.1052
5.0	1.00	.007765	.009403	.007996	.005584	.003294	.001599	.000578	.000107	.000013	3.1304	−1.2194
	.75	.007236	.008525	.007015	.004719	.002680	.001266	.000467	.000113	.000008	3.3972	−1.4033
	.50	.006658	.007574	.005966	.003812	.002053	.000943	.000379	.000139	.000038	3.7044	−1.6696
	.25	.006021	.006532	.004827	.002832	.001375	.000585	.000263	.000147	.000063	4.0590	−2.1812
	.05	.005466	.005630	.003842	.001978	.000758	.000203	.000057	.000052	.000036	4.3810	−3.4387

6.0	1.00	.006775	.007744	.006116	.003869	.001971	.000729	.000102	−.000090	−.000057	3.5219	−1.1898
	.75	.006328	.007026	.005357	.003258	.001602	.000597	.000129	−.000014	−.000017	3.7941	−1.3532
	.50	.005843	.006254	.004550	.002616	.001221	.000465	.000158	.000066	.000026	4.1025	−1.6101
	.25	.005316	.005420	.003683	.001926	.000797	.000286	.000139	.000108	.000055	4.4525	−2.1769
	.05	.004864	.004709	.002949	.001335	.000407	.000064	.000018	.000041	.000031	4.7665	−3.8572
8.0	1.00	.005410	.005556	.003805	.001957	.000678	.000023	−.000188	−.000161	−.000059	4.2173	−1.0948
	.75	.005068	.005041	.003318	.001636	.000563	.000072	−.000062	−.000047	−.000012	4.4963	−1.2538
	.50	.004704	.004497	.002306	.001295	.000428	.000095	.000035	.000045	.000027	4.8056	−1.5430
	.25	.004317	.003921	.002267	.000930	.000259	.000063	.000065	.000082	.000047	5.1496	−2.2673
	.05	.003989	.003440	.001820	.000623	.000093	−.000021	.000004	.000030	.000024	5.4538	−4.6791
10.0	1.00	.004496	.004180	.002489	.001016	.000173	−.000152	−.000187	−.000115	−.000036	4.8304	−1.0257
	.75	.004223	.003792	.002160	.000843	.000161	−.000057	−.000060	−.000016	.000002	5.1130	−1.2048
	.50	.003934	.003386	.001817	.000656	.000127	.000002	.000027	.000051	.000030	5.4225	−1.5448
	.25	.003631	.002963	.001461	.000455	.000063	.000010	.000049	.000068	.000039	5.7629	−2.4003
	.05	.003376	.002611	.001168	.000285	−.000010	−.000027	.000007	.000023	.000018	6.0608	−5.3303
12.0	1.00	.003836	.003243	.001677	.000518	−.000028	−.000167	−.000132	−.000065	−.000018	5.3840	−.9915
	.75	.003609	.002940	.001447	.000426	.000003	−.000072	−.000030	.000008	.000010	5.6690	−1.1940
	.50	.003372	.002627	.001210	.000325	.000011	−.000013	.000033	.000052	.000028	5.9788	−1.5752
	.25	.003124	.002302	.000966	.000213	−.000005	.000003	.000043	.000055	.000031	6.3168	−2.5159
	.05	.002917	.002034	.000765	.000116	−.000038	−.000018	.000009	.000017	.000013	6.6100	−5.8335
14.0	1.00	.003334	.002572	.001149	.000243	−.000101	−.000138	−.000082	−.000031	.000007	5.8922	−.9797
	.75	.003142	.002330	.000986	.000200	−.000054	−.000057	−.000006	.000020	.000012	6.1793	−1.2006
	.50	.002942	.002081	.000819	.000148	−.000029	−.000007	.000036	.000047	.000024	6.4894	−1.6087
	.25	.002733	.001825	.000647	.000087	−.000026	.000006	.000038	.000043	.000025	6.8254	−2.6038
	.05	.002560	.001614	.000506	.000031	−.000040	−.000009	.000009	.000012	.000010	7.1149	−6.2407

continued

Table 12.27 (continued) Bending moment variation and rotations D_2 and D_4 at bottom and top edges, respectively; triangular load of intensity q per unit area at bottom edge and zero at top; bottom edge hinged, top edge free

η	h_t/h_b	x/l										
		.1	.2	.3	.4	.5	.6	.7	.8	.9	D_2	D_4
16.0	1.00	.002937	.002073	.000794	.000089	–.000118	–.000103	–.000047	–.000012	–.000002	6.3646	–.9790
	.75	.002771	.001876	.000678	.000075	–.000068	–.000037	.000008	.000023	.000012	6.6533	–1.2119
	.50	.002599	.001675	.000558	.000052	–.000038	.000001	.000035	.000040	.000020	6.9636	–1.6364
	.25	.002421	.001468	.000435	.000021	–.000027	.000010	.000032	.000034	.000020	7.2980	–2.6699
	.05	.002273	.001298	.000334	–.000011	–.000033	–.000003	.000008	.000009	.000008	7.5847	–6.5829
18.0	1.00	.002616	.001691	.000549	.000004	–.000112	–.000073	–.000025	–.000003	–.000000	6.8076	–.9822
	.75	.002471	.001529	.000466	.000006	.000065	–.000021	.000015	.000023	.000011	7.0977	–1.2225
	.50	.002320	.001364	.000380	.000001	–.000034	.000008	.000033	.000033	.000017	7.4082	–1.6572
	.25	.002165	.001194	.000291	–.000012	–.000022	.000013	.000027	.000027	.000017	7.7413	–2.7216
	.05	,002036	.001056	.000218	–.000030	–.000025	.000001	.000007	.000007	.000007	8.0256	–6.8756
20.0	1.00	.002350	.001393	.000377	–.000042	–.000097	.000049	–.000011	.000001	.000000	7.2261	–.9860
	.75	.002221	.001259	.000318	.000030	–.000054	–.000009	.000018	.000020	.000009	7.5173	–1.2307
	.50	.002088	.001121	.000256	–.000024	–.000027	.000012	.000029	.000028	.000014	7.8280	–1.6726
	.25	.001951	.000980	.000192	–.000028	–.000015	.000014	.000022	.000022	.000015	8.1599	–2.7640
	.05	.001838	.000866	.000138	–.000037	–.000013	.000003	.000006	.000005	.000005	8.4422	–7.1280
24.0	1.00	.001934	.000966	.000167	–.000074	–.000064	–.000020	.000000	.000003	–.000000	8.0026	–.9909
	.75	.001830	.000870	.000139	–.000053	–.000032	.000004	.000017	.000015	.000006	8.2958	–1.2402
	.50	.001724	.000773	.000108	–.000039	–.000012	.000015	.000021	.000020	.000011	8.6066	–1.6933
	.25	.001614	.000673	.000074	–.000034	–.000004	.000013	.000015	.000015	.000011	8.9368	–2.8314
	.05	.001524	.000592	.000045	–.000035	.000007	.000004	.000004	.000004	.000004	9.2159	–7.5366

Notes: M = Coefficient × ql^2; D_2 or D_4 = Coefficient × $qr^2/(Eh_b l)$.

Table 12.28 Hoop force variation and shear at bottom edge; uniform load of intensity q per unit area; bottom edge hinged, top edge free

η	h_t/h_b	x/l											Shear
		0.0	*.1*	*.2*	*.3*	*.4*	*.5*	*.6*	*.7*	*.8*	*.9*	*1.0*	*at base*
.4	1.00	0.0000	.1566	.3118	.4647	.6150	.7627	.9081	1.0519	1.1944	1.3364	1.4783	.2458
	.75	0.0000	.1871	.3632	.5276	.6800	.8205	.9495	1.0676	1.1756	1.2741	1.3637	.2264
	.50	0.0000	.2348	.4437	.6261	.7818	.9111	1.0145	1.0926	1.1463	1.1763	1.1834	.1961
	.25	0.0000	.3205	.5882	.8030	.9649	1.0744	1.1321	1.1387	1.0946	1.0000	.8545	.1414
	.05	0.0000	.4632	.8291	1.0982	1.2714	1.3494	1.3330	1.2219	1.0139	.7032	.2770	.0497
.8	1.00	0.0000	.1746	.3437	.5044	.6552	.7967	.9299	1.0570	1.1800	1.3009	1.4212	.2342
	.75	0.0000	.2058	.3961	.5680	.7204	.8539	.9702	1.0715	1.1603	1.2392	1.3099	.2148
	.50	0.0000	.2524	.4739	.6622	.8170	.9390	1.0304	1.0937	1.1318	1.1473	1.1422	.1858
	.25	0.0000	.3293	.6022	.8179	.9770	1.0812	1.1330	1.1347	1.0880	.9933	.8491	.1376
	.05	0.0000	.4435	.7934	1.0510	1.2192	1.3015	1.3003	1.2155	1.0405	.7560	.3164	.0637
1.2	1.00	0.0000	.1992	.3873	.5580	.7093	.8418	.9585	1.0633	1.1605	1.2537	1.3457	.2187
	.75	0.0000	.2309	.4397	.6208	.7726	.8965	.9960	1.0757	1.1404	1.1947	1.2419	.1996
	.50	0.0000	.2750	.5124	.7076	.8602	.9726	1.0488	1.0940	1.1137	1.1125	1.0934	.1731
	.25	0.0000	.3413	.6210	.8372	.9917	1.0883	1.1321	1.1282	1.0796	.9864	.8451	.1328
	.05	0.0000	.4261	.7604	1.0051	1.1656	1.2488	1.2609	1.2033	1.0663	.8155	.3639	.0779
1.6	1.00	0.0000	.2262	.4345	.6154	.7661	.8885	.9875	1.0691	1.1396	1.2046	1.2679	.2021
	.75	0.0000	.2574	.4852	.6750	.8252	.9384	1.0206	1.0786	1.1199	1.1508	1.1755	.1841
	.50	0.0000	.2985	.5514	.7525	.9017	1.0033	1.0643	1.0926	1.0959	1.0803	1.0497	.1606
	.25	0.0000	.3548	.6417	.8577	1.0060	1.0935	1.1287	1.1197	1.0708	.9814	.8449	.1279
	.05	0.0000	.4171	.7411	.9744	1.1248	1.2040	1.2227	1.1861	1.0834	.8678	.4100	.0883

continued

Table 12.28 (continued) Hoop force variation and shear at bottom edge; uniform load of intensity q per unit area; bottom edge hinged, top edge free

η	h_t/h_b	x/l											Shear at base
		0.0	*.1*	*.2*	*.3*	*.4*	*.5*	*.6*	*.7*	*.8*	*.9*	*1.0*	
2.0	1.00	0.0000	.2526	.4800	.6696	.8188	.9308	1.0128	1.0731	1.1198	1.1600	1.1981	.1866
	.75	0.0000	.2827	.5280	.7248	.8720	.9745	1.0405	1.0796	1.1012	1.1127	1.1193	.1701
	.50	0.0000	.3206	.5875	.7927	.9371	1.0279	1.0750	1.0892	1.0801	1.0544	1.0160	.1496
	.25	0.0000	.3691	.6630	.8778	1.0188	1.0963	1.1230	1.1097	1.0623	.9788	.8484	.1232
	.05	0.0000	.4165	.7357	.9594	1.0987	1.1696	1.1386	1.1657	1.0913	.9097	.4516	.0945
3.0	1.00	0.0000	.3100	.5757	.7787	.9190	1.0062	1.0535	1.0745	1.0803	1.0793	1.0764	.1558
	.75	0.0000	.3369	.6160	.8217	.9569	1.0336	1.0677	1.0742	1.0651	1.0490	1.0302	.1433
	.50	0.0000	.3686	.6624	.8702	.9986	1.0634	1.0833	1.0749	1.0503	1.0155	.9718	.1291
	.25	0.0000	.4055	.7153	.9239	1.0435	1.0954	1.1024	1.0827	1.0440	.9793	.8671	.1127
	.05	0.0000	.4371	.7584	.9644	1.0743	1.1169	1.1223	1.1128	1.0865	.9777	.5354	.0980
4.0	1.00	0.0000	.3559	.6477	.8539	.9804	1.0453	1.0684	1.0670	1.0535	1.0353	1.0161	.1353
	.75	0.0000	.3806	.6823	.8876	1.0064	1.0604	1.0726	1.0620	1.0414	1.0177	.9931	.1258
	.50	0.0000	.4088	.7208	.9240	1.0333	1.0751	1.0763	1.0575	1.0309	.9997	.9613	.1153
	.25	0.0000	.4408	.7633	.9623	1.0595	1.0880	1.0800	1.0582	1.0302	.9843	.8891	.1039
	.05	0.0000	.4684	.7981	.9901	1.0734	1.0891	1.0779	1.0697	1.0694	1.0134	.5971	.0943
5.0	1.00	0.0000	.3945	.7043	.9071	1.0175	1.0627	1.0688	1.0556	1.0351	1.0130	.9909	.1210
	.75	0.0000	.4179	.7350	.9344	1.0354	1.0697	1.0668	1.0481	1.0255	1.0034	.9816	.1135
	.50	0.0000	.4440	.7687	.9633	1.0531	1.0755	1.0640	1.0414	1.0182	.9944	.9638	.1053
	.25	0.0000	.4734	.8053	.9927	1.0688	1.0782	1.0600	1.0389	1.0207	.9898	.9081	.0965
	.05	0.0000	.4995	.8365	1.0151	1.0760	1.0716	1.0488	1.0397	1.0526	1.0319	.6436	.0892

6.0	1.00	0.0000	.4284	.7507	.9464	1.0402	1.0684	1.0625	1.0436	1.0223	1.0021	.9329	.1105
	.75	0.0000	.4507	.7787	.9691	1.0525	1.0702	1.0571	1.0353	1.0148	.9974	.9810	.1043
	.50	0.0000	.4755	.8089	.9926	1.0639	1.0706	1.0511	1.0282	1.0100	.9934	.9699	.0976
	.25	0.0000	.5031	.8415	1.0161	1.0731	1.0678	1.0437	1.0248	1.0145	.9945	.9227	.0904
	.05	0.0000	.5278	.8697	1.0347	1.0769	1.0587	1.0300	1.0204	1.0389	1.0408	.6797	.0844
8.0	1.00	0.0000	.4864	.8237	.9995	1.0617	1.0640	1.0446	1.0233	1.0071	.9954	.9855	.0958
	.75	0.0000	.5072	.8473	1.0154	1.0668	1.0605	1.0371	1.0162	1.0030	.9951	.9881	.0911
	.50	0.0000	.5299	.8723	1.0312	1.0706	1.0557	1.0296	1.0110	1.0020	.9954	.9808	.0862
	.25	0.0000	.5549	.8988	1.0466	1.0722	1.0483	1.0211	1.0085	1.0083	1.0009	.9415	.0810
	.05	0.0000	.5769	.9216	1.0590	1.0719	1.0395	1.0098	1.0018	1.0205	1.0445	.7327	.0765
10.0	1.00	0.0000	.5352	.8783	1.0312	1.0669	1.0527	1.0284	1.0101	1.0001	.9953	.9923	.0858
	.75	0.0000	.5547	.8984	1.0422	1.0676	1.0471	1.0216	1.0054	.9985	.9963	.9943	.0820
	.50	0.0000	.5758	.9193	1.0526	1.0670	1.0406	1.0152	1.0026	.9997	.9978	.9869	.0781
	.25	0.0000	.5987	.9412	1.0624	1.0646	1.0326	1.0084	1.0016	1.0059	1.0041	.9524	.0740
	.05	0.0000	.6185	.9596	1.0700	1.0619	1.0254	1.0014	.9964	1.0103	1.0400	.7702	.0705
12.0	1.00	0.0000	.5773	.9202	1.0499	1.0643	1.0406	1.0165	1.0027	.9975	.9967	.9971	.0783
	.75	0.0000	.5957	.9374	1.0571	1.0624	1.0346	1.0111	1.0001	.9975	.9978	.9975	.0752
	.50	0.0000	.6155	.9550	1.0637	1.0593	1.0280	1.0064	.9991	.9996	.9994	.9899	.0720
	.25	0.0000	.6367	.9732	1.0694	1.0549	1.0207	1.0017	.9990	1.0047	1.0055	.9594	.0686
	.05	0.0000	.6547	.9882	1.0734	1.0508	1.0152	.9981	.9958	1.0047	1.0338	.7985	.0657
14.0	1.00	0.0000	.6143	.9529	1.0602	1.0583	1.0299	1.0084	.9990	.9970	.9980	.9995	.0725
	.75	0.0000	.6317	.9675	1.0647	1.0548	1.0242	1.0046	.9979	.9977	.9989	.9988	.0699
	.50	0.0000	.6503	.9825	1.0683	1.0503	1.0183	1.0013	.9980	1.0000	1.0003	.9914	.0671
	.25	0.0000	.6701	.9977	1.0711	1.0450	1.0123	.9985	.9984	1.0038	1.0059	.9645	.0642
	.05	0.0000	.6868	1.0100	1.0726	1.0402	1.0081	.9971	.9967	1.0016	1.0276	.8207	.0618

continued

Table 12.28 (continued) Hoop force variation and shear at bottom edge; uniform load of intensity q per unit area; bottom edge hinged, top edge free

		x/l											*Shear*
η	h_t/h_b	*0.0*	*.1*	*.2*	*.3*	*.4*	*.5*	*.6*	*.7*	*.8*	*.9*	*1.0*	*at base*
16.0	1.00	0.0000	.6472	.9786	1.0652	1.0509	1.0210	1.0032	.9974	.9974	.9989	1.0005	.0679
	.75	0.0000	.6638	.9912	1.0674	1.0465	1.0160	1.0007	.9973	.9984	.9996	.9992	.0656
	.50	0.0000	.6813	1.0038	1.0689	1.0415	1.0111	.9987	.9980	1.0004	1.0008	.9923	.0632
	.25	0.0000	.6999	1.0165	1.0696	1.0359	1.0064	.9973	.9985	1.0031	1.0058	.9685	.0607
	.05	0.0000	.7155	1.0267	1.0694	1.0311	1.0034	.9971	.9978	1.0000	1.0221	.8388	.0586
18.0	1.00	0.0000	.6768	.9990	1.0666	1.0432	1.0141	1.0001	.9971	.9981	.9995	1.0008	.0640
	.75	0.0000	.6926	1.0097	1.0672	1.0386	1.0099	.9985	.9975	.9991	1.0000	.9993	.0620
	.50	0.0000	.7092	1.0204	1.0670	1.0334	1.0059	.9975	.9983	1.0007	1.0010	.9930	.0599
	.25	0.0000	.7266	1.0310	1.0661	1.0279	1.0025	.9970	.9989	1.0025	1.0055	.9717	.0576
	.05	0.0000	.7413	1.0394	1.0647	1.0235	1.0004	.9975	.9987	.9992	1.0175	.8537	.0558
20.0	1.00	0.0000	.7035	1.0152	1.0656	1.0360	1.0088	.9983	.9973	.9987	.9999	1.0007	.0608
	.75	0.0000	.7186	1.0243	1.0650	1.0313	1.0054	.9975	.9979	.9995	1.0002	.9992	.0589
	.50	0.0000	.7344	1.0333	1.0636	1.0263	1.0024	.9972	.9988	1.0008	1.0011	.9936	.0570
	.25	0.0000	.7509	1.0422	1.0616	1.0212	1.0000	.9973	.9992	1.0019	1.0051	.9743	.0550
	.05	0.0000	.7647	1.0491	1.0594	1.0171	.9987	.9981	.9995	.9989	1.0138	.8663	.0534
24.0	1.00	0.0000	.7502	1.0382	1.0598	1.0236	1.0021	.9972	.9982	.9996	1.0001	1.0003	.0555
	.75	0.0000	.7640	1.0447	1.0575	1.0194	1.0001	.9973	.9989	1.0001	1.0002	.9991	.0540
	.50	0.0000	.7783	1.0509	1.0547	1.0153	.9986	.9976	.9995	1.0008	1.0010	.9946	.0524
	.25	0.0000	.7932	1.0569	1.0513	1.0112	.9977	.9982	.9997	1.0010	1.0043	.9783	.0508
	.05	0.0000	.8055	1.0615	1.0482	1.0080	.9974	.9990	.9999	.9989	1.0083	.8864	.0494

Notes: N = Coefficient × qr; V = Coefficient × ql.

Table 12.29 Bending moment variation and rotations D_2 and D_4 at bottom and top edges, respectively; uniform load of intensity q per unit area; bottom edge hinged, top edge free

		x/l										
η	h_t/h_b	.1	.2	.3	.4	.5	.6	.7	.8	.9	D_2	D_4
.4	1.00	.019835	.031231	.035745	.034905	.030212	.023145	.015157	.007687	.002162	1.5689	1.4181
	.75	.017949	.027761	.031194	.029895	.025386	.019073	.012246	.006087	.001677	1.9213	1.7419
	.50	.014998	.022322	.024063	.022042	.017819	.012686	.007677	.003574	.000915	2.4737	2.2786
	.25	.009684	.012530	.011216	.007889	.004168	.001149	–.000590	–.000983	–.000471	3.4662	3.4141
	.05	.000760	–.003927	–.010402	–.015973	–.018907	–.018423	–.014686	–.008808	–.002873	5.1182	7.1069
.8	1.00	.018709	.029159	.033039	.031955	.027415	.020835	.013550	.006831	.001911	1.7558	1.2023
	.75	.016816	.025678	.028486	.026958	.022619	.016805	.010681	.005262	.001438	2.1208	1.4747
	.50	.014003	.020505	.021719	.019528	.015479	.010797	.006396	.002911	.000727	2.6648	1.9815
	.25	.009318	.011884	.010425	.007099	.003499	.000668	–.000874	–.001108	–.000500	3.5658	3.4070
	.05	.002130	–.001381	–.007033	–.012247	–.015338	–.015483	–.012692	–.007818	–.002627	4.9030	11.9281
1.2	1.00	.017192	.026366	.029400	.027998	.023674	.017755	.011411	.005694	.001579	2.0134	.9191
	.75	.015342	.022974	.024980	.023170	.019063	.013902	.008684	.004212	.001135	2.3877	1.1409
	.50	.012766	.018251	.018826	.016442	.012628	.008511	.004855	.002120	.000505	2.9127	1.6401
	.25	.008858	.011079	.009458	.006159	.002731	.000142	–.001165	–.001225	–.000525	3.7025	3.4684
	.05	.003522	.001230	–.003531	–.008308	–.011492	–.012246	–.010447	–.006679	–.002337	4.7150	18.1501
1.6	1.00	.015583	.023412	.025564	.023845	.019763	.014547	.009192	.004519	.001237	2.2962	.6320
	.75	.013839	.020227	.021436	.019363	.015512	.011020	.006714	.003181	.000839	2.6719	.8234
	.50	.011558	.016064	.016042	.013503	.009942	.006382	.003437	.001399	.000304	3.1702	1.3507
	.25	.008389	.010270	.008511	.005273	.002046	–.000289	–.001372	–.001290	–.000533	3.8583	3.6231
	.05	.004547	.003190	–.000830	–.005174	–.008327	–.009489	–.008468	–.005638	–.002063	4.6231	24.7463

continued

Table 12.29 (continued) Bending moment variation and rotations D_2 and D_4 at bottom and top edges, respectively; uniform load of intensity q per unit area; bottom edge hinged, top edge free

		x/l										
η	h_t/h_b	.1	.2	.3	.4	.5	.6	.7	.8	.9	D_2	D_4
2.0	1.00	.014072	.020648	.021993	.020002	.016167	.011617	.007177	.003456	.000929	2.5748	.3802
	.75	.012477	.017750	.018263	.015984	.012389	.008509	.005013	.002298	.000587	2.9456	.5643
	.50	.010499	.014160	.013647	.011010	.007702	.004637	.002296	.000828	.000147	3.4157	1.1486
	.25	.007942	.009514	.007651	.004506	.001497	–.000590	–.001480	–.001299	–.000525	4.0237	3.8590
	.05	.005166	.004417	.000947	–.002997	–.006010	–.007369	–.006875	–.004765	–.001824	4.6270	31.2658
3.0	1.00	.011097	.015257	.015122	.012723	.009471	.006249	.003540	.001565	.000387	3.1895	–.0305
	.75	.009897	.013116	.012434	.009912	.006911	.004212	.002167	.000852	.000182	3.5404	.2002
	.50	.008532	.010688	.009398	.006744	.004024	.001902	.000590	.000014	–.000067	3.9574	.9744
	.25	.006958	.007892	.005897	.003067	.000617	–.000915	–.001444	–.001160	–.000458	4.4522	4.6264
	.05	.005547	.005371	.002684	–.000437	–.002869	–.004162	–.004243	–.003207	–.001367	4.8890	46.3551
4.0	1.00	.009121	.011748	.010781	.008287	.005548	.003228	.001572	.000576	.000112	3.6948	–.1926
	.75	.008217	.010174	.008874	.006380	.003897	.001985	.000780	.000187	.000006	4.0327	.1108
	.50	.007227	.008461	.006815	.004333	.002129	.000643	–.000095	–.000264	–.000128	4.4235	1.0558
	.25	.006142	.006593	.004577	.002100	.000162	–.000924	–.001221	–.000941	–.000372	4.8745	5.3884
	.05	.005238	.005046	.002722	.000211	–.001622	–.002584	–.002764	–.002238	–.001054	5.2758	59.4172
5.0	1.00	.007765	.009403	.007996	.005584	.003295	.001601	.000581	.000110	–.000010	4.1305	–.2210
	.75	.007054	.008202	.006603	.004268	.002233	.000866	.000154	–.000079	–.000056	4.4630	.1429
	.50	.006289	.006919	.005130	.002887	.001122	.000084	–.000323	–.000316	–.000127	4.8417	1.2067
	.25	.005469	.005554	.003573	.001430	–.000079	000828	–.000978	–.000736	–.000296	5.2735	6.0060
	.05	.004791	.004441	.002328	.000269	–.001082	–.001731	–.001880	–.001611	–.000836	5.6640	70.6885

6.0	1.00	.006775	.007744	.006116	.003869	.001972	.000730	.000105	–.000087	–.000055	4.5219	–.1917
	.75	.006193	.006790	.005060	.002935	.001281	.000306	–.000105	–.000162	–.000068	4.8524	.2123
	.50	.005573	.005782	.003957	.001966	.000563	–.000155	–.000364	–.000287	–.000108	5.2245	1.3479
	.25	.004914	.004721	.002808	.000961	–.000205	–.000707	–.000767	–.000572	–.000238	5.6451	6.4697
	.05	.004364	.003851	.001891	.000178	–.000815	–.001228	–.001324	–.001193	–.000677	6.0264	80.5658
8.0	1.00	.005410	.005555	.003805	.001957	.000679	.000024	–.000186	–.000159	–.000057	5.2173	–.0973
	.75	.004985	.004900	.003146	.001452	.000379	–.000101	–.000207	–.000145	–.000048	5.5459	.3374
	.50	.004537	.004216	.002466	.000931	.000059	–.000264	–.000285	–.000187	–.000067	5.9098	1.5308
	.25	.004066	.003504	.001768	.000399	–.000287	–.000491	–.000471	–.000356	–.000162	6.3153	7.0749
	.05	.003669	.002912	.001202	–.000020	–.000557	–.000697	–.000720	–.000703	–.000470	6.6768	97.3513
10.0	1.00	.004496	.004180	.002489	.001016	.000173	–.000151	–.000186	–.000112	–.000034	5.8304	–.0288
	.75	.004165	.003697	.002048	.000726	.000044	–.000169	–.000159	–.000086	–.000025	6.1573	.4033
	.50	.003818	.003197	.001596	.000427	–.000105	–.000228	–.000188	–.000114	–.000042	6.5156	1.6101
	.25	.003455	.002680	.001136	.000121	–.000274	–.000536	–.000299	–.000236	–.000119	6.9102	7.4563
	.05	.003150	.002251	.000759	–.000129	–.000417	–.000440	–.000434	–.000449	–.000343	7.2559	111.3552
12.0	1.00	.003836	.003243	.001677	.000517	–.000028	–.000166	–.000131	–.000063	–.000015	6.3839	.0047
	.75	.003566	.002872	.001369	.000346	–.000077	–.000150	–.000101	–.000045	–.000011	6.7096	.4265
	.50	.003286	.002490	.001056	.000168	–.000146	–.000171	–.000121	–.000073	–.000029	7.0636	1.6412
	.25	.002993	.002097	.000737	–.000017	–.000235	–.000232	–.000199	–.000168	–.000093	7.4500	7.7426
	.05	.002749	.001772	.000475	–.000172	–.000320	–.000299	–.000286	–.000306	–.000260	7.7848	123.3553
14.0	1.00	.003334	.002572	.001149	.000242	–.000101	–.000138	–.000082	–.000029	–.000005	6.8922	.0159
	.75	.003108	.002278	.000929	.000142	–.000112	–.000114	–.000060	–.000022	–.000005	7.2168	.4294
	.50	.002874	.001978	.000705	.000034	–.000142	–.000123	–.000078	–.000050	–.000023	7.5676	1.6562
	.25	.002630	.001669	.000477	–.000081	–.000191	–.000163	–.000139	–.000126	–.000076	7.9475	7.9803
	.05	.002429	.001416	.000291	–.000181	–.000247	–.000214	–.000202	–.000219	–.000202	8.2746	133.8149

continued

Table 12.29 (continued) Bending moment variation and rotations D_2 and D_4 at bottom and top edges, respectively; uniform load of intensity q per unit area; bottom edge hinged, top edge free

η	h_t/h_b	x/l									D_2	D_4
		.1	.2	.3	.4	.5	.6	.7	.8	.9		
16.0	1.00	.002937	.002073	.000794	.000089	–.000118	–.000103	–.000046	–.000011	.000000	7.3646	.0160
	.75	.002744	.001836	.000634	.000031	–.000112	–.000081	–.000034	–.000010	–.000002	7.6884	.4257
	.50	.002545	.001593	.000471	–.000034	–.000123	–.000086	–.000053	–.000037	–.000019	8.0365	1.6681
	.25	.002338	.001346	.000305	–.000107	–.000153	–.000117	–.000101	–.000098	–.000063	8.4114	8.1852
	.05	.002167	.001143	.000169	.000173	–.000192	–.000160	–.000151	–.000164	–.000161	8.7326	143.0531
18.0	1.00	.002616	.001691	.000549	.000003	–.000112	–.000073	–.000024	–.000002	.000002	7.8076	.0121
	.75	.002443	.001496	.000431	.000028	–.000099	–.000055	–.000019	–.000005	–.000002	8.1308	.4212
	.50	.002275	.001298	.000311	–.000066	–.000101	–.000061	–.000037	–.000030	–.000017	8.4767	1.6802
	.25	.002096	.001096	.000188	–.000113	–.000121	–.000087	–.000077	–.000078	–.000054	8.8475	8.3635
	.05	.001949	.000931	.000087	–.000158	–.000151	–.000123	–.000118	–.000127	–.000130	9.1639	151.3024
20.0	1.00	.002350	.001393	.000377	–.000042	–.000097	.000049	–.000011	.000002	.000002	8.2261	.0077
	.75	.002202	.001231	.000290	–.000057	–.000082	–.000037	–.000010	–.000003	–.000002	8.5486	.4182
	.50	.002050	.001067	.000200	–.000079	–.000081	–.000043	–.000028	–.000025	–.000015	8.8927	1.6928
	.25	.001894	.000899	.000108	–.000109	–.000095	.000066	–.000061	–.000064	–.000046	9.2602	8.5197
	.05	.001765	.000763	.000032	–.000140	–.000120	.000098	.000095	–.000101	–.000107	9.5726	158.7358
24.0	1.00	.001934	.000966	.000167	–.000074	–.000064	–.000020	.000000	.000003	.000001	9.0026	.0016
	.75	.001816	.000851	.000119	–.000072	–.000051	–.000015	–.000003	–.000002	–.000002	9.3243	.4166
	.50	.001696	.000734	.000069	–.000077	.000049	–.000023	–.000018	–.000018	–.000012	9.6654	1.7169
	.25	.001572	.000616	.000016	–.000090	–.000060	–.000042	–.000042	–.000045	–.000035	10.0276	8.7795
	.05	.001470	.000520	–.000028	–.000106	–.000078	–.000067	–.000066	–.000068	–.000075	10.3338	171.6498

Notes: M = Coefficient × ql^2; D_2 or D_4 = Coefficient × $qr^2/(Eh_b l)$.

Appendix A: Stiffness and fixed-end forces for circular and annular plates

The bending of circular and annular plates subjected to axisymmetrical load has been exhaustively treated by many authors.[1] The equations and tables presented in this appendix provide the fixed-end forces and the stiffness coefficients necessary for the analysis of circular or annular plates continuous with circular-cylindrical walls.

A.1 GOVERNING DIFFERENTIAL EQUATION

With the usual assumptions considered in the plate bending theory of thin elastic plates, the deflection of a circular plate of constant thickness is governed by the differential equation

$$\frac{d^4w}{dr^4}+\frac{2}{r}\frac{d^3w}{dr^3}-\frac{1}{r^2}\frac{d^2w}{dr^2}+\frac{1}{r^3}\frac{dw}{dr}=\frac{q12(1-v^2)}{Eh^3}, \tag{A.1}$$

where w and q are the deflection and load intensity at all points that lie on a circle of radius r; and E and v are the modulus of elasticity and Poisson's ratio of the plate material.

When q is constant, the solution of Equation (A.1) is

$$w=B_1+B_2\log\xi+B_3\xi^2+B_4\xi^2\log\xi+qr_B^4\frac{\xi^4}{64}\frac{12(1-v^2)}{Eh^3}, \tag{A.2}$$

where r_B is the radius at the outer edge; $\xi = r/r_B$; log indicates natural logarithm; and the B's are constants to be determined using the boundary conditions.

The principal moments in radial and tangential directions, M_r and M_t, respectively, are related to the deflection

$$M_r = -\frac{Eh^3}{12(1-\nu^2)}\left(\frac{d^2w}{dr^2} + \frac{\nu}{r}\frac{dw}{dr}\right), \tag{A.3}$$

$$M_t = -\frac{Eh^3}{12(1-\nu^2)}\left(\frac{1}{r}\frac{dw}{dr} + \nu\frac{d^2w}{dr^2}\right). \tag{A.4}$$

The shear on a circle of radius r is

$$V_r = \frac{Eh^3}{12(1-\nu^2)}\left(\frac{d^3w}{dr^3} + \frac{1}{r}\frac{d^2w}{dr^2} - \frac{1}{r^2}\frac{dw}{dr}\right). \tag{A.5}$$

A downward deflection w is here considered positive; M_r and M_t are positive when they produce tensile stress on the bottom surface of the plate; an upward force on the outer edge produces positive V_r.

A.2 STIFFNESS AND FIXED-END MOMENTS OF CIRCULAR PLATES

For a circular plate encastré at its outer edge and subjected to uniform load of intensity q, the equations of the previous section give (Figure A.1a)

$$M_r = \frac{qr_B^2}{16}\left[1+\nu-\xi^2(3+\nu)\right], \tag{A.6}$$

$$M_t = \frac{qr_B^2}{16}\left[1+\nu-\xi^2(1+3\nu)\right]. \tag{A.7}$$

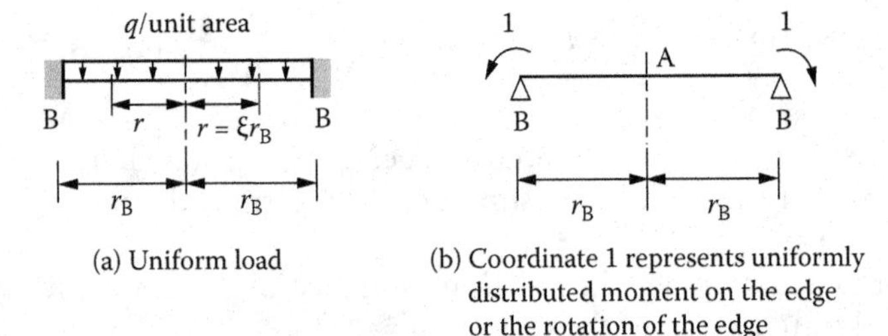

(a) Uniform load

(b) Coordinate 1 represents uniformly distributed moment on the edge or the rotation of the edge

Figure A.1 Circular plate of constant thickness.

At the fixed edge $\xi = 1$ and

$$M_r = -qr_B^2/8; \qquad M_t = -vqr_B^2/8. \tag{A.8}$$

A uniformly distributed moment of intensity unity applied in radial planes on the edge ($F_1 = 1$ in Figure A.1b) produces bending moments

$$M_r = M_t = -1 \tag{A.9}$$

and rotation at the edge

$$f_{11} = \frac{12(1-v^2)}{Eh^3}\frac{r_B}{(1+v)}, \tag{A.10}$$

where f_{11} represents the flexibility coefficient for the plate. Its inverse is the stiffness coefficient

$$S_{11} = \frac{Eh^3}{12(1-v^2)}\frac{(1+v)}{r_B}. \tag{A.11}$$

A.3 STIFFNESS OF ANNULAR PLATES

Figure A.2a represents a cross-section of an annular plate with two coordinates at each of the inner and outer edges. The coordinates represent the deflection w or the rotation dw/dr at the edges, or the corresponding forces $\{F\}$. A value F_i represents the intensity of a uniformly distributed force or moment at coordinate i. The forces $\{F\}$ and the displacement $\{D\}$ can be related

$$[S]\{D\} = \{F\}, \tag{A.12}$$

where $[S]$ is a square nonsymmetrical matrix representing the stiffness of the annular plate and will be derived later. The elements S_{ij} of this matrix are related by Betti's theorem

$$S_{ij} = S_{ji} \tag{A.13}$$

when coordinates i and j are on the same edge of the plate, and

$$r_A S_{ij} = r_B S_{ij} \tag{A.14}$$

or

$$S_{ij} = S_{ij}/\psi \tag{A.15}$$

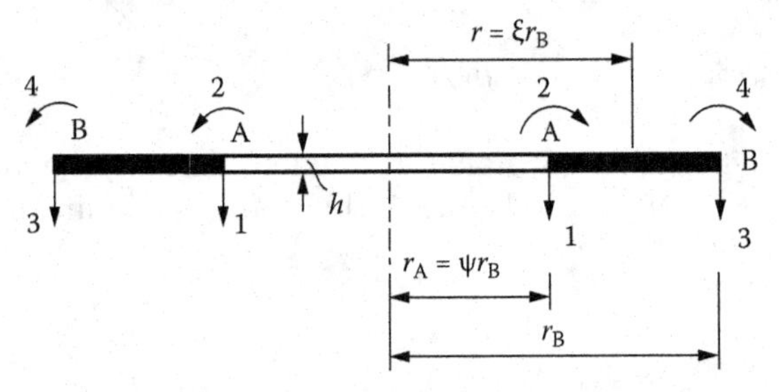

(a) Cross-section of an annular plate

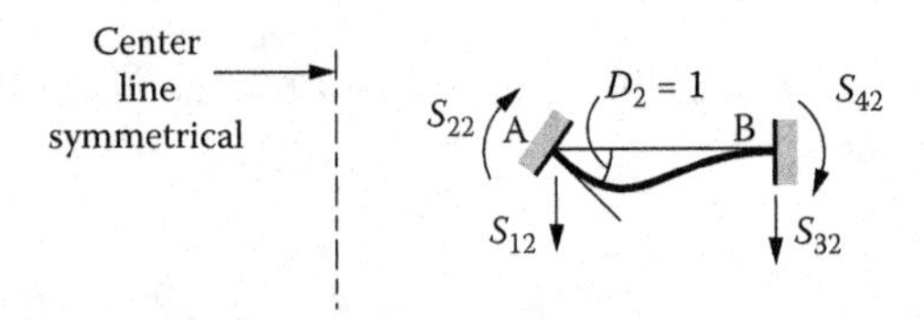

(b) Forces at the edges caused by displacement $D_2 = 1$

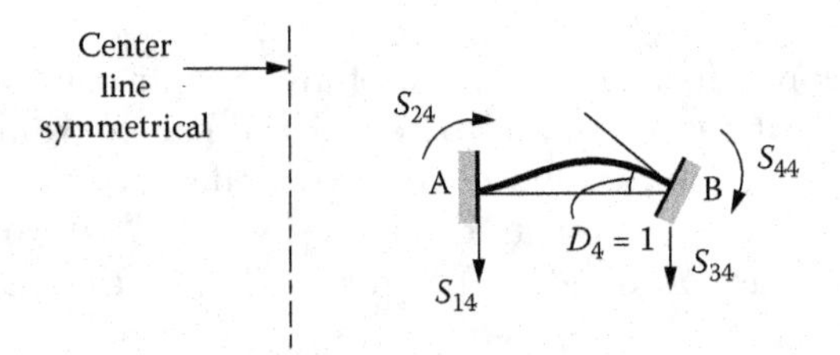

(c) Forces at the edges caused by displacement $D_4 = 1$

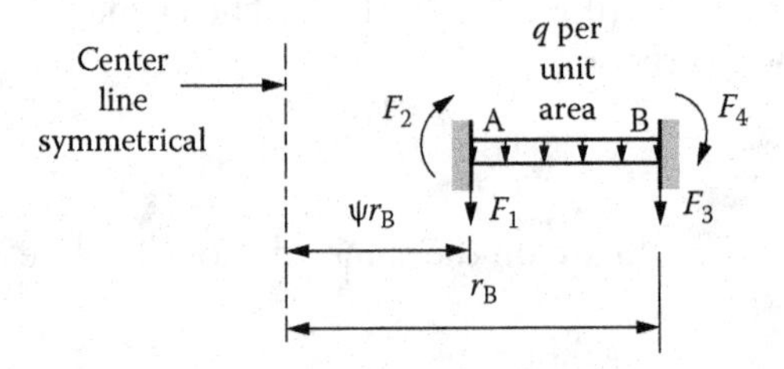

(d) Fixed-end forces caused by uniform load

Figure A.2 Derivation of stiffness coefficients and fixed-end forces for annular plates.

when coordinates i and j are at radii r_A/r_B, respectively; $\psi = r_A/r_B$. As an example of the application and proof of Equation (A.14), apply Betti's theorem[2] to the two systems of forces in Figure A.2b,c:

$$(2\pi r_A S_{24})D_2 = (2\pi r_B S_{42})D_4.$$

With $D_2 = D_4 = 1$, Equation (A.14) is proved.

In the absence of uniform load on the plate, q = 0, and the edge displacements {D} are related to the deflection w by Equation (A.2); thus

$$\{D\} = \begin{Bmatrix} (w)_{\xi=\psi} \\ \left(\frac{dw}{dr}\right)_{\xi=\psi} \\ (w)_{\xi=1} \\ \left(\frac{dw}{dr}\right)_{\xi=1} \end{Bmatrix} = \begin{bmatrix} 0 & \log\psi & \psi^2 & \psi^2\log\psi \\ 0 & \frac{1}{\psi r_B} & \frac{2\psi}{r_B} & \frac{\psi(2\log\psi+1)}{r_B} \\ 1 & 0 & 1 & 0 \\ 0 & \frac{1}{r_B} & \frac{2}{r_B} & \frac{1}{r_B} \end{bmatrix} \begin{Bmatrix} B_1 \\ B_2 \\ B_3 \\ B_4 \end{Bmatrix}, \quad \text{(A.16)}$$

or in shorter form,

$$\{D\} = [C]\{B\}, \quad \text{(A.17)}$$

where [C] is the 4 × 4 matrix in Equation (A.16).

The forces $\{F\}$ are related to the radial moment and shear at the edges, using Equations (A.2), (A.3), and (A.5):

$$\{F\} = \begin{Bmatrix} (V_r)_{\xi=\psi} \\ (M_r)_{\xi=\psi} \\ -(V_r)_{\xi=1} \\ -(M_r)_{\xi=1} \end{Bmatrix}$$

$$= \frac{Eh^3}{12(1-v^2)} \begin{bmatrix} 0 & 0 & 0 & \frac{4}{\psi r_B^3} \\ 0 & \frac{1-v}{\psi^2 r_B^2} & \frac{-2(1+v)}{r_B^2} & \frac{-2\log\psi(1+v)-3-v}{r_B^2} \\ 0 & 0 & 0 & -\frac{4}{r_B^3} \\ 0 & \frac{-(1-v)}{r_B^2} & \frac{2(1+v)}{r_B^2} & \frac{3+v}{r_B^2} \end{bmatrix} \{B\} \quad \text{(A.18)}$$

or

$$\{F\} = [L]\{B\}, \tag{A.19}$$

where $[L]$ is the 4×4 matrix in Equation (A.18) multiplied by $Eh^3/[12(1 - \upsilon^2)]$.
Solving for {B} from Equation (A.17) and substituting into Equation (A.19),

$$\{F\} = [L][C]^{-1}\{D\}. \tag{A.20}$$

Comparing Equations (A.12) and (A.20),

$$[S] = [L][C]^{-1}. \tag{A.21}$$

The variation of M_r, M_t, or V_r caused by edge displacements can be obtained by application of Equations (A.2) to (A. 5) after the constants $\{B\}$ have been determined.

Equation (A.21) is used to derive the stiffness coefficients in Table A.1 for annular plates of constant thickness having Poisson's ratio, $\upsilon = 1/6$.

The elements S_{22} and S_{44} of the stiffness matrix are, respectively, equal to the end-rotational stiffnesses S_{AB} and S_{BA} usable in the method of moment distribution, while the carry-over moments $t_{AB} = S_{42}$ and $t_{BA} = S_{24}$; the carry-over factors $C_{AB} = S_{42}/S_{22}$ and $C_{BA} = S_{24}/S_{44}$ (see Section 1.7).

Equations (A.13) and (A.15) are used to reduce the number of columns of Table A.1. The displacements $D_1 = 1$ produce the same end forces as $D_3 = 1$ but with reversed signs; thus

$$S_{13} = -S_{11}, \quad S_{23} = -S_{21}, \quad S_{33} = -S_{31}, \quad \text{and} \quad S_{43} = -S_{41}. \tag{A.22}$$

These relations are also used to reduce the size of Table A.1.

A.4 FIXED-END FORCES OF ANNULAR PLATES

For an annular plate with edges encastré and subjected to uniform load (Figure A.2d), the deflection w and the slope dw/dr are zero at $\xi = \psi$ and 1. Applying these four conditions, using Equation (A.2) gives

$$[C]\{B\} = -\frac{qr_B^4}{64}\begin{Bmatrix} \psi^4 \\ 4\psi^3/r_B \\ 1 \\ 4/r_B \end{Bmatrix}\frac{12(1-\upsilon^2)}{Eh^3}, \tag{A.23}$$

where [C] is defined by Equation (A.16).

Table A.1 Stiffness coefficients, end-rotational stiffnesses, and carry-over factors for annular plates (Figure A.2a,b,c) ($v = 1/6$)[a]

ψ	S_{11}	S_{21}	S_{41}	S_{22}	S_{42}	S_{44}
0.95	8445.9	209.30	202.26	6.9320	3.3418	6.7854
0.90	1086.1	53.302	49.688	3.5073	1.6265	3.3565
0.85	331.94	24.167	21.688	2.3687	1.0539	2.2134
0.80	144.88	13.891	11.973	1.8020	0.76696	1.6416
0.75	77.011	9.1003	7.5157	1.4645	0.59425	1.2984
0.70	46.454	6.4828	5.1154	1.2420	0.47858	1.0694
0.65	30.636	4.8975	3.6807	1.0858	0.39547	0.90576
0.60	21.613	3.8665	2.7577	0.97148	0.33264	0.78289
0.55	16.092	3.1606	2.1305	0.88583	0.28328	0.68721
0.50	12.536	2.6589	1.6858	0.82110	0.24328	0.61055
0.45	10.166	2.2932	1.3598	0.77272	0.21000	0.54773
0.40	8.5580	2.0225	1.1142	0.73818	0.18167	0.49527
0.35	7.4759	1.8221	0.92481	0.71657	0.15703	0.45080
0.30	6.7916	1.6769	0.77613	0.70867	0.13512	0.41260
0.25	6.4535	1.5790	0.65758	0.71779	0.11518	0.37945
0.20	6.4900	1.5271	0.56196	0.75218	0.09651	0.35044
0.15	7.0751	1.5291	0.48429	0.83295	0.07836	0.32494
0.10	8.8391	1.6139	0.42139	1.0263	0.05962	0.30263
0.05	15.112	1.8914	0.37212	1.6739	0.03798	0.28370
Multiplier	Eh^3/r_B^3	Eh^3/r_B^2	Eh^3/r_B^2	Eh^3/r_B	Eh^3/r_B	Eh^3/r_B

Notes: End-rotational stiffnesses, $S_{AB} = S_{22}$; $S_{BA} = S_{44}$. Carry-over factors, $CAB = S_{42}/S_{22}$; $CBA = S_{24}/S_{44} = S_{42}/(\psi S_{44})$.

[a] $S_{12} = S_{21}$; $S_{13} = S_{31}/\psi = -S_{33}/\psi = -S_{11}$; $S_{14} = S_{41}/\psi$; $S_{23} = S_{32}/\psi = -S_{21}$; $S_{24} = S_{42}/\psi$; $S_{34} = -S_{41}$.

Solution of this equation gives the constants $\{B\}$. Hence, the variation of M_r, M_t, and V_r can be determined (using Equations A.2 to A.5).

The fixed-end forces $\{F\}$ are obtained by appropriate differentiation of Equation (A.2) (see Equation A.18)

$$\{F\} = [L]\{B\} + q\begin{Bmatrix} \psi r_B/2 \\ -r_B^2\psi^2[(3+v)/16] \\ -r_B/2 \\ r_B^3[(3+v)/16] \end{Bmatrix}, \tag{A.24}$$

where [L] is defined by Equation (A.18).

Table A.2 Fixed-end forces in annular plates subjected to uniform load q per unit area (Figure A.2d) (v = 1/6)

ψ	F_1	F_2	F_3	F_4
0.95	−25.307	−0.21052	−24.708	0.20624
0.90	−51.296	−0.85168	−48.833	0.81653
0.85	−78.089	−1.9402	−72.375	1.8180
0.80	−105.83	−3.4966	−95.333	3.1977
0.75	−134.72	−5.5463	−117.71	4.9423
0.70	−165.01	−8.1210	−139.49	7.0383
0.65	−197.00	−11.261	−160.70	9.4720
0.60	−231.13	−15.019	−181.32	12.229
0.55	−268.00	−19.464	−201.35	15.296
0.50	−308.41	−24.689	−220.79	18.657
0.45	−353.57	−30.820	−239.64	22.298
0.40	−405.26	−38.041	−257.90	26.203
0.35	−466.30	−46.619	−275.54	30.355
0.30	−541.41	−56.969	−292.58	34.737
0.25	−639.11	−69.774	−308.97	39.328
0.20	−776.53	−86.247	−324.70	44.103
0.15	−993.87	−108.82	−339.67	49.024
0.10	−1412.7	−143.32	−353.73	54.016
0.05	−2647.2	−209.77	−366.39	58.886
Multiplier	$10^{-3}qr_B$	$10^{-3}qr_B^2$	$10^{-3}qr_B$	$10^{-3}qr_B^2$

The preceding equations are used to generate Table A.2, which gives the fixed-end forces for annular plates of constant thickness subjected to uniformly distributed load, with Poisson's ratio v = 1/6.

A.5 APPROXIMATE ANALYSIS FOR ANNULAR PLATES

In the case when y is close to unity ($0.6 \leq \psi \leq 1.0$), the tangential moment in an annular plate is small and the only important moment is in the radial direction. An approximate analysis can be obtained by considering an elemental radial beam with linearly varying width, equal to unity at the outer edge B and equal to ψ at edge A. This approximation is particularly useful in practical cases when the thickness is variable[3] and Table A.1 cannot be applied. It is to be noted that the bending moment and the shear of the elemental beam are, respectively, equal to $\xi\, M_r$ and $\xi\, V_r$ of the annular plate.

NOTES

1. See, for example, Timoshenko, S. P., and Wojnowsky-Krieger, S., 1959, *Theory of Plates and Shells*, 2nd edn, McGraw-Hill, New York; and Márkus, G., 1964, *Theorie und Berechnung Rotationssymmetrischer Bauwerke*, Werner-Verlag, Düsseldorf.
2. See Ghali, A., Neville, A. M., and Brown, T. G., 2009, *Structural Analysis: A Unified Classical and Matrix Approach*, 6th ed., Spon Press, London.
3. See Ghali, A., 1958, Analysis of cylindrical tanks with flat bases by moment distribution methods, *The Structural Engineer*, 36, no. 5, pp. 165–76.

Appendix B: Computer programs provided at http://www.crcpress.com/product/isbn/9781466571044

B.1 INTRODUCTION

At the Web site http://www.crcpress.com/product/isbn/9781466571044 two computer programs are provided as tools to assist in the analyses discussed in this book. One program is for the analysis of circular-cylindrical shells subjected to axially symmetrical loading; the same program can analyze beams on elastic foundation. The other program applies finite-element analysis to shells of revolution. For both programs, the loading on the structures is assumed axially symmetrical. A simple input file is required for running any of the two programs.

The names of the programs are

- CTW (Cylindrical Tank Walls)
- SOR (Shells of Revolution)

A folder is provided for each program containing files named:

Manual CTW.doc	Manual SOR.doc
CTW.EXE	SOR.EXE
EXAMPLE1.in	EXAMPLE.in
EXAMPLE2.in	
EX2_3.in	
EX8_3.in	
EX15_2.in	

For the analysis of a circular-cylindrical wall, the program CTW is preferred over the program SOR; the input data can be expediently generated for all load types expected to occur in the design. The program SOR applies the finite element method for any shell of revolution. The structure is idealized as an assemblage of shell elements, generally in the shape of a frustum of a cone, connected at circular nodal lines. The accuracy of the results increases by increasing the number of elements.

Four input file examples of cylindrical walls and one example of a beam on elastic foundation (EX15_2.IN) are provided for the computer program CTW.

B.2 COMPUTER PROGRAM CTW (CYLINDRICAL TANK WALLS)

The program CTW performs elastic analysis of circular-cylindrical walls of variable thickness subjected to axially symmetrical external applied loads or to temperature variations. Application examples are the vertical walls of circular storage tanks and silos.

The analysis gives the variation of the bending moments, M and M_ϕ in the vertical and the circumferential directions (force length/length) and the hoop force, N (force/length). The reactions at the edges are also given.

The top or bottom edges of the wall can rotate freely or the rotation can be completely restrained. The translation of the edges in the radial direction can be either free or elastically restrained. The latter represents the case when the wall slides on an elastomeric pad.

The wall thickness is to be specified at a minimum of three points; the computer assumes parabolic variation between the three points. By giving the thickness values at additional points, practically any variation of thickness can be analyzed. Any consistent system of units may be used.

B.2.1 Load types

The load can be of the following types.

B.2.1.1 Type 0

Type 0 is a concentrated line load of constant intensity, Q (force/length). An example is the load normal to the wall produced by a post-tensioned circumferential tendon. In this case, Q is equal to the force in the tendon divided by the radius r of the wall middle surface.

A distributed couple of intensity C (force length/length) applied at an edge or at any intermediate level is to be replaced by two line loads of intensities, Q and $-Q$; where $Q = C/(\Delta l)$, with(Δl) being an arbitrary small distance (suggested equal to one hundredth of the wall height).

B.2.1.2 Type 1

A distributed load whose intensity, q (force/length^2) is specified at an odd number of points, not less than three. The program assumes parabolic variation between each three consecutive points. A straight line variation is treated as a special case of a parabolic variation; thus three values of q must be specified.

B.2.1.3 Type 2

Temperature rise varying linearly between specified values in degrees, T_o and T_i at the outer and the inner faces of the wall, respectively.

B.2.1.4 Type 3

Prescribed radial outward translation, w_B and/or rotation θ_B at the bottom edge.

B.2.1.5 Type 4

Prescribed radial outward translation w_T and/or rotation θ_T at the top edge.

B.2.1.6 Types –1, –2, to –9

A distributed load whose intensity q = Multiplier × Z; where Z = Z(x) = one of nine shape functions defined in Section B.3. The multiplier is a real value specified in the input data.

Nine shape functions are selected to represent variations of load intensity; these are identified by the negative integers –1 to –9. The shapes can be combined to give distribution of circumferential prestressing forces, which produce an improved distribution of circumferential prestressing (see Chapter 8, Section 8.3).

The nine shape functions and their identifying integers are given next in terms of ξ, where $\xi = x/l$, with l being the wall height and x is the distance between the bottom edge and any point.

Function "–1": $Z = sin(\pi\xi)$
Function "–2": $Z = sin(2\pi\xi)$
Function "–3": $Z = (\exp(-\pi\xi))\, sin(\pi\xi)$
Function "–4": $Z = (\exp(-2\pi\xi))\, sin(2\pi\xi)$
Function "–5": $Z = (\exp(-\beta\xi))\, cos(\beta\xi)$

where $\beta = 3(1 - \mu^2)^{0.25}/(r\, h)$; μ = Poisson's ratio; and h = wall thickness at or close to the bottom end.

Function "–6": $Z = (\exp(-1.25\,\beta\xi))\, cos(1.25\,\beta\xi)$
Function "–7": $Z = (\exp(-\beta\xi))\, sin\,(\beta\xi)$

Function "–8": $Z = (\exp(-1.25\,\beta\xi))\,sin(1.25\,\beta\xi)$

Function "–9": $$Z = l\{1-\xi-\exp(-7.203\xi)[\cos(7.203\xi)+\sin(7.203\xi)]+ \left(\tfrac{1}{7.203}\right)\exp(-7.203\xi)\sin(7.203\xi)\}$$

All the shape functions are dimensionless except the last one, which has unit of length. This function is suggested by Brandom-Nielson (*ACI Structural Journal*, July/August 1985). With this function, the multiplier must have the unit force/length^2.

B.2.2 Method of analysis

The analysis is done by finite differences using 101 equally spaced nodes (see Chapter 3). However, the values of M, M_ϕ, and N_ϕ are calculated only at the top and bottom edges and at 19 equally spaced inner nodes.

B.2.3 Sign convention

The load intensity q is positive when acting outward. The bending moment, M or M_ϕ, is positive when it produces tension at the outer face of the wall. The hoop force is positive when tensile. A positive temperature value represents a temperature rise. A positive prescribed edge translation, w_B or w_T, represents an outward radial deflection. The edge rotation, θ_B or θ_T, is positive when the derivative (dw/dx) is positive, with w being the outward radial deflection and x the distance between the bottom edge and any point. A positive value of the reaction at an edge indicates an inward line load normal to the wall surface at one of the edges.

B.2.4 Input data

To run the program, type: CTW. This must be preceded by the preparation of an input file of any name, containing 10 or 11 lines or sets of lines:

1. A title not exceeding 76 characters.
2. Modulus of elasticity, E, and Poisson's ratio, μ, and thermal expansion coefficient, α.
3. Radius of middle wall surface, r, and wall height, l.
4. A set of lines indicating the thickness variation. Each line consists of three x values and the corresponding wall thicknesses, h. Repeat this line as many times as necessary to describe the thickness variation over the whole wall height. The first x value on the first line of the set must be 0.0 and the last x value on the last line must be equal to the wall height, l. Also, the third x value on any line must be the same as the first value of the subsequent line.

5. Number of load cases, NLC (not exceeding 8) followed by the parameter "IRESULTS", which is equal to 0 or 1. When the integer 1 is selected, the results will be printed at 21 nodes to be specified below by the user. When IRESULTS = 0, the results will be printed at 21 equally spaced points.
6. The integer 0 or 1, followed by a value of a variable SSB, representing the conditions at the bottom edge. The integers 1 and 0 are indicators of free and prevented rotation, respectively. The value SSB (force/length^2) is the stiffness of springs restraining the radial translation. The stiffness is the force per unit length of the periphery per unit radial translation.
7. A line containing the integer 0 or 1 and the value SST, representing the conditions at the top edge. When the radial translation is prevented at the bottom or top edge, enter a large value for SSB or SST; entering the value 0.0 will indicate that the edge is free to slide. Note that the boundary conditions specified by this data line and the preceding one apply to all the loading cases except a single case of prescribed edge displacement, if any.
8. A set of lines of arbitrary number giving the load data. The last line of the set starts with an integer greater than NLC (see item 9). Each line of the set gives

   LCASE, LTYPE, *VAR1*, *VAR2*

   where LCASE is the load case number; LTYPE is the integer 0, 1, 2, 3, 4, or –1 to –9, representing one of the load types defined in Section 2; *VAR1* and *VAR2* are real numbers whose meanings differ with the load type:

   Type 0
   - *VAR1* = the x value of the point at which the line load is applied.
   - *VAR2* = the line load intensity, Q (force/length).

   Type 1
   - *VAR1* = the x value of a point.
   - *VAR2* = the load intensity, $q(x)$ (force/length^2). Note that this load type requires a minimum of three data lines, specifying the values of $q(x)$ at an odd number of points, whose x-coordinates must be listed in ascending order.

   Type 2
   - *VAR1* = the temperature rise, T_O at the outer face.
   - *VAR2* = the temperature rise, T_I at the inner face.

   Types 3 and 4
   - *VAR1* = w_B or w_T, the prescribed radial translation at bottom or top edge.
   - *VAR2* = θ_B or θ_T, the prescribed rotation at bottom or top edge.

The value of $VAR1$ or $VAR2$ can be any real number other than 0.0. The value 0.0 is reserved to indicate a free displacement. The following are two data line examples and their meaning:

```
8  3  1.0  1.0E-6
8  4  0.0  0.0
```

The first line indicates that for the case of loading 8, $w_B = 1.0$, while θ_B is restricted to the negligible value 1.0E-6. The second line indicates that the same load case, the translation and rotation are free to occur at the top edge. Absence of this data line has no effect in this case; the free displacement condition will be assumed when no prescribed values are given. Note that only one case of prescribed displacements at the edges can be analyzed in one run. For the other load cases analyzed in the same run, the edge conditions are as specified earlier by the parameters IBB, SSB, IBT, and SST.

Types –1 to –9
 $VAR1$ = the constant by which the shape function –1, –2, or –9 is to be multiplied to give the load intensity, q. In this case, $VAR2$ is a dummy real value, which is not to be omitted.

9. One line to indicate the end of the load data set. It consists of an integer greater than the number of load cases, followed by a dummy integer and two dummy real values.
10. A set of lines listing 21 integers in ascending order, to select 21 nodes, out of 101, where the results are required. This set is to be omitted when IRESULTS = 0.
11. A set of lines indicating any number of load combinations. Each line starts with a load combination number followed by a set of load factors whose number equals NLC. The load factors, in the order in which they are listed, will be used as multipliers to the respective load cases. The last line of the set starts with the integer 99, followed by NLC dummy real values, which must not be omitted. The results of a load combination give the effect of all the cases combined, with each case being multiplied by its respective load factor.

B.2.5 Beam on elastic foundation

The program CTW can be used for analysis of a beam on elastic foundation only when its moment of inertia, I, of the cross-section and foundation modulus, K (force/length^2), are constants. This is possible because the beam has the same deflection and bending moment as for a strip of unit width in an analogous circular cylinder. The strip runs in the direction of the cylinder axis. To use program CTW for the analysis of a beam on elastic foundation, enter as input the elasticity modulus, $(E)_{\text{cylinder}}$ equal to that of the beam; Poisson's ratio for the cylinder = 0 and the length of the cylinder, $(l)_{\text{cylinder}}$ = the length of the beam, but

$$(r\, h)_{\text{cylinder}} = \sqrt{}\,(12\, E\, I/k)_{\text{beam}}.$$

An arbitrary value can be chosen for $(h)_{\text{cylinder}}$, then $(h)_{\text{cylinder}}$ calculated to satisfy this equation. The symbols C, Q, and q for the beam will, respectively, mean applied couple, transverse force, and load intensity. Their units will be (force length), force, and (force/length), respectively. The symbols SSB and SST will have the units (force/length). No thermal loading can be analyzed. The bending moment obtained by the analysis of the cylinder will be the same as the beam's bending moment, but the circumpherential bending moment calculated for the cylinder has no meaning. The hoop force, $(N_\phi)_{\text{cylinder}}$, can be used to calculate the beam's deflection or the elastic foundation reaction by the equations

$$\text{Beam deflection} = \left(\frac{rN_\phi}{Eh}\right) \text{cylinder;}$$

$$\text{foundation reaction} = k\left(\frac{rN_\phi}{Eh}\right) \text{cylinder}$$

B.2.6 Input data examples

The input data is to be prepared in a file with any name. Four example input data files are included at the companion Web site for this book. The data of each line in an input file is followed by explanatory words. To prepare the input file for a new problem, it is recommended to copy one of the input files in a new file and change the numbers to suit the problem in hand. An example input file is

```
Prestressing and hydrostatic pressure; units: N and m

32.e9   0.1667   10.e-6  E, μ, α
30. 10.                  r, l
0.  5.  10. 0.3 0.3 0.3 {x1, x2, x3}, thicknesses{h1, h2, h3}
2   0                   NLC; IRESULTS
0   1.e10                  Edge conditions at bottom
1   0.0                 Edge conditions at top
1   1 0.   -140.e3        Load case no., load type, x, q
1   1 5.   -79.e3
1   1 10.  -18.e3
```

```
2   1  0.   100.e3
2   1  5.   50.e3
2   1  10. 0.0
9   0  0.   0.0       Termination of load data
1   0.85    1.0       Combination 1: 0.85 (Case 1) + 1.0 (Case 2)
99  0.0 0.0          Termination of load combinations
```

Notes: The analysis is for a tank of radius 30 m and height 10 m and constant wall thickness = 0.3 m, subjected in Case 1 to circumferential prestressing of linearly varying intensity: –140.0 and –18.0 kN/m^2 at bottom and top, respectively. The hydrostatic pressure in Case 2 is 100 and 0.0 kN/m^2 at bottom and top, respectively. Case 1 is to be combined with load with load factors 0.85 and 1.0, respectively.

B.3 COMPUTER PROGRAM SOR (SHELLS OF REVOLUTION)

SOR analyzes shells of revolution by the finite element method. The element used has the shape of a frustum of a cone of constant thickness. The idealized structure is defined by (x, r) coordinates of nodes situated on a radial half section passing through the axis of revolution. The x-axis is vertical, pointing upward; it coincides with the axis of revolution. The analysis gives the nodal displacements, the reactions, and four stress resultants at midheight of each element. The stress resultants are N_s, N_ϕ, M_s, M_ϕ; where N is normal force per unit length; M is bending moment per unit length; the subscript s refers to direction of a meridian; and the subscript ϕ refers to circumferential (hoop) direction. To run the program, type SOR and press Enter; this must be preceded by the preparation of an input file named SOR.IN. The program will execute the analysis and give the input and the results in a file named SOR.OUT.

B.3.1 Load types

The loading can be any, or a combination, of the following types.

1. Two concentrated forces and a couple, uniformly distributed over a circumpherential nodal line. Enter load intensities q_1 (vertical upward), q_2 (radial outward), and q_3 (radial moment, represented by a clockwise arrow on the right-hand half sectional elevation).
2. Uniform load on the surface of an element, whose intensities per unit area are q_t and q_n, where q_t is the tangential load directed along a meridian, from the element's starting node to its end node (Chapter 5, Figure 5.3c); and q_n is normal to element surface pointing in the direction of local

coordinates 2* and 5* (Chapter 5, Figure 5.3d). The self-weight of the structure can be determined by the computer. The command for this loading, which is a part of loading type (b), is indicated later.

3. Temperature rise in elements; in which case, the element number the temperatures T_i and T_o at the inner and outer surfaces are given. It is assumed that the rise of temperature varies linearly between T_i and T_o.

B.3.2 Sign convention

Positive displacement *u* is vertical upward; positive displacement *w* is horizontal radial outward; and positive rotation theta is represented by a clockwise arrow on the right-hand half of a sectional-elevation of the structure. *N* is positive when tensile. *M* is positive when producing tension at the outer surface of a conical element. When the two nodes defining an element have the same *x*-coordinate, the element has the shape of annular plane; positive moment produces tension at the bottom face, provided that r_1 is less than r_2. The subscripts 1 and 2 refer to the first and second nodes of the element. The reaction components are determined in the directions of the global axes.

B.3.3 Input data instructions

The data, in free format, must be in a file named SOR.IN. The data is composed of single lines or sets of lines as follows:

- Single line of title, not exceeding 72 characters.
- Single line of four integers: number of nodes, number of elements, number of joints with prescribed displacement(s), and number of load cases.
- Single line of three real values for elasticity modulus, *E*, Poisson's ratio, μ, and coefficient of thermal expansion, α.
- A set of lines whose number equals the number of nodes. Each line is composed of a node number followed by its *x* and *r* coordinates.
- A set of lines whose number equals the number of elements. Each line starts with three integers: a member number, its starting node, and its end node, followed by a real value, the thickness, *t*.
- A set of lines whose number equals the number of nodal lines with prescribed displacements. Each line starts with a node number and three restraint indicators: 0, 1, or 2, followed by three real values of prescribed displacements. The restraint indicators 1 and 0 mean free and prescribed displacement components, respectively. When a displacement component is free, a dummy value must be filled in for the prescribed displacement (any real number). When the indicator is 2 it indicates spring support; the corresponding real number on the same line is the spring stiffness (force/length^2) or (force-length/length).

- A set of any number of lines for loading of type (a). Each line is composed of two integers, a load case number and a node number, followed by three real values of nodal forces q_1, q_2, and q_3.The last line of the set must start with an integer greater than the number of load cases followed by a dummy integer and three dummy real numbers. The last line must not be omitted even when there is no loading of type (a).
- A set of any number of lines for loading of type (b). Each line is composed of two integers, a load case number and element number, followed by two real values of intensities (force/length^2) of tangential and normal loads, q_t and q_n. If the analysis is for the effect of self weight of the structure is required, include in this group a line composed of load case number, the integer zero, the weight per unit volume, and zero. The last line of this set must start with an integer greater than the number of load cases, followed by a dummy integer and two dummy real values. The last line must not be omitted even when there is no loading of type (b).
- A set of any number of lines for loading of type (c). Each line is composed of two integers, a load case number and an element number, followed by two real values of temperature rise at the inner and outer surfaces. The last line of this set must start with an integer greater than the number of load cases followed by a dummy integer and six dummy real values. The last line must not be omitted even when there is no loading of type (c).

B.3.4 Input data example

Cylindrical wall with ring-shaped base, Figure 6.12, Ghali's book 3rd ed.

```
18 17  2  1    NJ,NE,NSJ,NLC
  1.  0.2  0.  Elasticity modulus, Poisson's ratio, thermal exp. coeff.
1  0.00    84.0 Node no., x and r coords.
2  0.00    85.0
3  0.00    86.0
4  0.00    87.0
5  0.00    88.0
6  0.00    89.0
7  0.00    90.0
8  3.00    90.0
9  6.00    90.0
10 9.00    90.0
11 12.00   90.0
12 15.00   90.0
13 18.00   90.0
14 21.00   90.0
15 24.00   90.0
16 27.00   90.0
17 30.00   90.0
18 0.3   90.0
1  1  2  1.0    Element no., JS, JE, thickness
2  2  3  1.0
```

```
3  3  4  1.0
4  4  5  1.0
5  5  6  1.0
6  6  7  1.0
7  7  18 1.0
8  8  9  1.0
9  9  10 1.0
10 10 11 1.0
11 11 12 1.0
12 12 13 1.0
13 13 14 1.0
14 14 15 1.0
15 15 16 1.0
16 16 17 1.0
17 18 8  1.0
1  0  0  1  0. 0. 0. Node no., restr. indics., prsc. displs. (vl., hl. & rotn.)
7  0  1  1  0. 0. 0.
1  1  0. 0. 0.      Load case 1, node, F1 (up), F2 (horiz. out), F3 (radial moment)
10 0  0. 0. 0.      Dummy, end of data of loaded nodes
1  7  .0 -12.94 .0 Load case 1, element no., qt(tangential), qn(normal) (force/length2)
1  8  .0 -11.2 .0
1  9  .0 -10.0 .0
1  10 .0 -8.8  .0
1  11 .0 -7.6  .0
1  12 .0 -6.4  .0
1  13 .0 -5.2  .0
1  14 .0 -4.0  .0
1  15 .0 -2.8  .0
1  16 .0 -1.6  .0
1  17 .0 -12.34   .0
10 0  0. 0.  Dummy, end of data of loaded elements
1  1  0. 0.  Load case 1, element no., Ti, To (temp. inner& outer face)
10  0 0. 0.  Dummy end thermal load data
```

Notes: The analysis is for the effect of circumferential prestressing of wall, producing inward radial pressure of varying intensity: $13p$ to p at bottom and top, respectively. The vertical displacement of the base is prevented only at its inner and outer edges (nodes 1 and 7). The results will include N_ϕ in terms of ph and M in terms of ph^2. Node 18 is inserted near the bottom of the wall, so that the output will include the values of N_ϕ and M at a section closer to the bottom.

Index

D

E

F

R

S

T

W

Printed in the United States
by Baker & Taylor Publisher Services

Printed in the United States
by Baker & Taylor Publisher Services